Fundamental measurement problems in engineering, mechanics, manufacturing, and physics are now being solved by powerful optical methods.

This book presents a lucid, up-to-date discussion of these optical methods. Beginning from a firm base in modern optics, the book proceeds through relevant theory of interference and diffraction and integrates this theory with descriptions of laboratory techniques and apparatus. Among the techniques discussed are classical interferometry, photoelasticity, geometric moire, moire interferometry, holography, holographic interferometry, laser speckle interferometry, and video-based speckle methods.

By providing a firm base in the physical principles and at the same time allowing the reader to perform meaningful experiments related to the topic being studied, the book offers a unique user-oriented approach that will appeal to students, researchers, and practicing engineers.

OPTICAL METHODS OF ENGINEERING ANALYSIS

OPTICAL METHODS
OF ENGINEERING ANALYSIS

GARY L. CLOUD
Michigan State University

CAMBRIDGE
UNIVERSITY PRESS

PUBLISHED BY THE PRESS SYNDICATE OF THE UNIVERSITY OF CAMBRIDGE
The Pitt Building, Trumpington Street, Cambridge CB2 1RP

CAMBRIDGE UNIVERSITY PRESS
The Edinburgh Building, Cambridge CB2 2RU, United Kingdom
40 West 20th Street, New York, NY 10011-4211, USA
10 Stamford Road, Oakleigh, Melbourne 3166, Australia

© Cambridge University Press 1998

This book is in copyright. Subject to statutory exception
and to the provisions of relevant collective licensing agreements,
no reproduction of any part may take place without
the written permission of Cambridge University Press.

First published 1994
First paperback edition 1998

Printed in the United States of America

Library of Congress Cataloging-in-Publication Data is available.

A catalog record for this book is available from the British Library.

ISBN 0-521-45087-X hardback
ISBN 0-521-63642-6 paperback

UNIVERSITY
OF SHEFFIELD
LIBRARY   D

Contents

Part II. Photoelasticity

Part III. Geometrical moire

Part VII. Speckle methods

Acknowledgments

Most of the material normally consigned to a preface has been incorporated into Chapter 1, which describes the purposes, approach, and compass of the book. I emphasize here only the notions that this is intended to be an educational book that is designed to meet a long-standing need, and that it utilizes an instructional paradigm that integrates theory and practice. I hope very much that the work will be useful.

Heartfelt appreciation is extended to Ms. Florence Padgett, Physical Sciences Editor at Cambridge University Press. Her encouragement, support, and wisdom have been important resources for this novice. I warmly thank Ms. Katharita Lamoza, Production Editor, for diligence, acumen, and sensitivity while transforming the manuscript to finished product. Her dedication and expertise bespeak the true professional who loves books.

I am glad to be able, at last, to offer praise and gratitude to mentor Dr. Clyde E. Work, formerly of Michigan Technological University and recently retired as Dean at Western New England College, for giving me a start in experimental mechanics and in the academic world. Identical warm sentiments are extended to mentor Dr. Jerzy T. Pindera, Emeritus Professor of the University of Waterloo, who inspired me by example and instruction to keep going in the right direction. Mentor Dr. John R. Powell is honored for wise counsel and unflagging assistance in the quest for creative direction.

Deserving of commendation are the many students who suffered the early handwritten versions of this work and actually viewed them as worthy of serious study and comment. I am deeply grateful to Dr. Pedro Herrera Franco, Dr. Richard Haskell, Dr. Robert Jones, Dr. Catherine Wykes, Dr. Katherine Creath, and Mr. Xiao-Lu Chen for allowing me to use some of their fine written materials and illustrations on Doppler, moire, and speckle interferometries. Dr. David Grummon, Mr. Henry Wede, and Mr. Richard Schalek have helped immeasurably by converting to computer graphics many of the drawings accumulated over the years. To my wife, Elizabeth Anne, and to my daughters, Sarah Elizabeth and Abigail Anne, I offer love along with my appreciation of their support and patience.

This book is dedicated to my mother, Mrs. Eliene P. Cloud, and to the memory of my father, Mr. Kenneth M. Cloud. They made it possible to get a decent education and so to lead a fulfilling life.

1

Introduction and orientation

This chapter states the goals of the book, traces some reasons for its existence, and describes the best ways to use it. Some of the material that appears here would normally be found in an Author's Preface and would, because of its position of exile outside the main text, suffer the fate of being unread. Given the character of this treatise and its somewhat odd but purposeful organization, it seems best to give this commentary the status of a chapter.

1.1 Objectives

The author perceives that a strong need exists for a book about optical methods of experimental engineering analysis, a book that begins from a firm base in the sciences of physics and modern classical optics, proceeds through careful discussion of relevant theory, and continues through descriptions of laboratory techniques and apparatus that are complete enough to help practicing experimental analysts solve their special measurement problems.

This book on optics, interferometry, and optical methods in engineering measurement is primarily a teaching tool, designed to meet that need. It is not intended to be a research monograph, although it contains many examples drawn from research applications. It is not an encyclopedia of results, nor is it a handbook on optical techniques. It grew from lecture notes prepared during the past 25 years for graduate and undergraduate courses in experimental mechanics. These courses are taken by graduate students and seniors who have a variety of educational and professional experiences in several science and engineering disciplines. The preparation of printed class notes began because of the difficulty of teaching without a single textbook that covers optical methods of measurement from the old-fashioned and still-useful photoelasticity through the newer techniques of moire interferometry and electronic speckle pattern interferometry. Instructional literature about the new methods that is useful and understandable for the typical engineering student or strain analyst is still scarce.

Any major piece of writing is, in the end, colored by compromises between purposes that often conflict. It is also reflective, inevitably, of the author's individual experience. In this case, goals concerning utility, completeness, accuracy, length, and efficiency in the teaching environment have dictated development of a particular pedagogical paradigm. Because the resulting organization might seem strange to persons familiar with more typical presentations, such as handbooks or other textbooks, more needs to be said about the goals, the constraints, and the result.

One intent has been to bring into a single binding as much theory and practical information in optical methods of engineering analysis as would be needed by the typical graduating engineer, the graduate student who might utilize some photomechanics in his research, and the industrial stress and motion analyst.

Another important goal is that of scientific accuracy. Mathematical models of optical phenomena tend to be complex and often not especially useful in general form. On the other hand, some commonly accepted simplified models are neither accurate nor physically meaningful. Further, this book is not intended to be another optics textbook, of which we already have several fine renditions.

The major divisions of the text are intended to be as independently useful as possible. The chapters on holography, for example, might be studied without reference to any other part of the book. The sections on moire methods could stand by themselves, and so on.

In opposition to the goals just outlined was a desire to keep the book short and simple enough to be useful as a text in a 30-class-hour course and to be popular with the applied stress analyst as a quick source of answers to immediate problems. Because the contents of optical methods courses vary greatly depending on institution and professor, there is more here than one can comfortably cover in a single course. The idea is to offer some options while keeping the overall coverage adequate.

1.2 Approach and scope

A conventional approach would probably involve several systematic introductory chapters on optics theory. Some weeks of study of these chapters would be required before finally getting to applications, which are the optical measurement techniques themselves. Laboratory experience with the optical methods would start later yet. The result of this approach is that the term or semester is half over before meaningful laboratory experiments can be performed.

The chosen approach, which meets the goals and solves the laboratory scheduling problem, is to integrate theory as much as possible with the development of the optical methods. The theory and the practical use of the theory in the form of important classes of techniques are offered in a series of parcels. Presented at the beginning are just enough optical concepts and theory to understand basic interferometry and handle the mathematics. The concepts are then utilized, first in learning about some classical two-beam interferometers and subsequently in detailed study of one of the oldest of interferometric methods in engineering: photoelasticity. The student is then in a position to do meaningful experiments on these topics while the increment of theory needed to understand the next technique, geometric moire, is presented. This cycle repeats through the entire text.

This methodology seems to be efficient in teaching theory, technique, and laboratory practice in a limited time. In the author's experience, part of the effectiveness arises from the fact that much transfer of learning and reinforcement is incorporated.

Teaching, say, photoelasticity in this way serves several purposes at once. It emphasizes and illustrates all the important concepts of interferometry. The theory is a simple paradigm for the mathematical development of more complex techniques. Photoelasticity is an important and valuable interferometric technique in engineering analysis. The approach gets the students meaningfully involved in the laboratory early on. They are then in a position to go on to more theory and new methods.

The inclusion of much of the material in Chapter 6 on photoelasticity methods has been problematical because of length limitations and perceived distortion of balance in the book. The rationale is that many of the topics are useful in other areas of applied optical interferometry. Additionally, many conventional courses on experimental mechanics concentrate on photoelastic interferometry. In the end, this chapter represents a compromise that seems satisfactory.

Length limitations have affected the product. The book contains no discussion of continuum mechanics or elasticity; some viscoelasticity concepts are outlined in the chapter on properties of model materials because the concepts are important there. Some omissions reflect the author's own ignorance, but he hopes that most of them represent a willful choice that advances the cause of utility. The lack of at least one chapter on shearography represents the major exclusion. The concept of wavefront shearing is generic in that it is used to enhance other forms of interferometry for specific purposes. A useful but balanced treatment of shearography is left for a second edition, should there be one. Similar comments can be made about the exclusion of grid measurement methods and caustics techniques.

The original plan called for an appendix of scripts for a series of illustrative laboratory experiments. These have not been included for several reasons. A major one is that the book grew too long. Second, quite detailed instructions for most of the processes, such as making holograms, optical spatial filtering, obtaining moire fringes, and reducing fringe data, have been incorporated into the text where appropriate. Reiteration of these details seemed redundant. Finally, many of the procedures will be specific to the equipment indigeneous to a given laboratory; a script that covered all eventualities would have been very long and complex. The instructor or student should be able, with minimum investment of effort, to develop a set of experiments that are based on the instructions in the text but are modified specifically for the apparatus in hand.

In the theory developments, the preferred pattern is to start from basics with the simplest sound treatment of the mathematical and physical foundations of the various methods studied. Physical argument rather than mathematics is sometimes exploited to illustrate and explain. Such an approach is not a valid scientific method, but it serves well in teaching concepts. The danger is that it can lead to oversimplified thinking. This trap has been avoided in most instances by following up the conceptualization with rigorous discussion and theory.

The desire that each major section of the book stand by itself is fulfilled to a reasonable degree. It is possible, for example, to read only the section on moire interferometry and make sense of it. It would be better, though, to read the basic optics and moire material in Chapters 2 and 7 first. Even better would be to include Chapter 10. Some maps of this sort are included later in this chapter.

Finally, some comments about references and bibliography are appropriate. The handling of the references supports the primary purpose, which is education. Utility in teaching has been the only fundamental criterion for inclusion of a reference. Only basic papers and reference books on any one technique or piece of theory are cited in the bibliography. These will lead into the body of research literature on a given topic, should the reader want to pursue it. Exceptions are where controversy is perceived or where the origins of a particular treatment seem clearly connected. When there is a choice, published books are cited in preference to proceedings, theses, and reports. There is no intent to slight or promote any person's work, to assign criticism or credit, or to act as arbiter of turf disputes. This approach to the bibliography requires much judgment on the part of the author. Lapses are inevitable; one hopes that they are not incredible!

Given the broad coverage of the book, the development of a consistent notation that is appropriate for all the methods has been incompletely accomplished. It has not seemed wise to carry such a movement to the extreme. Historically conventional notation for photoelasticity is not entirely compatible with that used for moire interferometry, for example. For this reason, a list of symbols is not included; rather, the meanings are made clear by defining them in the local context.

1.3 Motivations

The last three decades or so have produced many important advances in the discovery and understanding of optical phenomena, in the refinement of optical techniques of measurement, and in the utilization of interferometry to solve problems in physics and engineering. Indeed, many advances in optics theory and technology have been stimulated by the demands of engineering research.

From another viewpoint, one observes that the science of optics unifies and expands diverse important areas of experimental research. Fundamental problems in rapidly developing areas – such as fluid mechanics, rock mechanics, biomechanics, fracture mechanics, materials science, manufacturing technology, nondestructive inspection, and glacier mechanics – are being solved by new methods of optical metrology. Examples of these methods include speckle interferometry, acousto-optic holography, infrared photo-elasticity, holographic interferometry, white-light speckle photography, advanced moire methods, electronic speckle, laser Doppler velocimetry, and optical data processing.

It is a fact that much of the development and many of the applications of advanced optical techniques of experimental mechanics have been carried out by dedicated and capable individuals or small groups in government, university, and industrial laboratories. On the other hand, many of the measurement applications are pursued by people whose backgrounds are in engineering and who have little formal training in optics. The discovery of new phenomena and the invention of new methods tend to be concentrated in physics laboratories, and much of this development of optical techniques is undertaken by specialists who might have a rather narrow concept of applications. Problems-oriented experimental engineering persons tend to be less aware of new choices and conservative in selection of experimental methods.

The need for a book of this type has been perceived by teachers and practitioners in engineering research for many years. It has also been obvious that such a treatise would be very difficult to write, partly because the sheer volume of material is forbidding and partly because of the breadth of background required of the author.

This textbook is a sincere attempt to provide this long-needed bridge, which will tie experimental engineering analysis firmly to its parent sciences while meeting the responsibility of providing training in old and new optical methods of experimental analysis.

1.4 Suggestions for using the book

This text was prepared for both the academic and the industrial laboratory environments, and there are several patterns by which it can be effectively used.

The general sequence to be remembered is that an increment of theory is presented and then a method that utilizes that theory is developed. Thus, the student can perform a laboratory experiment involving that method while learning the next increment of theory.

1.4.1 As a textbook in a graduate course

Consider first the use of the text in a typical graduate course in optical methods, which probably comprises thirty lecture hours and about the same number of laboratory hours. In that case, the first week of instruction deals with Chapters 1 and 2, with only cursory treatment, if at all, of the section on matrix methods. Chapter 3 can be examined to any depth desired, depending on purpose and student interest. If the emphasis is on mechanics-type analysis, Chapter 3 is given only about one-half hour of lecture, which is sufficient to place interferometry in its historical context. At the beginning of the second week, the students are in position to do a meaningful experiment on geometric optics and basic interferometry using Young's fringes, Newton's rings, or basic Michelson interferometry. While this is going on, photoelasticity theory is discussed, with emphasis on the fact that it serves as a paradigm for other interferometric methods while also being very useful in engineering analysis. The class should be well into Chapter 4 at the end of week 2, and they can begin experiments on photoelasticity as Chapter 5 is begun.

At this point, some decisions need to be made about the direction of the course. If skill in photoelasticity is to be attained, then considerable time with the broad range of material in Chapter 6 is necessary. Otherwise, some selected topics from this chapter can be discussed at less-than-thorough depth to round out the overall understanding of the topic and to provide some information that will be useful later in the course. Calibration of materials, model similarity and scaling laws, compensation methods (now called phase measurement techniques in other contexts), and the concepts behind reflection and three-dimensional photoelasticity can be covered, for example.

The next topic is geometric moire, which is discussed while the students are finishing the photoelasticity experiments. Here again, the depth of treatment can be adjusted to the goals of the course. Study of only Chapter 7 is sufficient to aid in understanding the more sophisticated moire techniques. If some skill is to be developed and an experiment on geometric moire is to be conducted, then Chapter 8 and/or Chapter 9 will be useful.

Attention is then turned to the very important topic of diffraction by an aperture. Chapter 10 should not be slighted, no matter what is to follow in the course, because this material is fundamental to much of the subsequent development. Chapter 10 is the second of the twin pillars of the book, the other being Chapter 2. They are separated by many pages and by about a month's study, for good reasons having to do with scheduling the laboratory experiments, developing the requisite mathematical skills, maintaining student interest, and teaching effectively.

From here on, the sequence and depth can be adjusted to suit goals and interests. The material on intermediate sensitivity moire, Chapter 11, builds on geometric moire and is a good practical application of diffraction theory. It is a useful improvement on geometric moire, and it is quite easy to set up a laboratory experiment to illustrate the method.

Alternatively, one can skip to moire interferometry, holographic interferometry, or speckle methods at this point. There is some sequential dependence among these topics, but it has been held to a minimum. If utility in engineering analysis is the major concern, then at least the basics of moire interferometry should be covered. This would comprise most of the first seven sections of Chapter 13 and the two short Chapters 14 and 15. A one-dimensional version of a moire interferometer with minimum adjustments can be set up for the laboratory experiment on this method. An effective alternative is, for this one instance, to rely on a laboratory demonstration with a television camera to show the fringes.

Most students are keen on making holograms and doing simple holographic interferometry, so this topic will probably be an essential part of the course. From the pedagogical viewpoint, discussion of this subject effectively rounds out the understanding of interferometry. The experiments are quite easy to set up. Given typical laboratory budgets, the instructor must be careful to limit film and plate usage, but much practical experience is gained through these experiments.

The final topics in a course with practical orientation are contained in Chapters 18 and 19 on the speckle phenomenon and speckle photography. As with holography, an experiment on speckle photography is easy to set up and effective in terms of teaching and time limitations. Speckle correlation and electronic speckle can be discussed as the closing topics while the students are finishing laboratory experiments. If electronic speckle apparatus is at hand, a demonstration of it is useful because it illustrates an important direction of future development.

1.4.2 As text for a survey course

This material has been used successfully as one of the source books for a senior-level introductory course that covers all the basic methods of strain analysis and motion measurement. It is augmented by materials that cover strain gages and accelerometers. Roughly fifteen class periods are used for discussion of optical techniques. The development of in-depth skill is not a goal. The emphasis is on fundamental concepts and capabilities.

In this type of course, Chapters 1 and 2 are covered but in less depth than before and with emphasis on the concepts of interferometry. Chapter 4 is presented, and either a demonstration of photoelasticity is conducted or else a simple "canned" experiment is performed by the students. Only sections 4–6 and 9 of Chapter 5 are discussed, with emphasis on the procedure rather than the mathematics. The general ideas of stress separation, reflection photoelasticity, and three-dimensional photoelasticity are presented in about two class hours.

The first halves of Chapters 7 and 8 are then discussed. Again, either a demonstration or a simple experiment on geometric moire is offered.

One class hour is then devoted to discussing the concept of diffraction at an aperture and the lens as a Fourier analyzer, with examples.

At this point, the class is in position to cover all or any of the remaining topics at the depth desired for the time left. Coverage of the first nine sections of Chapter 16 on holographic interferometry, with the mathematics de-emphasized in favor of physical concepts, is recommended. This is supported by another demonstration or a basic holography experiment.

The course closes with study of the first halves of Chapters 18 and 19. Demonstrations of speckle photography and, perhaps, electronic speckle provide an effective and forward-looking end.

1.4.3 As a laboratory source book

This book should be useful for the laboratory manager or technician who needs a speedy solution to a certain measurement problem. Review of the strengths and weaknesses of the various methods will indicate which techniques might be most appropriate for a given application. The user can then concentrate on only the section that seems to offer the best probability for solving the problem. One must remember that the technique must be chosen and modified to fit the problem, and that the reverse leads to trouble.

The major divisions of the book are written so that they can stand by themselves to a significant degree. Part V on moire interferometry or Part VI on holographic interferometry, to name two examples, can be studied without reference to other sections, assuming that the reader has some minimal background in optics. Otherwise, preliminary study of Chapters 2 and 10 is advised. A technician who needs to quickly set up a holographic interferometry experiment could even begin with a reading of only Chapter 17 and expect success in the laboratory.

1.5 Closure

The author's fervent wish is that this book will prove useful. Considerable effort has been expended in editing for accuracy, and contemplation of mistakes is distasteful. Given the breadth of coverage, however, it seems inevitable that the work must contain some errors. The author hopes that the mistakes will be neither foolish nor serious. Users of the book are cordially invited to offer constructive comments and corrections, which might lead to an improved second edition.

PART I
Optics and interferometry

2

Light and interference

This chapter provides background, defines concepts, and furnishes some of the mathematical tools for understanding and using optical methods of measurement. Interference of light waves and diffraction at an aperture, the two cornerstones of interferometry, are introduced.

2.1 Nature of light

The exact nature of light is a profound question, which seems to be not yet completely answered. Fortunately, one does not need to know exactly what light is in order to understand how it behaves and to utilize it. There are two convenient ways to describe the propagation of light and its interactions with materials. Neither system is sufficient alone, nor are they completely sufficient together. At least, the two systems are not contradictory. The two approaches are outlined next.

2.1.1 Electromagnetic wave model

For most practical purposes, light (including infrared, ultraviolet, and radiowaves) can be considered to be energy in the form of electromagnetic waves. At any time, the wave train of radiation can be completely described by two vectors that are perpendicular to the direction of travel of the ray and perpendicular to each other. These vectors are

$$\mathbf{E} = \text{the electric vector}$$
$$\mathbf{H} = \text{the magnetic vector}$$

Starting with the fundamental quantity known as "charge" and Coulomb's law, we can develop rationally the rules governing the behavior of electromagnetic waves and their interaction with matter. The resulting equations are Maxwell's equations. These equations and solutions for various cases form the science of electromagnetic field theory.

As shall be seen, field theory accounts for the transmission, propagation, and reflection of light. It has, however, serious shortcomings including:

1. It predicts infinite energy at small wavelength.
2. It lacks a clear expression of what light actually is, except to say that it behaves like energy in the form of waves.
3. It fails to account adequately for such readily observable phenomena as photo-emission and photoresistivity.

2.1.2 *Quantum model*

Some of the deficiencies of electromagnetic theory are eliminated by considering light to be composed of "bundles" or "photons" whose characteristics are predicted through statistical mechanics. For most work in photomechanics, quantum mechanics does not need to be used to explain observed behavior.

Having two systems for describing our basic tool can be annoying. Neither system is incorrect, but neither is sufficient by itself. The limitation is one of the human senses and understanding. It is something like being blind and deaf on alternate days. There is still room for research in this area by another Maxwell.

It may be helpful to correlate wavelength with the applicable theory, as follows:

Longwaves (radio, etc.)	Electromagnetic theory (waves)
Shortwaves (x-ray, gamma waves, etc.)	Quantum theory (photons)
Light (ultraviolet, visible, infrared)	Both, depending on effect observed

2.2 Maxwell's equations

Electromagnetic field theory is adequate for describing all aspects of interferometry except the recording of fringe data by a detector or a photographic emulsion. The basic physical models and mathematical developments in this book start with the assumption that optical energy exists in the form of electromagnetic waves.

The set of mathematical laws that describe the propagation of light waves and their interactions with materials or objects are known as Maxwell's equations, named after James Clerke Maxwell (1831–1879) (Borne and Wolf 1975; Garbuny 1965). Maxwell did not by himself invent the entire set of parameters and relationships that are named after him. He collected the work of others, including Coulomb and Faraday, saw that their individual contributions could be related to one another, added some elegant developments of his own, and ended up with the beautifully insightful set of equations.

Many users of optics and optical methods do not ever bother to look at

Maxwell's equations, and their success does not seem to be diminished by the omission. They just start with the wave equation and do not worry about its antecedents. Because these laws are so fundamental in interferometry, and because optical methods of measurement are used on an ever-expanding variety of materials and structures, they are deserving of the investment of some space and time.

Strictly speaking, Maxwell's equations encompass four relationships among field parameters. They are augmented with four material or constitutive equations that account for the fact that material response to electromagnetic radiation depends on certain material properties. In effect, these material equations define the important material properties in an empirical way.

To present the equations in the most orderly and compact way, it is well to first define the symbols that represent the field parameters and the material coefficients. In addition to **E** and **H**, defined in section 2.1, which define or represent the electromagnetic field, the field quantities are

$$\mathbf{D} = \text{electric displacement vector}$$
$$\mathbf{J} = \text{current density vector}$$
$$\mathbf{B} = \text{magnetic induction vector}$$

The material properties are represented as the following scalar parameters in an isotropic medium:

$$\sigma = \text{specific conductivity}$$
$$\epsilon = \text{dielectric constant}$$
$$\mu = \text{magnetic permeability}$$
$$\rho = \text{electric charge density}$$

If the medium is not isotropic, then the material properties become tensors.

The four field equations of Maxwell are, using ∇ to represent the operator

$$\frac{d}{dx}\mathbf{i} + \frac{d}{dy}\mathbf{j} + \frac{d}{dz}\mathbf{k}$$

and a dot above the vector to indicate d/dt,

$$\nabla \times \mathbf{H} - \frac{1}{c}\dot{\mathbf{D}} = \frac{4\pi}{c}\mathbf{J} \tag{2.1}$$

$$\nabla \times \mathbf{E} + \frac{1}{c}\dot{\mathbf{B}} = 0 \tag{2.2}$$

$$\nabla \cdot \mathbf{D} = 4\pi\rho \tag{2.3}$$

$$\nabla \cdot \mathbf{B} = 0 \tag{2.4}$$

Equation 2.3 is recognized as the law of Gauss defining charge density. Equation 2.4 states that there are no magnetic sources or sinks; magnets always have two poles. Also, c turns out to be the speed of light.

The material constitutive equations, which actually define the material parameters, are, for the isotropic medium,

$$\mathbf{J} = \sigma\mathbf{E} \tag{2.5}$$

$$\mathbf{D} = \epsilon\mathbf{E} \tag{2.6}$$

$$\mathbf{B} = \mu\mathbf{H} \tag{2.7}$$

Note that these material equations will be tensor relationships of similar form for materials that are optically anisotropic. Such materials include birefringent crystals and photoelasticity models.

Maxwell's equations can be greatly simplified for specific ideal classes of materials. Optical interferometry is usually concerned with materials, such as glass and air, that approximate homogeneous transparent dielectrics. In that case, the specific conductivity and the current density are zero. Also, the dielectric coefficient and the magnetic permeability do not vary through the field. With these restrictions in place, it is fairly easy, though tedious, to show that the electric vector and the magnetic induction vector satisfy the basic wave equation. This development is worth doing once, so it is outlined here for the electric vector (only) in an isotropic medium. The steps parallel those in the standard reference works (Born and Wolf 1975; Coker and Filon 1957; Garbuny 1965); the first two of these texts also deal with optically anisotropic materials. Boldface type is used to represent vector quantities in this book.

Substitute equation 2.7 into 2.2 and take the curl,

$$\nabla \times (\nabla \times \mathbf{E}) = -\frac{\mu}{c}\nabla \times \dot{\mathbf{H}} \tag{2.8}$$

Differentiate equation 2.1 with respect to time,

$$\dot{\nabla} \times \mathbf{H} + \nabla \times \dot{\mathbf{H}} - \frac{1}{c}\ddot{\mathbf{D}} = \frac{4\pi}{c}\dot{\mathbf{J}} \tag{2.9}$$

Combining these two equations to eliminate $\nabla \times \dot{\mathbf{H}}$ gives, after recognizing that $\dot{\nabla} \times \mathbf{H} = 0$ and $d\mathbf{J}/dt = 0$,

$$-\frac{c}{\mu}\nabla \times (\nabla \times \mathbf{E}) - \frac{1}{c}\ddot{\mathbf{D}} = 0 \tag{2.10}$$

Now differentiate equation 2.6 twice, introduce it into equation 2.10, then tidy it a bit to get

$$\nabla \times \left(\frac{1}{\mu}\nabla \times \mathbf{E}\right) + \frac{\epsilon}{c^2}\ddot{\mathbf{E}} = 0 \tag{2.11}$$

Use now the identities

$$\nabla \times m\mathbf{V} = m(\nabla \times \mathbf{V}) + \nabla m \times \mathbf{V}$$

and (2.12)

$$\nabla \times (\nabla \times \mathbf{V}) = \nabla(\nabla \cdot \mathbf{V}) - \nabla^2 \mathbf{V}$$

so that the first expression of equation 2.11 can be modified as follows,

$$\nabla \times \left(\frac{1}{\mu}\nabla \times \mathbf{E}\right) = \frac{1}{\mu}(\nabla \times (\nabla \times \mathbf{E})) + \nabla\left(\frac{1}{\mu}\right) \times (\nabla \times \mathbf{E}) \qquad (2.13)$$

$$\frac{1}{\mu}(\nabla \times (\nabla \times \mathbf{E})) = \frac{1}{\mu}(\nabla(\nabla \cdot \mathbf{E}) - \nabla^2 \mathbf{E}) \qquad (2.14)$$

When these results are used in equation 2.11, the result is

$$\frac{1}{\mu}\nabla(\nabla \cdot \mathbf{E}) - \frac{1}{\mu}(\nabla^2 \mathbf{E}) + \nabla\left(\frac{1}{\mu}\right) \times (\nabla \times \mathbf{E}) - \frac{\epsilon}{c^2}\ddot{\mathbf{E}} = 0 \qquad (2.15)$$

or

$$\nabla^2 \mathbf{E} - \frac{\mu\epsilon}{c^2}\ddot{\mathbf{E}} + \nabla \ln \mu \times (\nabla \times \mathbf{E}) + \nabla(\nabla \cdot \mathbf{E}) = 0 \qquad (2.16)$$

Use again material equation 2.6 and the identity

$$\nabla \cdot m\mathbf{V} = m\nabla \cdot \mathbf{V} + \mathbf{V} \cdot \nabla m \qquad (2.17)$$

to write

$$\nabla \cdot \mathbf{D} = \nabla \cdot \epsilon\mathbf{E} = \epsilon\nabla \cdot \mathbf{E} + \mathbf{E} \cdot \nabla\epsilon \qquad (2.18)$$

But equation 2.3 implies that $\nabla \cdot \mathbf{D} = 0$ if $\rho = 0$ as assumed, so

$$\nabla \cdot \mathbf{E} = \frac{1}{\epsilon}\mathbf{E} \cdot \nabla\epsilon \qquad (2.19)$$

When equation 2.19 is substituted into equation 2.16, the result is

$$\nabla^2 \mathbf{E} - \frac{\mu\epsilon}{c^2}\ddot{\mathbf{E}} + (\nabla \ln \mu) \times (\nabla \times \mathbf{E}) + \nabla\left(\frac{1}{\epsilon}\mathbf{E} \cdot \nabla\epsilon\right) = 0 \qquad (2.20)$$

The last term may be manipulated further, depending on the situation. For the case at hand, recall the assumption that ϵ and μ are uniform in the space being studied. The last two expressions of the equation vanish to leave the vector wave equation

$$\nabla^2 \mathbf{E} - \frac{\mu\epsilon}{c^2}\ddot{\mathbf{E}} = 0 \qquad (2.21)$$

A parallel development gives a similar result for **H**.

2.3 Wave equation and simple harmonic wave

In a nonconducting medium that is free of electric charge, Maxwell's equations require that the electric vector $\mathbf{E}(t)$ satisfy the wave equation, which was derived in the previous section and which is rewritten here in slightly modified form:

$$\nabla^2 \mathbf{E} = K\mu \frac{d^2 \mathbf{E}}{dt^2} \tag{2.22}$$

where $K\mu = 1/v^2$
              v is the speed of propagation of the wave
              K is the dielectric coefficient of the medium $= \epsilon/c^2$
              μ is the magnetic permeability of the medium

The general solution is, for the one-dimensional case with the propagation direction z,

$$\mathbf{E} = \mathbf{f}(z - vt) + \mathbf{g}(z + vt) \tag{2.23}$$

The simplest solution is a harmonic plane wave: that is, a sine or cosine wave (which one is not important) such as,

$$\mathbf{E} = \mathbf{A} \cos\left[\frac{2\pi}{\lambda}(z - vt)\right] \tag{2.24}$$

where $\mathbf{A}$ is a vector specifying amplitude and plane of wave
              λ is the wavelength

This wave is the basic element in analyzing optical interferometric systems. It is monochromatic (single wavelength) as well as plane polarized (also called linearly polarized). There are other ways to describe this wave mathematically; particularly useful is the complex algebra form, which will be introduced presently.

The one-dimensional solution is easily generalized for a wave propagating along some axis specified by unit vector $\boldsymbol{\alpha}$ having direction cosines l, m, n in Cartesian coordinates.

$$\boldsymbol{\alpha} = l\mathbf{i} + m\mathbf{j} + n\mathbf{k} \tag{2.25}$$

The wave traveling in the $\boldsymbol{\alpha}$ direction has the following equation,

$$\mathbf{E} = \mathbf{A} \cos\left\{\frac{2\pi}{\lambda}[vt - (lx + my + nz)]\right\} \tag{2.26}$$

If a general position vector is

$$\mathbf{r} = x\mathbf{i} + y\mathbf{j} + z\mathbf{k} \tag{2.27}$$

then equation 2.26 can be written as

$$\mathbf{E} = \mathbf{A} \cos(\omega t - \mathbf{k} \cdot \mathbf{r}) \tag{2.28}$$

where $\quad \mathbf{k} = k\boldsymbol{\alpha} = \dfrac{2\pi\boldsymbol{\alpha}}{\lambda} = \text{vector wave number}$

$\omega = \dfrac{2\pi v}{\lambda} = \text{angular frequency of radiation} = 2\pi v$

$v = \text{optical frequency, Hz}$

In practice, this single basic wave is rarely encountered, and creating one in the laboratory is difficult. More realistic waves consist of more than one wavelength and of various polarizations. Fortunately, these complicated waves can be treated simply as superpositions of more than one simple wave.

A straightforward analysis shows that the propagation velocity v and frequency v of the simple wave are related as follows,

$$v = \lambda v \tag{2.29}$$

Frequency is independent of the medium through which the wave is passing. That is,

$$v = \frac{v}{\lambda} = \text{constant, for any given wave} \tag{2.30}$$

2.4 Electromagnetic spectrum

The interactions between light and a transmitting medium are highly dependent on frequency and wavelength. Let us establish the units used for frequency and wavelength and then see how the electromagnetic spectrum is divided on the basis of utilization.

The wavelength of light is best expressed in the following units:

$1 \text{ micron} = 1 \text{ } \mu\text{m} = 10^{-6} \text{ meter}$
$1 \text{ millimicron} = 1 \text{ m}\mu = 1 \text{ nanometer} = 1 \text{ nm} = 10^{-9} \text{ meter}$
$1 \text{ angstrom} = 1 \text{ Å} = 10^{-10} \text{ meter}$

The first two units (μm and nm) will be used in this book because they give numbers of convenient size for working with visible and infrared wavelengths. Note that some authors shorten μm to μ.

If the unit of frequency is taken to be the hertz (Hz) = 1 cycle per second, then equation 2.29 indicates that

$$v \text{ (Hz)} = \frac{v \text{ } (\mu\text{m/sec})}{\lambda \text{ } (\mu\text{m})}$$

If the vacuum wavelength λ_{vac} is used, then the velocity of light in a vacuum must also be employed: $v_{\text{vac}} = c = 3 \times 10^5 \text{ km/sec} = 3 \times 10^{14} \text{ } \mu\text{m/sec}$.

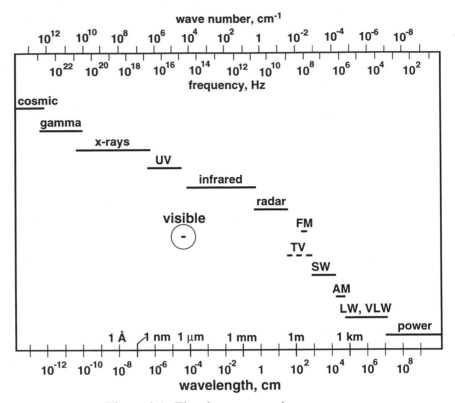

Figure 2.1. The electromagnetic spectrum.

A unit of frequency much used by spectroscopists is the reciprocal centimeter (cm^{-1}), also called the wave number:

$$\text{wave number} = v = \frac{1}{\lambda \, (cm)}$$

Figure 2.1 is a chart of the electromagnetic spectrum showing the various divisions. Note the relatively small span of wavelengths covered by the visible spectrum. Regions of special interest in photomechanics are, approximately,

visible	0.4–0.7 μm
near infrared	0.7–2 μm
ultraviolet	1–400 nm

2.5 Polarization

The polarization of radiation is an important consideration in optical interferometry. For one thing, interference between light waves is possible only if they are identically polarized. Several types of polarization can be found. In general, the polarization state can be described on the basis of what is happening to the electric vector.

If the electric vector for a given wave or group of waves has a fixed orientation in space and time, then the light is said to be linearly polarized. This polarization state is important in interferometry, and a device to convert radiation to linear polarization is often inserted into optical instruments so that full interference can be achieved.

If the electric vector has constant magnitude but is known to rotate in time or space, then the radiation is said to be circularly polarized. One way to visualize this state is that a snapshot of the electric vector of the wave at a given instant would show that its tip traces out a circular helix. The other visualization is that over a period of time, the electric vectors crossing a given cross section define a circle. This state of polarization and methods for creating it are described in detail in Chapter 4, as it is very important in interferometry. Perfectly polarized radiation, linear or circular, is not easy to create. Typically the cross section is elongated into an ellipse, and such a state is termed elliptical polarization.

Radiation in which the electric vectors of the several component waves do not have any preferred orientation in time or space is said to be randomly polarized or unpolarized. True random polarization is not all that common because emission, reflection, and scattering processes usually cause some ordering of the electric vector.

2.6 Mathematical approaches

In the preceding sections, only the simplest representation of the monochromatic polarized light wave has been utilized. This is a good place to outline alternative approaches for representing electromagnetic radiation and for performing calculations to describe the behavior of that radiation.

But first, the rationalization for using more than one mathematical approach in a single book should be explained. A goal is to make this work useful to engineers, scientists, and practicing technicians who have a variety of backgrounds. It seems best to start with the trigonometric formulation of the wave, to use that representation to explain the basic concepts of interference and diffraction, and to carry it through the theory of photoelastic interferometry. The trigonometric function approach is physically meaningful to anyone who has studied the most elementary vibrations; and it is adequate, and satisfyingly simple, for describing a surprising number of optics phenomena.

One finds that representations in the form of trigonometric functions become quite unwieldy when the optical processes become geometrically complicated, as in off-axis holography, or when the theory requires extensive manipulation, as in general diffraction development. As these topics are approached, the complex variable representation of the electromagnetic wave is introduced. By that time, physical insight has been fully associated with the mathematics. For those already accustomed to using complex variables, the trigonometric function equations of the first few chapters are easily converted to complex form.

In addition to the trigonometric and the complex variable representations utilized herein, other systems of mathematics could be used constructively for at least some of the development. These include, in particular, two matrix methods whereby prediction of the effects of optical elements on a simple harmonic wave are reduced to a series of matrix multiplications.

One of the surprising and useful aspects of certain optical processes is that they turn out to be physical renditions of Fourier transforms. For example, any lens can be seen as a Fourier transforming device. Even if this property of a lens is not of obvious importance in the device being considered, such as a camera lens, the performance of that lens is given in terms of a resolution capability or optical transfer function that originates in the physical Fourier analysis of spatial signals. To properly describe diffraction, as in moire analysis or in reconstructing a holographic image, knowledge of elementary Fourier transforms is crucial.

The basic concepts and manipulations belonging to these various branches of mathematics are now given, except for the trigonometric approach (which has already been treated adequately) and for basic Fourier transforms (which will be discussed in the context of diffraction theory in Chapter 10).

2.6.1 Complex variables and some useful complex algebra

Representation of the simple harmonic wave of electromagnetic energy in terms of trigonometric functions is physically meaningful and adequate for many purposes, though that approach tends to become cumbersome when describing the function of more complicated interferometric systems.

In optics and electronics, the complex number representation of the harmonic wave is customarily used, and it will be used in portions of this book. By using the identities giving the trigonometric functions in terms of complex exponentials, equations 2.26 and 2.28 are easily converted to complex numbers, yielding

$$\mathbf{E} = \text{Re } \mathbf{A}e^{i[\omega t - \mathbf{k} \cdot \mathbf{r}]} \tag{2.31}$$

where Re means "real part of" and i is $\sqrt{-1}$ (designated by j in many books). As a useful tip, it is worth mentioning that, in texts and papers, the Re designation is omitted; it is supposed to be understood by the reader wherever it is needed.

It is a simple matter to include in the preceding equations for the elementary wave a phase-shift term ϕ, giving, for example,

$$\mathbf{E} = \text{Re } \mathbf{A}e^{i(\omega t - \mathbf{k} \cdot \mathbf{r} + \phi)} \tag{2.32}$$

Equation 2.32 can also be written as a product

$$\mathbf{E} = \text{Re}[\mathbf{A}e^{i\phi}e^{-i\mathbf{k}\cdot\mathbf{r}}e^{i\omega t}] \tag{2.33}$$

As a further notational simplification, the terms containing time, which represent an oscillation at optical frequency, are usually dropped. We have no way of even detecting this oscillation; our eyes and other detectors measure only some sort of average value. In most optics derivations, the terms containing the space variables and the amplitude and phase data are the only ones of interest. This information is contained in the so-called complex amplitude $\mathbf{U}$, which results from dropping the optical frequency terms from equation 2.33:

$$\mathbf{U} = \text{Re}[\mathbf{A}e^{i\phi}e^{-i\mathbf{k}\cdot\mathbf{r}}] \tag{2.34}$$

Often, the polarization state is of no consequence or can be assumed uniform, and the vector designation is dropped along with the real-part symbol to leave the following customary form for the complex amplitude,

$$U = Ae^{i\phi}e^{-i\mathbf{k}\cdot\mathbf{r}} \tag{2.35}$$

As an example, the complex amplitude for the wave traveling in the z-direction would be, in the notation just introduced,

$$E = Ae^{i\phi}e^{-ikz} \tag{2.36}$$

It is important to realize that equations such as these represent a whole family of waves moving in a specified direction with common velocity and having a common amplitude and phase relationship. That is, equation 2.32 represents a monochromatic plane wave group traveling in the given direction and having uniform intensity. Variations of phase and intensity over the extent of the group can be incorporated into the complex amplitude.

If the amplitude or intensity of the light changes within the volume occupied by the group; then the amplitude A might be a function of position. For such a ray moving along the z-axis, the complex amplitude would be

$$U = A(x, y, z)e^{i\phi}e^{-ikz} \tag{2.37}$$

It can also happen, as in many instances of whole-field interferometry, that the phase is a function of location in the ray of light. The complex amplitude would include a phase term that depends on position. For such a group moving in the z-direction, the complex amplitude then becomes

$$U = A(x, y, z)e^{i\phi(x, y, z)}e^{-ikz} \tag{2.38}$$

The general equations for a bundle of waves traveling in the direction specified by unit vector α are

$$\mathbf{E} = \mathrm{Re}[\mathbf{U}e^{i\omega t}] \qquad (2.39)$$

where

$$\mathbf{U} = \mathrm{Re}[\mathbf{A}(\mathbf{r})e^{i\phi(\mathbf{r})}e^{-i\mathbf{k}\cdot\mathbf{r}}] \qquad (2.40)$$

and recall that

$$\mathbf{r} = x\mathbf{i} + y\mathbf{j} + z\mathbf{k} = \text{general position vector}$$

$$\mathbf{k} = k\alpha = \left(\frac{2\pi}{\lambda}\right)\alpha$$

$$\alpha = l\mathbf{i} + m\mathbf{j} + n\mathbf{k}$$

An algebraic detail that proves useful in optics derivations with complex amplitudes is that twice the real part of the complex conjugate of a complex number can be represented as the sum of the number and its complex conjugate. The factor of 2 is often just absorbed in the other constants of the equation. In complex algebra symbolism, using an asterisk to represent complex conjugate,

$$z + z^* = 2\,\mathrm{Re}(z) \qquad (2.41)$$

also,

$$z - z^* = 2\,\mathrm{Im}(z)$$

For much optical work, data collection such as photography involves measurement of irradiance, commonly called intensity. The irradiance of a wave is defined as the time average (optical frequency oscillation averaged out) of the square of the amplitude:

$$I(\mathbf{r}) = 2\langle \mathbf{E}^2(\mathbf{r}, t)\rangle \qquad (2.42)$$

$$= \lim_{T\to\infty} \frac{1}{2}T \int_{-T}^{T} E^2(r, t)\,dt \qquad (2.43)$$

where $\langle\ \rangle$ represents the average over many oscillations at optical frequencies. For a simple harmonic wave, the irradiance is

$$I(\mathbf{r}) = 2A^2(\mathbf{r})\langle\cos^2(\omega t - \mathbf{k}\cdot\mathbf{r})\rangle \qquad (2.44)$$

but the time average of the $\cos^2$ is just $\frac{1}{2}$, so

$$I(\mathbf{r}) = A^2(\mathbf{r}) \qquad (2.45)$$

This result states that the irradiance distribution is just the square of the amplitude distribution.

Another bit of useful complex algebra is established by using equations 2.41 and 2.39 to write

$$E(\mathbf{r}, t) = U_1 e^{i\omega t} + U_1^* e^{-i\omega t} \qquad (2.46)$$

Utilize equation 2.42 to obtain

$$I(\mathbf{r}) = \langle U_1^2 e^{i2\omega t} + U_1^{*2} e^{-i2\omega t} + U_1 U_1^* \rangle \qquad (2.47)$$

but

$$\langle e^{\pm i2\omega t} \rangle = 0$$

so

$$I(\mathbf{r}) = U_1 U_1^* = |U_1|^2 \qquad (2.48)$$

The result is that the local irradiance of a wave that may contain a distribution of phase differences is just the square of the modulus of the local complex amplitude.

2.6.2 *Matrix methods*

For certain optical derivations, especially those involving a sequence of optical devices such as polarizers and lens elements, one or another of two computational schemes that reduce the necessary calculation to a sequence of matrix operations are advantageous (Shurcliff 1962). Both techniques represent the state of polarization as a vector. The transformation to a different state of polarization is represented by multiplication of the vector by a matrix. We shall introduce these ideas and look at a few examples without getting too deeply involved. The disadvantage of the matrix approach is that insight into the physical nature of the processes tends to be lost.

Mueller calculus

The Mueller calculus was developed from a purely phenomenological point of view, and it does not rest on electromagnetic theory. The basic light vector contains four parameters that describe the light beam. It is called the Stokes vector, represented as a column vector, and its elements are

$$\{I, M, C, S\} \qquad (2.49)$$

where I is intensity
         M is preference for horizontal polarization
         C is preference for 45° polarization
         S is preference for right-handed circular polarization

examples   vertically polarized $= \{1, -1, 0, 0\}$
            horizontally polarized $= \{1, 1, 0, 0\}$

The optical element is represented by a 4 × 4 matrix. For example, a 45°

half-wave retarder acting on horizontally polarized light to obtain vertical polarization would be represented by

$$
\begin{bmatrix} 1 & 0 & 0 & 0 \\ 0 & -1 & 0 & 0 \\ 0 & 0 & 1 & 0 \\ 0 & 0 & 0 & -1 \end{bmatrix} \begin{bmatrix} 1 \\ 1 \\ 0 \\ 0 \end{bmatrix} = \begin{bmatrix} 1 \\ -1 \\ 0 \\ 0 \end{bmatrix}
$$

half-wave   horz.   vert.
                 pol.   pol.

The 4 × 4 matrix is, basically, derived from experiment for many types of optical elements. Usually, these can be looked up in books and charts. Although this approach does not rely on electromagnetic theory, it may be interpreted to give electromagnetic parameters, as follows:

$$
I = \langle A_x^2 + A_y^2 \rangle
$$

$$
M = \langle A_x^2 - A_y^2 \rangle
$$

$$
C = \langle 2 A_x A_y \cos \gamma \rangle
$$

$$
S = \langle 2 A_x A_y \sin \gamma \rangle \tag{2.50}
$$

where A_x, A_y are amplitudes of x, y components
             γ is the phase angle between components, related to relative retardation
             $\langle \rangle$ is the time average over several waves, RMS level

Although it is premature to do so, let us now write the Mueller calculus representation of a linear polariscope with a photoelastic model in place. Multiplication is from right to left.

$$
\frac{1}{2} \begin{bmatrix} 1 & -1 & 0 & 0 \\ -1 & 1 & 0 & 0 \\ 0 & 0 & 0 & 0 \\ 0 & 0 & 0 & 0 \end{bmatrix} \begin{bmatrix} 1 & 0 & 0 & 0 \\ 0 & D^2 - E^2 - G^2 & 2DE & -2EG \\ 0 & 2DE & 2D^2 + E^2 + G^2 & 2DG \\ 0 & 2EG & -2DG & 2G^2 - 1 \end{bmatrix}
$$

analyzer   model

$$
\times \begin{bmatrix} 1 \\ 1 \\ 0 \\ 0 \end{bmatrix} = \begin{bmatrix} I \\ M \\ C \\ S \end{bmatrix} \tag{2.51}
$$

horz. emerging
pol.   light

where $\quad D = M \sin \delta/2 \qquad E = C \sin \delta/2$
$\qquad\qquad F = S \sin \delta/2 \qquad G = \cos \delta/2$
$\qquad\qquad \delta$ is retardation in radians $= [(n_1 - n_2)/n_0]d/\lambda$
$\qquad\qquad M, C, S$ are the second, third, and fourth Stokes parameters of the
$\qquad\qquad$ normalized fast electric vector of the given retarder

The result is, for the Stokes vector of the emerging light,

$$\begin{bmatrix} 1 - D^2 - E^2 - G^2 \\ -1 + D^2 + E^2 + G^2 \\ 0 \\ 0 \end{bmatrix} \tag{2.52}$$

The first parameter is the intensity or irradiance, which is what is observed in photoelasticity:

$$I = -1 - M^2 \sin^2 \frac{\delta}{2} - C^2 \sin^2 \frac{\delta}{2} - \cos^2 \frac{\delta}{2} \tag{2.53}$$

This problem will be pursued no further now. It should transform to the result given in Chapter 4 after considerable algebra.

Jones calculus

This approach is similar except that it utilizes 2×2 matrices that are complex. It does depend on electromagnetic theory for its rationale. A major advantage is that Jones calculus preserves all absolute phase information. In the Jones calculus approach, the beam of light is represented by its electric vector **E**, which is written in matrix form in terms of its components as follows,

$$E = \begin{bmatrix} E_x \\ E_y \end{bmatrix} = [E_x, E_y] \tag{2.54}$$

Thus, linearly polarized light with a horizontal polarization axis might be represented by the vector

$$\begin{bmatrix} A \cos \dfrac{2\pi}{\lambda} (z - vt) \\ 0 \end{bmatrix} \tag{2.55}$$

Usually, the vector is normalized so that the nonzero term in the polarized light matrix is just 1,

$$\begin{bmatrix} 1 \\ 0 \end{bmatrix}$$

Note that absolute phase data are not contained in normalized forms. The effect of an optic element such as a polarizer or model is represented by matrix multiplication. We need know only the "transfer function" matrix for the various optic elements. These functions can be determined from study of the electromagnetic equations for simple systems. Actually, they have been already established in general form for most conceivable optic elements. One just looks them up on a chart (Shurcliff 1962).

Now, let us again, albeit prematurely, see how this works out for the photoelasticity experiment. The transfer function for any optical element with two principal axes – that is, any birefringent plate – is

$$[T] = \begin{bmatrix} \cos^2 \phi + e^{-i\delta} \sin^2 \phi & (1 - e^{-i\delta}) \cos \phi \sin \phi \\ (1 - e^{-i\delta}) \cos \phi \sin \phi & \sin^2 \phi + e^{-i\delta} \cos^2 \phi \end{bmatrix} \qquad (2.56)$$

where ϕ is the angle between principal axis I of the optic element and the reference x-axis;

δ is the relative retardation of the element

Surprisingly, the preceding matrix includes polarizers. For example, a horizontal polarizer is obtained by taking $\phi = 0$, $\delta = \infty$ (one component of E doesn't get through at all) to give

$$[T]_{\text{pol.}} = \begin{bmatrix} 1 & 0 \\ 0 & 0 \end{bmatrix} \qquad (2.57)$$

The photoelastic model is a retarding plate, and its matrix is just the general matrix given by equation 2.56, which can be represented in symbolic form by

$$[T]_{\text{model}} = \begin{bmatrix} m_{11} & m_{12} \\ m_{21} & m_{22} \end{bmatrix} \qquad (2.58)$$

The incident light is pretty much arbitrary. We don't really have a matrix for unpolarized light, so we just take

$$E_{\text{in}} = \begin{bmatrix} 1 \\ 1 \end{bmatrix}$$

Assembling these matrices gives the matrix representation of the linear polariscope as

$$\begin{bmatrix} E_x \\ E_y \end{bmatrix} = \begin{bmatrix} 0 & 0 \\ 0 & 1 \end{bmatrix} \begin{bmatrix} m_{11} & m_{12} \\ m_{21} & m_{22} \end{bmatrix} \begin{bmatrix} 1 & 0 \\ 0 & 0 \end{bmatrix} \begin{bmatrix} 1 \\ 1 \end{bmatrix} \qquad (2.59)$$

emerging

The matrix multiplications are easily carried out and are left as an exercise. The raw results are already similar to those given in Chapter 4.

2.7 Wavefronts: plane, spherical, and warped

The complex amplitude, $U = A(x, y, z)e^{i\phi}$ may be looked on as the equation of the wavefront of a beam of light. The $e^{i\phi}$ describes the change of phase across the beam whereas the $A(x, y, z)$ gives the amplitude distribution. Alternatively, the phase term can be viewed as the equation of a "surface of equal phase" or wavefront. Figure 2.2 is a simple illustration of the wavefront concept. The equiphase surface ideas can be utilized in the development of the "warped wavefront" explanations of photoelasticity, holography, and other optical processes (Post 1970). It is especially useful in enhancing visualization.

Two wavefront configurations are of special interest in optics calculations. The first of these is the *plane wavefront*, usually called a plane wave, in which there is no phase variation through the beam cross section ($\phi = \text{const} = 0$). The plane wave, which is illustrated in Figure 2.3, is easily handled mathematically, and it is easy to generate and reproduce to a reasonable degree of accuracy in the laboratory.

The second special wavefront is that which results when light radiates in all directions from a true point source. It is called a spherical wave because the surface of uniform phase is a sphere centered at the source, and all the waves propagate along radii. See Figure 2.4 for a cross-sectional representation.

The general equation for a spherical wavefront is

$$E(r, t) = A(r)\cos(\omega t - kr + \phi_0) \tag{2.60}$$

or in complex form,

$$E(r, t) = U(r, \phi)e^{i\omega t} \tag{2.61}$$

where $\quad U = \text{complex amplitude} = A(r)e^{-ikr}e^{i\phi_0}$
$$= A(r)e^{i(\phi_0 - kr)}$$

One reason for the importance of spherical waves is that they are generated by passing a parallel beam of light (plane wave) through a lens. Figure 2.5 shows how two lenses are used to convert from a narrow plane to a spherical wavefront and back to a plane wavefront of larger beam diameter. This technique is employed in expanding a laser beam to a larger diameter beam with plane wavefront. Often, the second lens is eliminated and the spherical wavefront is utilized.

The two-dimensional analog of the spherical wavefront is the cylindrical wavefront. The preceding two figures accurately depict the cross sections of the cylindrical case if one remembers that the wavefront pictured in Figure 2.4 is symmetrical about a line rather than a point. The system in Figure 2.5 is, for the cylindrical case, symmetric about a plane rather than the optic z-axis.

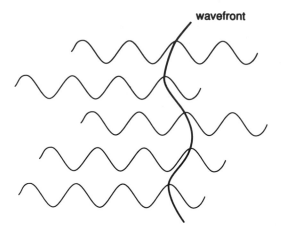

Figure 2.2. The warped
wavefront concept.

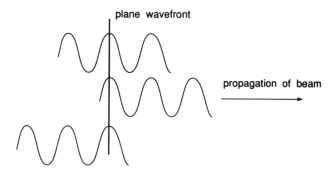

Figure 2.3. Simple plane
wave.

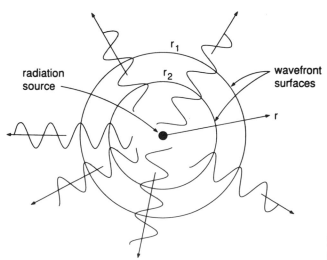

Figure 2.4. Cross section of
a spherical wave.

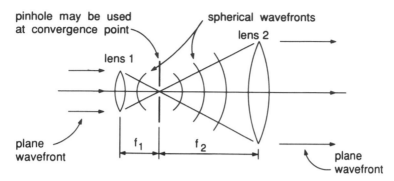

f₁ and f₂ are focal lengths of lenses

Figure 2.5. Spherical and plane wavefront conversion.

2.8 Interference of collinear waves

The development of insight into the nature of interferometric processes is begun by considering the interaction between two waves that travel along the same axis. The waves are assumed to be "coherent," which means they are able to interfere. For simplicity, assume also that the waves travel along the z-axis, as represented in Figure 2.6. Because polarization is the same for the two coherent waves, only the magnitudes of the electric vectors need to be considered. Likewise, the wavelengths and wave velocities are identical. The scalar representations are, in the elementary notation introduced earlier,

$$E_1 = A_1 \cos\left[\frac{2\pi}{\lambda}(z - vt)\right]$$

$$E_2 = A_2 \cos\left[\frac{2\pi}{\lambda}(z - vt - r)\right]$$

(2.62)

The extra parameter r in the second wave is introduced because that wave lags behind the other one; that is, the two are out of phase. To shorten the development, assume that the amplitudes are equal and call it just A. The combination of the two waves is E_s and is given by

$$E_s = A\left\{\cos\left[\frac{2\pi}{\lambda}(z - vt)\right] + \cos\left[\frac{2\pi}{\lambda}(z - vt - r)\right]\right\}$$

(2.63)

At this point, use the identity

$$\cos A + \cos B = 2\cos\left(\frac{A + B}{2}\right)\cos\left(\frac{A - B}{2}\right)$$

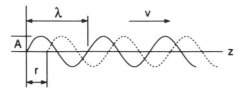

Figure 2.6. Interference of
two collinear waves.

The result is

$$E_s = 2A \cos \frac{\pi r}{\lambda} \cos\left[\frac{2\pi}{\lambda}\left(z - vt - \frac{r}{2}\right)\right]$$ (2.64)

This result represents a wave whose amplitude is $2A \cos(\pi r/\lambda)$. The amplitude depends on the degree to which the two waves are out of phase. The irradiance is the square of this amplitude,

$$I_s = 4A^2 \cos^2 \frac{\pi r}{\lambda}$$ (2.65)

This function can be plotted as a function of r. We are most interested in the values of r for which maxima or minima are attained. If $r/\lambda = n$, where $n = 0, 1, 2, \ldots$, then the irradiance is maximum, and the interaction between the two waves is called constructive interference. The irradiance is minimum, zero in this ideal case, for $r/\lambda = 1/2, 3/2, \ldots$; this situation is called destructive interference. It causes the dark fringes to be formed in whole-field interferometry.

If the two waves are not of equal amplitude, then the waves do not completely cancel one another, and the minimum irradiance is greater than zero. In that case, the fringe visibility or fringe contrast suffers. Fringe visibility is an important issue in interferometry.

An interesting question is "What happens to the energy flux in interferometry?" Is the energy destroyed when two waves interfere destructively?

2.9 Interference of two plane waves

A process that is fundamental to the creation of a hologram, and equally important in many other interferometric situations, is the interference of two beams of light that might impinge on a film plate or other detector. In interferometry, one of these rays of light usually has a plane or spherical wavefront whereas the other has a complicated, warped wavefront. Consideration of the special case of interference of two plane waves guides one's thinking and provides insight into the more complex general cases. At least, the observable manifestations of the interferometric phenomena are easily correlated with theory for this simple situation. The result is a physical understanding of holography and interferometry that cuts through the many confusing phenomenological complexities.

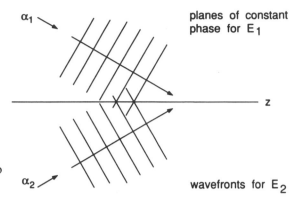

Figure 2.7. Geometry for the oblique meeting of two plane waves.

For educational purposes, part of the theory will be presented simultaneously in trigonometric and complex forms. Familiarity with both mathematical approaches to describing interference will be useful in reading this book. For the moment, the reader can just concentrate on either form of the equations if having both forms at hand seems confusing.

Consider two wave groups of the same wave length that are capable of interfering with one another (that is, they are identically polarized and "coherent") and that are propagating along axes that are defined by vectors α_1 and α_2. The geometry is pictured in Figure 2.7. The magnitudes of the electric vectors of the two waves are

$$
\begin{aligned}
E_1 &= A_1 \cos(\omega t - \mathbf{k}_1 \cdot \mathbf{r} + \phi_{01}) = \mathrm{Re}(U_1 e^{i\omega t}) \\
E_2 &= A_1 \cos(\omega t - \mathbf{k}_2 \cdot \mathbf{r} + \phi_{02}) = \mathrm{Re}(U_2 e^{i\omega t})
\end{aligned}
\tag{2.66}
$$

where ϕ_{01} and ϕ_{02} are initial phases at $t = 0$ and $\mathbf{r} = 0$; $\mathbf{k}_1$ and $\mathbf{k}_2$ are the propagation vectors that depend on α_1 and α_2. Define two phase functions

$$
\begin{aligned}
\phi_1(\mathbf{r}) &= \mathbf{k}_1 \cdot \mathbf{r} - \phi_{01} \\
\phi_2(\mathbf{r}) &= \mathbf{k}_2 \cdot \mathbf{r} - \phi_{02}
\end{aligned}
\tag{2.67}
$$

At a particular point in space given by position vector $\mathbf{r}$, the sum E_s of the two waves is found by simple superposition.

$$
\begin{aligned}
E_s &= E_1 + E_2 \\
&= A_1 \cos(\omega t - \phi_1) + A_2 \cos(\omega t - \phi_2) \\
&= \tfrac{1}{2} U_1 e^{i\omega t} + \tfrac{1}{2} U_1^* e^{-i\omega t} + \tfrac{1}{2} U_2 e^{i\omega t} + \tfrac{1}{2} U_2^* e^{-i\omega t} \\
&= [A_1 \cos\phi_1 + A_2 \cos\phi_2] \cos\omega t \\
&\quad + [A_1 \sin\phi_1 + A_2 \sin\phi_2] \sin\omega t
\end{aligned}
\tag{2.68}
$$

The irradiance distribution of E_s is now calculated, using trigonometric notation for illustration and convenience of physical interpretation. Recall

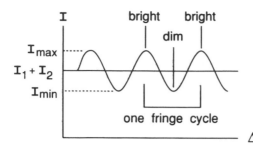

Figure 2.8. Irradiance distribution showing interference fringe formation.

that $\langle\ \rangle$ designates time average, used to get rid of the undetectable optical frequency oscillation. After some algebra, the result is

$$I_s = 2A_s^2(\mathbf{r})\langle\cos^2(\omega t - \phi_s(\mathbf{r}))\rangle \tag{2.69}$$

where

$$A_s^2(r) = A_1^2 + A_2^2 + 2A_1A_2\cos[(\mathbf{k}_1 - \mathbf{k}_2)\cdot\mathbf{r} - (\phi_{02} - \phi_{01})]$$

and

$$\phi_s(r) = \arctan\frac{A_1\sin(\mathbf{k}_1\cdot\mathbf{r} - \phi_{01}) + A_2\sin(\mathbf{k}_2\cdot\mathbf{r} - \phi_{02})}{A_1\cos(\mathbf{k}_1\cdot\mathbf{r} - \phi_{02}) + A_2\cos(\mathbf{k}_2\cdot\mathbf{r} - \phi_{02})}$$

Remember that the time average of the $\cos^2$ function is just $\frac{1}{2}$, so

$$\begin{aligned} I_s &= A_s^2(\mathbf{r}) \\ &= A_1^2 + A_2^2 + 2A_1A_2\cos[(\mathbf{k}_2 - \mathbf{k}_1)\cdot\mathbf{r} - (\phi_{02} - \phi_{01})] \end{aligned} \tag{2.70}$$

The equation for I_s means more if it is noted that A_1^2 and A_2^2 are just the irradiances I_1 and I_2 of the incoming beams. Making this substitution in equation 2.70 gives the final result,

$$I_s(r) = I_1 + I_2 + 2\sqrt{I_1I_2}\cos[(\mathbf{k}_2 - \mathbf{k}_1)\cdot\mathbf{r} - \phi_0] \tag{2.71}$$

where

$$\phi_0 = \phi_{02} - \phi_{01}$$

An important point is that the irradiance of the sum is not simply the sum of I_1 and I_2; it contains an interference term dependent on the incidence angles, wavelength, and initial phases as well as on the position in space $\mathbf{r}$ where irradiance is being observed. Because irradiance varies with location, the existence of interference bands can be expected. That they do indeed exist is easily demonstrated.

Let Δ replace the argument $[(\mathbf{k}_2 - \mathbf{k}_1)\cdot\mathbf{r} - \phi_0]$ in equation 2.71, which becomes

$$I_s = I_1 + I_2 + 2\sqrt{I_1I_2}\cos\Delta \tag{2.72}$$

Figure 2.8 is a qualitative graph of I_s as a function of Δ. This result indicates that the resultant irradiance varies sinusoidally in space; the maxima and minima are visible as interference bands or fringes. The spacing and location

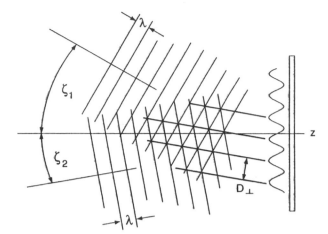

Figure 2.9. Formation of
interference bands as two
beams having plane
wavefronts meet
obliquely.

of the fringes are not yet determined. The minimum irradiance is not always
necessarily zero, and the "visibility" or "contrast" of the fringes is open to
question.

The question about the ease of seeing the interference bands can be
answered in part by introducing a parameter known as "fringe visibility,"
which is much used in interferometry.

$$\text{Fringe visibility} = V = \frac{I_{\max} - I_{\min}}{I_{\max} + I_{\min}} = \frac{\text{irradiance variation}}{2 \,(\text{intensity average})} \qquad (2.73)$$

For the interference of two beams, the maximum and minimum intensities
are extracted from equation 2.72 to give

$$V = \frac{(I_1 + I_2 + 2\sqrt{I_1 I_2}) - (I_1 + I_2 - 2\sqrt{I_1 I_2})}{(I_1 + I_2 + 2\sqrt{I_1 I_2}) + (I_1 + I_2 - 2\sqrt{I_1 I_2})} = \frac{2\sqrt{I_1 I_2}}{I_1 + I_2} \qquad (2.74)$$

If the two incident beams are of equal irradiance, then

$$V = \frac{4I}{4I} = 1$$

which means that the irradiance goes from 0 to $4I$ with an average of $2I$.
This is an ideal case with maximum visibility. The ideal is approached by
some forms of interferometry, including photoelasticity.

The next step is to determine the spacing of the interference bands and
their locations in space. After some mathematical manipulation, which is not
especially profound, the result is geometrically very simple. Figure 2.9 shows
in cross section how the two incident waves combine to form a stationary
system of parallel interference bands in space. If the two beams of light fall
on a screen placed as shown in this figure, the irradiance distribution is the

sinusoidal one in Figure 2.8. As a technical note, observe that here is a good way of producing dense sinusoidal grills for moire or spatial filtering investigations. Also, the fringe formation process pictured here exactly corresponds to the creation of moire fringes by superimposing two gratings (see Chapter 7). The lines of the gratings are analogous to the wave-fronts.

The perpendicular distance between adjacent interference fringes is calculated from Figure 2.9 to be

$$D_{\perp} = \frac{\lambda}{2 \sin \dfrac{\psi}{2}} \tag{2.75}$$

where $\psi = \zeta_1 + \zeta_2$ is the angle between the propagation axes of the two beams of light.

Equation 2.75 is useful in calculating the resolution requirements for photographic media that might be used to record the interference pattern, as well as for finding out just how microscopic the intelligence on the film is. For a He–Ne laser, $\lambda = 633$ nm. Take $\psi = 90°$, which is a limiting value. With these parameters,

$$\text{fringe frequency} = \frac{1}{\text{spacing}} = \frac{2\left(\sin \dfrac{\psi}{2}\right)}{\lambda}$$

$$= \frac{2 \sin 45}{633 \times 10^{-6} \text{ mm}}$$

$$\cong 2200 \text{ lines/mm} \tag{2.76}$$

This resolution is not within the capability of most photographic emulsions, and it is orders of magnitude higher than can be attained with electronic media and xerography.

The resolution requirement is not as severe if the angle between the beams is reduced. Even so, the resolution capability of films used in holography and similar applications should be somewhere between 500 lines/mm (corresponding to $\psi = 12°$) and 2000 lines/mm.

Another interesting and useful case is when the angle between the two beams is near either zero or 180°. This problem reduces to the case of two collinear waves in the preceding section. A standing wave is set up, and its amplitude depends on the phase difference. The fringe spatial frequency is $2/\lambda$; that is, the fringes are one-half wavelength apart. More important, the fringes are perpendicular to the wave axes so as to lie in the plane of the screen or film. This situation is used for making reflection holograms of the sort seen on magazine covers, credit cards, and jewelry.

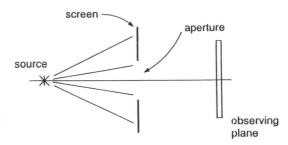

Figure 2.10. The basic
diffraction problem.

2.10 Concept of diffraction at an aperture

Two-beam interference and diffraction are the twin pillars of interferometry. Sufficient detail about interference has been presented so that some interferometric methods can now be understood. A full treatment of the diffraction problem is put off until Chapter 10, as it is not prerequisite to learning about the basic techniques of photoelasticity and moire. A short presentation of the concepts involved is, however, useful at this point.

The basic diffraction problem is pictured in Figure 2.10 where an aperture, such as a simple hole in an opaque screen, is illuminated. Required is a prediction of the nature of the illumination on the opposite side of the screen. The aperture might contain an optical signal, such as a transparency, or an optical element, such as a lens. Obviously, this problem encompasses almost all optical processing systems, whether a camera, a telescope, a flashlight, or the eye.

The optical field downstream from the aperture depends to a degree on whether the incident radiation is incoherent, as in typical imaging systems, or coherent, as in interferometric devices. For our purposes, the illumination is taken as coherent.

The solution of this difficult boundary value problem is conceptually very meaningful. The amplitude at some distance from the aperture turns out to be proportional to the Fourier transform of the spatial aperture function. That is, the light distribution indicates the spatial frequency content of whatever is in the aperture. This effect is easy to see if an experiment is conducted with an arrangement such as that shown in Figure 2.11. In this experiment, the aperture contains some sort of grating having a specific spatial frequency expressed as a given number of lines per millimeter. A lens is also placed in the aperture for convenience of observation. In the back focal plane of the lens will appear a series of bright dots, one dot on-axis and the rest symmetrically displaced along a line through the central dot. The focal plane is the Fourier transform plane. Distance in this plane is indicative of spatial frequency (the reciprocal of the line spacings) in the aperture plane. Thus, the degree to which the bright dots are off-axis indicates the spatial frequency of the input transparency.

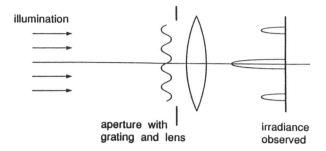

illumination

aperture with
grating and lens

irradiance
observed

Figure 2.11. Diffraction to create Fourier transform of a grating transparency.

In general, diffraction is a Fourier spectrum analyzer for spatial signals. The implications and applications of this fact are many, as will become apparent presently.

2.11 The generic interferometer

Interferometry is a method whereby the difference between the distances traveled by two waves is measured through interference. A conceptual model of this process is pictured in Figure 2.12.

Because interference between two waves is to be used, it is necessary that they be coherent, meaning that they are able to interfere with one another. The conditions for coherence imply, in the practical sense, that the two waves have come simultaneously from the same source. The first component in the optical system beyond the source is, therefore, some sort of beam divider. This device, called a beam splitter in optical interferometry, can divide the amplitude of a single wave or wave group, or it can divide the wavefront of a broad group of waves. The divider starts the two related waves along their separate paths. The sketch suggests that the two paths differ in length. One of them is often considered to be a reference path, and it can even have a known length, although this is not necessary. A path can involve a wave being sent along an optic fiber or through a volume of refractive material. If the radiation is microwave, then it can be a waveguide or free space. It might just be a system of mirrors and lenses for optical energy.

Eventually the two beams arrive at the ends of their paths. Because of the difference in the distance traveled, the waves are no longer in phase. The phase difference is a measure of the path length difference (PLD), and comparison of the two waves should yield this quantity. The trouble is that detectors, whether photocells, voltmeters, photographic film, microwave detectors, or the eye, are not sensitive to absolute phase. They are all "square-law detectors," which respond to the square of amplitude or "intensity." It is necessary, therefore, to convert the phase difference between the two beams to amplitude information. This task is accomplished by causing the two waves to interfere. To facilitate interference, the waves must be

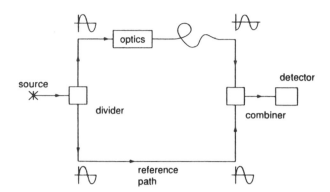

Figure 2.12. The generic interferometry model.

combined or mixed by a device that reverses the action of the beam divider. After mixing, the resultant wave, whose amplitude depends on the phase information, is sent to the detector. The detector indicates whether the two waves are in-phase, out-of-phase, or somewhere between these extremes.

An important fact is that absolute phase relationships are lost. One cannot ascertain by a single interferometry experiment whether two waves are out-of-phase by one wavelength or by many wavelengths. If absolute path length difference is required, then additional measures must be taken.

References

Born, M., and Wolf, E. (1975). *Principles of Optics*, 5th ed. Oxford: Pergamon Press.

Coker, E. G., and Filon, M. A. (1957). *A Treatise on Photoelasticity*. Cambridge University Press.

Garbuny, M. (1965). *Optical Physics*. New York: Academic Press.

Post, D. (1970). Optical analysis of photoelastic polariscopes. *Experimental Mechanics*, 10, 1: 15–23.

Shurcliff, W. A. (1962). *Polarized Light*. Cambridge, MA: Harvard University Press.

3

Classical interferometry

Some examples of so-called classical interferometric techniques are described in this brief chapter. Also included is a discussion of laser Doppler interferometry, which is fundamentally different from the other methods discussed in this book, but which has matured into an extremely useful approach for dynamic measurement.

3.1 Newton's rings

One of the oldest and most easily observed of interferometric phenomena is the formation of interference fringes in thin films. Apparently, they were first described scientifically by Boyle and Hooke, but they are named after Newton because he first analyzed their properties. This type of interference is responsible for the colors observed in an oil slick. It causes troublesome spurious fringes when one photoplate is contact copied onto another and when glass cover plates are used to protect your favorite 35-mm slide. On the other hand, Newton's rings provide an easy way of checking for full contact between two surfaces, as when an optical flat is used to check the flatness of a finely lapped surface. A similar process accounts for the functioning of dielectric interference filters that are used to isolate a narrow band of wavelengths from a beam of light. In fracture mechanics research, Newton's rings have been used for measuring the opening of cracks in transparent materials.

Analysis of the formation of Newton's rings in the general case (Born and Wolf 1975; Tolansky 1973) is not as simple as might be thought. Only the most basic case is discussed here, as the purpose is primarily one of example.

This type of interferometry is classed as amplitude division because each wave train of light is divided into two parts that are subsequently recombined. The light in one part of the field is not caused to interfere with the light in another part of the field.

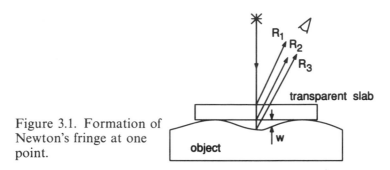

Figure 3.1. Formation of
Newton's fringe at one
point.

Figure 3.1 shows the basic optical setup for analyzing the formation
of Newton's rings at one point with near-normal incidence and near-
normal observation of the interference. The ray paths are separated for
clarity.

A slab of transparent material is placed in close contact with a surface that
is at least partially reflective. Some of the incident light is reflected from the
first surface of the transparent slab to form wave R_1, some is reflected from
the second surface to form R_2, and a portion travels on through the slab and
the space between the two objects to be reflected or scattered from the object
surface to form wave R_3. If the object surface is only gently convoluted, it
will reflect the incident light along an axis close to the line of R_2. In practice,
this reflected wave might be slightly scattered so as to form a cone that
encompasses the other reflected waves. R_2 and R_3 are in condition to interfere.
The product of this interference depends on the path length difference, which
in turn depends on the distance w between the two plates at the location
being examined, the inclinations of the rays, and the relative refractive index
n of the medium that fills the space. If the path lengths differ by $\lambda/2$, then the
interference is destructive and the patch being examined will appear dark.

Now, suppose that a great many such points are examined simultaneously,
as suggested by Figure 3.2. This arrangement is more complicated in that a
partial mirror is used to facilitate both illumination and viewing along the
normal. A field lens is also employed so that the whole field can be viewed
or photographed at once. The eye or the camera is placed at the focus of the
field lens. A color filter is introduced to limit the span of wavelengths. A point
source and a collimating lens create the needed collimated illumination.

What was true for the single point that was interrogated, as in Figure 3.1,
is now true for every point in the entire field. If the interference between the
reflected waves R_2 and R_3 is constructive, then an illuminated image of that
point is formed by the eye. If the interference is destructive, then that point
is dark in the image. Superimposed on the whole image is a system of
interference fringes that are loci of constant path-length difference. A pattern
of fringes that is typical of those that might be produced by such an
experiment is shown in Figure 3.3.

UNIVERSITY
OF SHEFFIELD
LIBRARY D

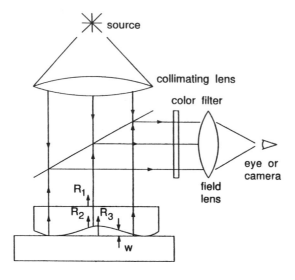

Figure 3.2. Whole-field
formation of Newton's rings.

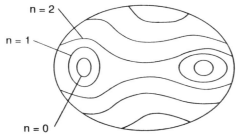

Figure 3.3. Typical form of
Newton's rings.

A zero-order fringe, indicating no path length difference, occurs at points
of contact between the two surfaces. One would expect the zero fringe to
appear light; however, it is dark because of a phase change that occurs at
reflection from the object surface if the object is made of a dielectric such as
glass. That is to say, the appearance of the zero-order fringe is material-
dependent. In any case, one starts with the zero fringe and numbers the rest
of the fringes consecutively, recognizing that the gradient can be positive or
negative. Some prior knowledge of the general shapes of the surfaces is helpful
in numbering the fringes, which is not a trivial matter. Once the fringe order
is established for any particular point in the field, then the path length
difference is known for that point, and the separation is given by

$$w = \frac{N\lambda}{2n} \tag{3.1}$$

where w is the gap to be measured
             N is the fringe order
             λ is the wavelength of light·
             n is the refractive index in the gap

The factor of 2 is necessary because the light reflected from the object travels twice through the gap.

If the incident light and/or the viewing axis are not normal to the object, then the equation relating gap to fringe order is modified to account for the angles.

Note that several other waves will be gathered by the viewing system. These waves, caused by the partial reflections that occur whenever a wave meets a surface, will contaminate the image and cause a reduction of fringe contrast.

3.2 Young's fringes

In 1802, Thomas Young described an experiment that had great impact on the physical sciences (Born and Wolf 1975; Tolansky 1973). He illuminated a screen containing two small apertures with light from a small source, and he found a pattern of interference fringes in the space beyond the screen. Among other implications, this experiment gave evidence of the wave nature of light. The fringes created by similar experiments are called Young's fringes. They are important in many areas of interferometry for several reasons.

Figure 3.4 shows an experimental setup that reproduces Young's experiment. For convenience, only the symmetrical case is considered; the light source is at O on the optical axis, and the two apertures are equidistant from the axis and separated by distance p. The light source can be an ordinary small-filament lamp, although better results are obtained with a concentrated-arc source. A color filter to limit the wavelength band is useful. With incandescent or arc sources, the aperture pinholes should be small, and the viewing screen should be in darkness, because the fringes will be dim. An expanded laser beam gives the best fringes, even though they are contaminated by laser speckle. One wonders what the early, brilliant researchers such as Young could have done had they had lasers.

Figure 3.5 is a photograph of Young's fringes, which were created by illuminating a pair of apertures with laser light.

Notice that this interferometer operates on the basis of division of wavefront. Light waves from widely separated parts of the illuminated field must interfere with one another. This requirement for coherency over the extent of the cone of illumination is more severe than is the case for interferometry based on amplitude division. Still, a small source and small pinholes give sufficient coherence for fringes to be observed. The long coherence length of the laser is not really needed.

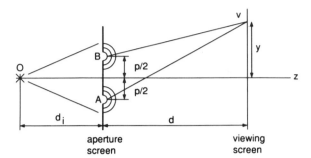

Figure 3.4. Apparatus to observe Young's fringes.

Figure 3.5. Example of Young's fringes formed by illuminating two pinholes with laser light.

There are several ways to describe the formation of fringes in Young's experiment. It is a prime demonstration of general diffraction theory, as discussed in Chapter 10. A more elementary approach, which is sufficient for the time being, is to merely calculate the path length difference between the waves that come from each aperture and arrive at a general point *V* in the viewing screen. This is basically the Huygens' construction, which gets around the diffraction problem by assuming that an illuminated small aperture emits a spherical wavefront. Otherwise there is no reason to believe that light traveling from the source *O* to the apertures at *A* and *B* will be bent and scattered so as to illuminate the entire screen. Such an approach explains in a heuristic way many of the phenomena that are observed in experiments involving diffraction.

The path length difference, PLD, is calculated using the Pythagorean theorem and the distances identified in Figure 3.4. Remember that only the axially symmetric case is considered, so the distances from the source to the apertures are equal.

$$PLD = AV - BV$$

$$= \left[d^2 + \left(y + \frac{p}{2} \right)^2 \right]^{1/2} - \left[d^2 + \left(y - \frac{p}{2} \right)^2 \right]^{1/2} \qquad (3.2)$$

$$= d \left\{ 1 + \left[\frac{y + p/2}{d} \right]^2 \right\}^{1/2} - d \left\{ 1 + \left[\frac{y - p/2}{d} \right]^2 \right\}^{1/2} \qquad (3.3)$$

The preceding equation can be written in terms of angular deviations; but a simpler alternative, which is sufficient for current purposes, is to impose an approximation. In the typical experiment, y and p are much smaller than distance d, so $(y + p/2)/d \ll 1$. Keeping only the first terms of the series expansion gives $(1 \pm \epsilon^2)^{1/2} \cong 1 \pm \frac{1}{2}\epsilon^2$. Utilization of this common approximation converts equation 3.3 to the simple form

$$PLD = d \left\{ 1 + \frac{1}{2} \left[\frac{y + \dfrac{p}{2}}{d} \right]^2 \right\} - d \left\{ 1 + \frac{1}{2} \left[\frac{y - \dfrac{p}{2}}{d} \right]^2 \right\} \qquad (3.4)$$

$$= \frac{yp}{d} \qquad (3.5)$$

Constructive interference occurs whenever the path length difference equals the wavelength λ. Equating this value to the *PLD* in equation 3.4 and inverting the result indicates that there is a series of bright fringes wherever

$$y = \frac{N\lambda d}{p} \qquad \text{where } N = 0, 1, 2, \ldots \qquad (3.6)$$

Dark fringes of destructive interference occur between the bright fringes. There is no dependence on the out-of-plane coordinate in this two-dimensional analysis. In fact, if the x-coordinate of the observing point were included, it would drop out for small values. The end result is that the interference pattern is approximated by a system of parallel interference fringes of equal brightness that have spacing $\lambda d/p$.

This simple analysis is supported by experiment and by more rigorous diffraction theory, except that the brightness of the pattern is found to fall off rapidly away from the optic axis. The reason is that the Huygens' construction ignores obliquity factors.

One example of the direct application of Young's interference in the measurement of deformation is afforded by the "interferometric strain gage" (Sharpe 1968). In this technique, tiny indentations are created in the specimen using a prismatic diamond indentor. When illuminated by coherent light, the indentations seem to act as pinhole sources, except that the diffraction pattern is biased by reflection effects caused by the planar sides of the indents in a manner similar to that observed with a blazed diffraction grating. An interference pattern is observed in the far field. Because the strain of the specimen affects the spacing of the indentations, it is evident as a change of spacing in the interference fringes. This approach is amenable to electronic sensing of the fringes and subsequent computerized data reduction. It is quite nondestructive and noncontacting, and it has been used at very high temperatures.

The primary use of Young's fringes in interferometric measurement of deformation is in interrogation of laser speckle photographs. This subject is discussed in the chapters on speckle methods.

3.3 Michelson interferometry

The Michelson interferometer is an instrument that, in one form or another, has yielded deep and lasting benefits to modern physics, optics, and metrology (Born and Wolf 1975; Tolansky 1973). With it, Michelson established the experimental basis of relativity, he defined the meter in terms of the wavelength of light, he provided the basis for many methods of optical testing, and so on. Several of the methods of optical measurement discussed in this book have their origins in the Michelson interferometer; these include, for example, speckle correlation interferometry, laser Doppler velocimetry, and holographic interferometry. Understanding the basics of the Michelson instrument facilitates learning about its modern offspring. Setting up a Michelson interferometer is often the easiest way to assess the stability of an optical bench or to determine the coherence capability of a laser or other light source.

The Michelson interferometer is an instrument that employs division of amplitude. The basic configuration is shown in Figure 3.6. In this version, collimated radiation is projected upon a precision beam splitter (BS). A portion is directed to the flat, fixed mirror M_1, where it is reflected back through the splitter to the observation screen. The remaining light passes to a mirror or test object M_2, where it is reflected back to the splitter and redirected to the screen. Of course, each interaction with the beam splitter causes another amplitude division; only the important divisions and reflections are shown. Also, for clarity, the various beams are shown slightly separated in the sketch.

Now, if the system is adjusted to be perfectly square, two separate but parallel plane wavefronts will arrive at the screen. The path length difference will be

$$PLD = (OA + AO + OC) - (OB + BO + OC) = 2(OA - OB) \qquad (3.7)$$

As usual, if the PLD is a multiple of the wavelength of light, then the screen will be light. Consider what happens when the test object represented by mirror M_2 translates axially. The irradiance at the screen will be alternately

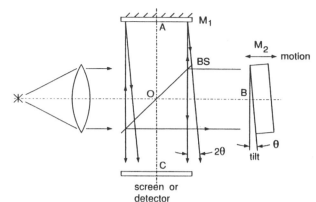

Figure 3.6. Classic configuration of the Michelson interferometer.

bright and dark. The same will happen if a test cell is placed on the arm *OB* and the refractive index of the contents of the cell is increased or decreased.

If instead of translating, the object M_2 is tilted by angle θ, the reflected beam will undergo a deviation of 2θ relative to the undisturbed beam, as shown exaggerated in the figure. The two beams will interfere to produce a system of parallel interference fringes according to the scheme discussed in section 2.9. The fringe spacing can be interpreted to give the tilt of M_2. If the tilted mirror is given some translation, then the result will be to cause the tilt fringes to move across the screen. If a point detector is placed at the screen location, then an oscillating signal proportional to the changing irradiance will be produced.

If the test object is not a simple flat mirror, then, clearly, the interference fringes will not be parallel. Instead, they will be loci of points of equal path difference. That is, they will form a contour map of the surface. If both mirrors are not flat, then the contour map will be a function of the difference between the shapes of the mirrors. The similarity to Newton's fringe methods is recognizable. This approach, in refined form, is the basis of several techniques for establishing precisely the figure or shape of optical components such as lenses and mirrors.

Much more can be said about variations and applications of Michelson interferometry; these topics can be found in the voluminous literature on the subject. A useful summary of the most important developments is provided by Tolansky (1973). The basic concepts and applications just described are sufficient for this text.

3.4 Laser Doppler interferometry

The Doppler effect provides the basis of a form of interferometry that is unique in that the interference fringes, or rather their motions, indicate the change in optical frequency of the light that is emitted by a moving source,

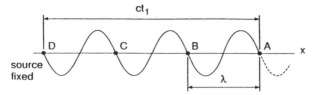

Figure 3.7. Space diagram of a harmonic wave with fixed source.

reflected from a moving test object, or received by a moving observer. This frequency shift is best observed by means of a Michelson interferometer. Laser Doppler interferometry is becoming widely used in investigating the dynamic response of solid materials, fluids, and structures; and sophisticated instruments based on this principle are now on the market. It is remote and noncontacting, requiring only optical access to the specimen. Its velocity response and frequency response are exceptional.

The following is a basic discussion of the laser Doppler technique. It is derived from the excellent development by Haskell (1971), and it is used with permission. Relativistic effects that lead to different results for moving source and moving observer are not considered because they are not important in most practical applications of interferometry.

The Doppler effect is the increase or decrease in the measured frequency of a wave depending on whether the source (or observer) is moving toward or away from the observer (or source). The effect has its origin in the fact that the velocity of the wave is independent of the velocity of the source.

Before learning how to detect and utilize the Doppler effect, the relationship between motion and frequency shift must be understood. Figure 3.7 shows a space diagram that provides a basis for understanding.

In the figure the source is at rest. At time $t = 0$, wave point A is emitted by the source. After a time t_1, wave point A has traveled the distance ct_1, and wave point D is just then being emitted by the source. Figure 3.7 illustrates the case in which only three cycles are emitted in the time t_1. In general, if n is the number of cycles emitted in the time t_1, then from the space diagram it follows that

$$n\lambda = ct_1$$

from which,

$$n = \frac{ct_1}{\lambda} = vt_1 \tag{3.8}$$

where $v = c/\lambda$ is the frequency of the wave.

Now suppose that the source is moving to the right with a constant velocity v. Then A will still leave the source at $t = 0$, B will still leave the source at $t = \frac{1}{3}t_1$, C will still leave the source at time $t = \frac{2}{3}t_1$, and D will just be leaving

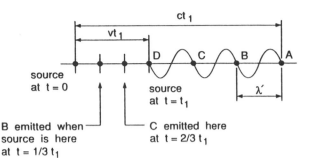

Figure 3.8. Space diagram of a harmonic wave with moving source.

B emitted when source is here at $t = 1/3\, t_1$

C emitted here at $t = 2/3\, t_1$

the source at time $t = t_1$. Of course, all this time the source has been moving to the right, so at time $t = t_1$ the space diagram will look like Figure 3.8.

This diagram shows that the wavelength λ' of the wave has decreased because the entire three cycles must fit into a shorter space. The frequency $v' = c/\lambda'$ as measured by a fixed observer would therefore increase. Note that the number of cycles emitted during the time t_1 is the same as when the source was not moving (three cycles, in this case). In general, if n cycles are emitted in time t_1, then from Figure 3.8 one can write,

$$n\lambda' = ct_1 - vt_1 \tag{3.9}$$

from which,

$$n = \frac{(c - v)t_1}{\lambda'}$$

$$= \left(\frac{v'}{c}\right)(c - v)t_1$$

$$= v'\left(1 - \frac{v}{c}\right)t_1 \tag{3.10}$$

Because n is the same whether or not the source is moving, one can equate equations 3.8 and 3.10 to obtain

$$v'\left(1 - \frac{v}{c}\right)t_1 = vt_1$$

from which,

$$v' = \frac{v}{1 - v/c} \tag{3.11}$$

Equation 3.11 gives the new frequency v' as would be measured by a fixed observer when the source is moving toward the observer with the velocity v.

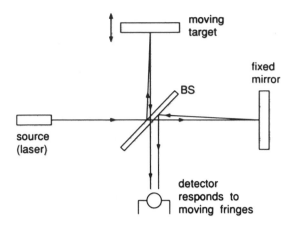

Figure 3.9. Interferometer for measuring Doppler shift of frequency.

In this case, $v' > v$. If the source is moving away from the observer, the v is replaced by $-v$ in equation 3.11 to give

$$v' = \frac{v}{1 + v/c} \qquad (3.12)$$

In general, the frequency observed by a moving observer who is intercepting waves emanating from a fixed source is different from that calculated for a moving source and a fixed observer. The explanation for this paradox is found in relativity theory. The difference is not apparent for sources or observers that are moving at speeds much less than the speed of light, so the moving observer problem will not be examined here.

Clearly, if the new frequency can be measured, perhaps by comparison with the at-rest frequency, then the velocity of the source or observer can be inferred. One way to determine the frequency shift is to use the Michelson interferometer. Consider the basic interferometer shown in Figure 3.9. Rather than using broad collimated beams, only a single wave train or a small pencil of waves, as from a laser, is employed. One of the target mirrors or just a scattering surface in the interferometer is attached to the object whose velocity is to be determined. Because the target object is moving, the source appears to be moving toward or away from the observer. This source motion causes a Doppler shift in the frequency of the light beam traveling in that arm of the interferometer. The two light beams, which add at the output of the interferometer, have two different frequencies v and v'. The sum of these two output fields at a given point can then be written as

$$E_s = A_1 \cos(2\pi vt + \phi_1) + A_2 \cos(2\pi v't + \phi_2)$$

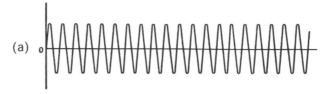

(a)

(b)

Figure 3.10. Laser Doppler output for two types of target motion: (a) target moving with constant velocity; (b) vibrating target (bias frequency added).

The irradiance will be

$$I = |E_s|^2$$

$$I = A_1^2 \cos^2(2\pi vt + \phi_1) + A_2^2 \cos^2(\pi v't + \phi_2)$$

$$+ 2A_1 A_2 \cos(2\pi vt + \phi_1) \cos(2\pi v't + \phi_2) \qquad (3.13)$$

Use the identity $\cos A \cos B = \frac{1}{2}[\cos(A + B) + \cos(A - B)]$. Then equation 3.13 becomes

$$I = A_1^2 \cos^2(2\pi vt + \phi_1) + A_2^2 \cos^2(2\pi v't + \phi_2)$$

$$+ A_1 A_2 \cos[2\pi(v' + v)t + \phi] + A_1 A_2 \cos[2\pi(v' - v)t + \phi] \qquad (3.14)$$

where $\phi = \phi_2 - \phi_1$.

The first three terms in equation 3.14 represent an irradiance that oscillates at frequencies equal to or greater than v or v'. If the radiation used is light, then these frequencies are too large to be detected. Only a time-average irradiance, which is essentially a constant, would be produced by these expressions. The last term, however, gives an output that is proportional to the cosine of the difference between the frequencies v and v'. This is the so-called beat frequency. For ordinary velocities of the moving target, the original frequency and the shifted frequency are close enough together so that the difference frequency can be observed with ordinary detectors. In particular, suppose that the target has some constant velocity, and its motion is observed with this Doppler version of the Michelson interferometer. A photodiode detector measures the irradiance at the output. If the detector can respond to frequencies $\Delta v = v' - v$, then the output as a function of time is another harmonic wave, as in Figure 3.10a.

The *frequency* of this output is the indicator of target velocity. This point should not be forgotten because there is always the temptation

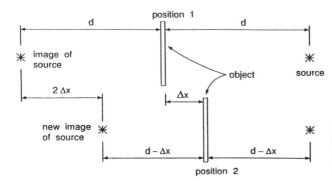

Figure 3.11. Relation between apparent source motion and object motion for Doppler interferometry.

to think that the amplitude of the output is proportional to the velocity. If the target is itself vibrating, then the interferometer detector gives an output that is frequency-modulated, as in Figure 3.10b. At any instant, the frequency is proportional to the instantaneous velocity of the target.

A question that remains to be answered has to do with the relationship between the apparent source velocity as indicated by the Doppler shift and the actual velocity of the target. We expect that they are not the same because the light traverses the object arm of the Michelson interferometer twice. For convenience, the target has been temporarily viewed as a moving source, when actually it is a moving object that reflects the light from a stationary source. Figure 3.11 guides one's thinking by showing the virtual images as created by a mirror for two positions of the mirror.

Suppose the target object moves a distance Δx. We see that the image of the source moves a distance $2\,\Delta x$. If the velocity of the object is $v_o = \Delta x/\Delta t$, then the velocity of the image of the source is

$$v = \frac{2\,\Delta x}{\Delta t} = 2v_o \tag{3.15}$$

The image is seen to move at twice the velocity of the mirror. This image velocity is what is detected by the Michelson Doppler interferometer.

An alternative approach to understanding the Michelson interferometer, as it is used to measure velocity by detecting the Doppler frequency shift, is to calculate the rate at which traveling interference fringes cross the detector. Recall that if the mirror is stationary, a set of stationary fringes appear at the output. It is helpful to think of the mirror as tilted slightly so that a pattern of parallel fringes is created. Moving the mirror an amount Δx causes a path length difference between the two arms of the interferometer of an amount $2\,\Delta x$. This *PLD* causes a phase shift difference of $(4\pi/\lambda)\,\Delta x$. When this phase difference is equal to 2π; that is, when $\Delta x = \lambda/2$, the output fringes shift by one fringe. The quantity $\Delta x/(\lambda/2)$ is the number of fringes that move across a detector when the mirror moves a distance Δx. If this motion occurs in a time Δt, then the output of the detector will vary sinusoidally at a frequency given by

$$\frac{\Delta x/(\lambda/2)}{\Delta t} = \frac{2\,\Delta x}{\lambda\,\Delta t} = \left(\frac{v}{c}\right)2v_{\text{o}} = \frac{vv}{c} \tag{3.16}$$

Comparison with results obtained earlier shows that this result is just the Doppler frequency shift $v' - v$. Remember that v is twice the object velocity in both cases.

One problem with the basic optical Doppler technique as just described is that it is not possible to distinguish the sign of the target velocity. A velocity away from the detector gives the same frequency shift in the output signal as a velocity toward the detector. This difficulty, and artifacts related to the processing of low-frequency signals, is eliminated by introducing a bias frequency shift. The frequency shifts that are related to target velocity are then manifested as a modulation of the bias frequency signal, which is now acting simply as an FM carrier wave.

The bias frequency shift is created by introducing into one arm of the optical path a frequency-shifting device. A constant velocity of the reference mirror cannot be used for long, of course. One possibility is merely to vibrate the reference mirror in the Michelson interferometer by using an electro-magnetic exciter with, say, a square-wave input, but this method introduces a frequency modulation itself; and it is also difficult to obtain a large enough bias. A practical method is to place in the optical path a rotating diffraction grating. The angular deviation of the diffracted portion of the beam coupled with the rotation produces a constant frequency shift. The preferred approach is to use an electro-optic device such as a Bragg cell to create the bias frequency deviation.

References

Born, M., and Wolf, E. (1975). *Principles of Optics*, 5th ed. Oxford: Pergamon Press.

Haskell, R. E. (1971). "Introduction to Coherent Optics." Unpublished manuscript, Oakland University, Rochester, MI.

Sharpe, W. N., Jr. (1968). The interferometric strain gage. *Experimental Mechanics*, 8, 4: 164–70.

Tolansky, S. (1973). *An Introduction to Interferometry*, 2nd ed. New York: Halsted Press.

PART II
Photoelasticity

4

Photoelasticity theory

One of the oldest and most useful forms of interferometric measurement for engineering purposes is photoelasticity, which involves the observation of fringe patterns for determination of stress-induced birefringence. It is important as a measurement technique. Further, it provides an instructive paradigm of applied interferometry. This chapter presents in some detail the fundamental theory of the photoelastic technique.

4.1 Photoelasticity as interferometry

For practical and instructional reasons it is important to recognize photoelasticity as a classic interferometric technique. The path length difference to be measured in the specimen depends on local direction-dependent variations in the refractive index; these variations are usually induced by stress. The surface of the photoelastic model itself acts as the beam splitter because it divides the incident light into orthogonally polarized components. These components travel through the same thickness of material, but the path lengths differ because of the difference of refractive index. Thus, the components exhibit a relative phase difference when they exit the specimen. The phase difference is converted to amplitude information through interference as the two components are recombined at the downstream polarizer, called the analyzer. Because the beam splitting divides a single wave train or a small pencil of waves, photoelasticity is of the amplitude-division class of techniques. It is also a common path interferometer since the two orthogonally polarized waves follow identical geometric paths through the whole instrument. These facts, plus the fact that the path lengths differ by only 20 or so wavelengths, mean that the coherence requirements are not stringent, and ordinary light sources are suitable. Also, vibrations do not have much effect on common path interferometers, so they are easy to use in noisy environments.

4.2 Refraction

The simplest form of electromagnetic radiation was shown to be a harmonic plane wave, which is described by the equation

$$\mathbf{E} = \mathbf{A} \cos \frac{2\pi}{\lambda} (z - vt) \tag{4.1}$$

The speed of propagation along the z-axis is dependent only on the material properties contained in the wave equation

$$v = \left(\frac{1}{k\mu}\right)^{1/2} \tag{4.2}$$

It is convenient to use the speed of light in a vacuum as a reference velocity for expressing the speed of light in other materials. For this reason, the "absolute index of refraction" of a material is defined as

$$n_1 = \frac{c}{v_1} \tag{4.3}$$

where v_1 is the speed of light in the material
             c is the speed of light in a vacuum

Often, it is desirable to relate the speed of light in an object to the speed of light in the medium surrounding the object. For this purpose, a "relative index of refraction" is defined to be

$$n_{21} = \frac{n_2}{n_1} = \frac{v_1}{v_2} \tag{4.4}$$

In equation 4.4, the first subscript refers to the object and the second refers to the immersion medium. Many references do not differentiate clearly between relative and absolute indexes of refraction. Little harm is done in most instances, such as when the immersion medium is air.

The index of refraction of a material depends on wavelength in a complicated way. These relationships can be understood through the use of simple models of the interaction of radiation and material. One such model uses a damped harmonic oscillator to represent an electron bound to a motionless (at optical frequencies) nucleus. The electromagnetic radiation forces a vibration of the electron. The percentage of radiant energy not absorbed by the oscillator represents transmittance. It turns out that index of refraction can be computed in terms of the oscillator parameters. This subject is fascinating and it is a fertile area of research.

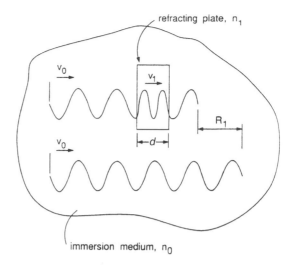

Figure 4.1. Absolute retardation by a refracting slab in an immersion medium.

4.3 Absolute retardation

Consider a ray of light passing through a plane parallel plate of thickness d and refractive index n_1 that is immersed in a medium whose index is n_0. The distance by which the refracted rays is retarded is easily calculated with the aid of Figure 4.1. By definition, the velocity of light in the refractive plate is $v_1 = c/n_1$. The time for the wave to traverse the plate is $t = d/v_1 = d/(c/n_1)$. During this same time period, an undisturbed wave will travel the distance $v_0 d/(c/n_1) = d_0$. The ray that travels a distance d through the plate thus lags behind an undisturbed ray by the amount

$$R_1 = d_0 - d = \left(\frac{v_0 d}{c/n_1}\right) - d$$

$$= \left(\frac{v_0 n_1}{c} - 1\right) d$$

Using now the definition of absolute index of refraction, one gets

$$R_1 = \left(\frac{n_1 - n_0}{n_0}\right) d \tag{4.5}$$

In terms of the relative refractive index, the result is

$$R_1 = (n_{10} - 1) d \tag{4.6}$$

R_1 is called the absolute retardation of the light by the refracting plate.

4.4 Birefringence

In certain materials, the refractive index varies with the directions of polarization and the propagation axis. Light polarized in one plane will

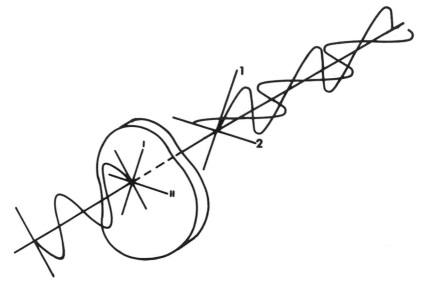

Figure 4.2. Effect of doubly refracting material on transmitted polarized light.

propagate at a speed that is different from the speed of a wave polarized in another plane. Photoelasticity depends entirely on the interaction of polarized light with these optically anisotropic materials.

Wave propagation in an anisotropic material is an area of study that can lead to some surprising results. A profound analysis is not needed for understanding photoelasticity; we can get by, for a time, with some general observations and qualitative knowledge of the phenomena involved.

Suppose one takes a light ray of some arbitrary polarization and passes it at normal incidence (the incidence angle is important) through a plate of optically anisotropic material. Careful examination of the light coming out of the plate will show that it always consists of two separate plane-polarized components. (Imagine the consternation of early investigators when faced with this behavior of light.) The electric vectors (in the planes of polarization) of the emerging components will be found to be mutually perpendicular. Furthermore, the components will be out of phase, indicating that they passed through the plate at different speeds. Figure 4.2 pictures this interaction of light and "doubly refracting" materials.

Experiments such as those just described lead to the following important generalizations (Jessop and Harris 1949):

1. A birefringent material divides the entering wave into orthogonal components. The inclinations of the planes of polarization of the emerging waves depend uniquely on the nature of the birefringence along the path of the wave through the plate. These directions of polarization are called the *principal directions*; the associated axes in the material are called the *principal axes of refractive index*.

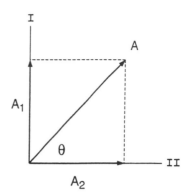

Figure 4.3. Resolution of
amplitude into components.

2. The amplitudes of the two components are found by simple resolution of the electric vector into components along the principal directions, assuming negligible reflection and absorption losses. Referring to Figure 4.3, where A is the amplitude of the light entering the plate and I and II are the principal directions, we see that the component A_1 has amplitude $A \sin \theta$, and so on.

3. The speeds of travel of the two components are different and can be described in terms of two indexes of refraction associated with the two principal directions. These values, called the *principal values of refractive index*, are designated as n_1 and n_2.

For most photoelastic and photoholelastic analyses, the optical behavior of anisotropic materials needs to be studied no further. Some additional comments and observations may prove illuminating and useful to those who must undertake three-dimensional analysis and absolute retardation experiments.

The conclusions drawn from experiments and summarized in the preceding list suggest that index of refraction is analogous to stress at a point. In fact, refractive index, like stress at a point, is a second-rank tensor quantity (Born and Wolf 1975; Coker and Filon 1957). Often, it is written in terms of the dielectric tensor. Now, if $\mathbf{E}$ is the electric vector and $\mathbf{D}$ is the so-called electric displacement for the material (depends on polarization of electrons in molecules), then an electromagnetic material law can be written in terms of components as

$$D_k = \epsilon_{kl} E_l \qquad (4.7)$$

As with other tensor parameters, principal values and principal axes can be defined and found. If $\epsilon_{11}, \epsilon_{22}, \epsilon_{33}$ are the principal values of the dielectric tensor, then the principal values of refractive index are related to them by an extension of equation 4.2 with an added simplification resulting from the fact that the magnetic permeability of a dielectric is close to unity,

$$n_1^2 = \epsilon_{11}, \qquad n_2^2 = \epsilon_{22}, \qquad n_3^2 = \epsilon_{33} \qquad (4.8)$$

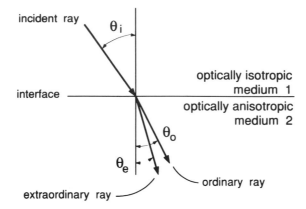

Figure 4.4 Effect of birefringent material on transmitted light with oblique incidence.

Although it is easy to use the principal values of the refractive index or dielectric tensor, the situation becomes more complicated when these quantities are transformed to a set of axes different from the principal axes. The physical meaning of the off-diagonal terms in the matrix becomes somewhat mysterious – they are analogous to shear stress or product of inertia. We have difficulty visualizing in simple terms what happens to a ray of light as it passes through a three-dimensional blob of material along some axis other than a principal axis.

The picture can be cleared somewhat by thinking about a simple experiment. Let a ray of light fall at *oblique incidence* on a birefringent plate. We expect the incident ray to be divided into two orthogonally polarized components, and this does happen. The peculiar thing is that the components travel through the plate along divergent paths, as is shown schematically in Figure 4.4. The two rays are referred to as "ordinary" (meaning *as expected* by early investigators) and "extraordinary" (meaning *unexpected*). Polarizing devices such as the Nicol prism operate by separating the two rays. Microscopes and other small-field instruments still use polarizers of this type, whereas dichroic polarizers such as those sold by the Polaroid Corporation are commonly used for large-field instruments.

Now, an experimenter who conducted a series of experiments such as the one just described would be led to make a three-dimensional plot of the refractive indexes for the ordinary and extraordinary rays for different angles of incidence. The result is an ellipsoidal representation of the refractive index tensor (Burger 1987; Jessop and Harris 1949). It is known as the "Fresnel ellipsoid." Figure 4.5 shows an example. It represents the refractive properties at a specific point in an anisotropic solid by an ellipsoid whose major axes correspond to the principal axes and principal values of refractive index at that point.

Finally, observe that if light passes through the body along some inclined axis *l* containing point A, then the body at A acts as a two-dimensional plate.

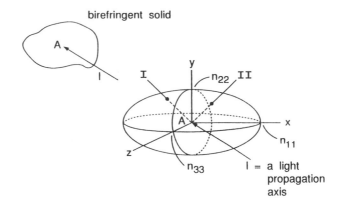

Figure 4.5. Refractive index ellipsoid for a point in a birefringent body.

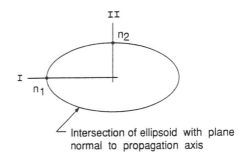

Figure 4.6. Refractive index ellipse for light incident along *l*-axis.

The refractive properties of the plate are given by the *intersection of the ellipsoid and the plane* normal to the axis *l*. Try to visualize this intersection: it will be an ellipse such as that shown in Figure 4.6. In general, the major and minor axes of this *ellipse*, shown as axes I and II in the figure, are different from the major axes of the *ellipsoid*. That is, the principal directions and principal values in the plane of the wavefront are different from *the* (absolute) principal values and principal axes for the material.

The light traveling along *l* will be divided at A into components polarized along the I and II axes. One can do this physically by sawing a slice out of the original body perpendicular to *l* and interpreting the birefringence in that slice according to the geometrical reasoning just described. This idea is refined and used in three-dimensional photoelasticity.

The whole picture is much simpler if the light is projected along one of the principal axes at a point in the body. Then, the principal axes in the plane of the wavefront correspond to two of the principal axes of the ellipsoid. It is worth noting that, if the indexes of refraction vary through the solid, then there is a unique Fresnel ellipsoid for every point.

A meaningful discussion of the mathematics and some physical interpretation for the three-dimensional situation are given by Born and Wolf (1975).

Note also that the human eye, which is not sensitive to phase or polarization, cannot detect that the light emerging from a birefringent plate has been affected in any way.

4.5 Relative retardation

The "relative retardation" or phase difference between the two components emerging from a birefringent plate can be computed with the aid of Figure 4.1. If n_1 and n_2 are the two principal indexes of refraction of the plate and n_0 is the index for the immersion fluid, then the components undergo the following absolute retardations,

$$R_1 = \left(\frac{n_1 - n_0}{n_0}\right)d$$
$$R_2 = \left(\frac{n_2 - n_0}{n_0}\right)d \tag{4.9}$$

The phase difference, expressed as a distance, is merely the difference between the absolute retardations. This *relative retardation* is

$$R = R_1 - R_2 = \left(\frac{n_1 - n_2}{n_0}\right)d \tag{4.10}$$

Often, R is conveniently expressed as a fraction, m, or multiple of wavelength,

$$m = \frac{R}{\lambda} = \left(\frac{n_1 - n_2}{n_0}\right)\frac{d}{\lambda} \tag{4.11}$$

The number m is called the "fringe order."

In terms of relative indexes of refraction, equation 4.10 becomes

$$R = (n_{10} - n_{20})d \tag{4.12}$$

4.6 Birefringence in materials: basic relations

Here we discuss some of the relationships among stress, strain, and birefringence for various materials (Coker and Filon 1967; Dally and Riley 1991; Frocht 1941; Wolf 1961).

In certain materials, the birefringence occurs naturally. Examples include calcite (Iceland spar, $CaO–CO_2$) and sodium nitrate ($NaNO_3$). These materials are used in the construction of birefringence-type polarizers such as the Nicol prism. In another large class of materials, the birefringence can be induced by stress or deformation. Examples are glass, many plastics, some elastomers, semiconductors, and certain fluids.

The relationships between birefringence and the composition and structure of materials are not completely understood, although some theoretical models have been proposed. Even though the theoretical basis is not yet well established, the birefringence in materials can be related through experiment to the stress and strain history of the material. To a first approximation, the correspondence is surprisingly simple.

Let σ_1 and σ_2 be principal stresses at a point in a two-dimensional plate of "momentarily linearly elastic" material (to be defined). Experiments show the following:

1. The principal axes of refraction correspond to the principal stress axes (same as principal strain axes for this case).
2. Each principal index of refraction is a linear function of the two stress components.

The absolute retardation can be expressed in terms of the stresses through the use of the *absolute photoelastic coefficients* C_1 and C_2, which may themselves be functions of stress, strain, time, and radiation wavelengths. The defining equations are, remembering that d = thickness of the plate,

$$R_1 = (C_1\sigma_1 + C_2\sigma_2)d$$
$$R_2 = (C_2\sigma_1 + C_1\sigma_2)d \tag{4.13}$$

The relative retardation is the difference between the absolute retardations of the two components,

$$R = R_1 - R_2 = (C_1 - C_2)(\sigma_1 - \sigma_2)d \tag{4.14}$$

Previously, we had

$$R = \left(\frac{n_1 - n_2}{n_0}\right)d \tag{4.15}$$

Comparing equations 4.14 and 4.15, we find it reasonable, but not necessary, to write

$$\frac{n_1}{n_0} = C_1(\sigma_1 - \sigma_2)$$

$$\frac{n_2}{n_0} = C_2(\sigma_1 - \sigma_2) \tag{4.16}$$

The absolute photoelastic coefficients are rarely used. Rather, we define an empirical coefficient C_σ, called the stress–optic coefficient, which is related to the absolute coefficients by

$$C_\sigma = C_1 - C_2 \tag{4.17}$$

Thus, the relative retardation takes the simple form,

$$R = C_\sigma(\sigma_1 - \sigma_2)d \tag{4.18}$$

Alternatively, everything can be formulated in terms of strains, to obtain

$$R = C_\epsilon(\epsilon_1 - \epsilon_2)d \tag{4.19}$$

where C_ϵ is the strain–optic coefficient.

Material stress–birefringence relationships will be discussed more fully in a later chapter, but attention should be drawn to some important qualifications on the preceding equations.

1. If equations 4.18 and 4.19 are to be equivalent, then stress axes and strain axes (principal axes) must be coincident. If photoelasticity is to be used to study certain plastic, viscoelastic, or viscous behaviors, then the analysis is much more complex. In fact, relatively few of these problems have been solved.
2. The restriction here to momentarily linearly elastic behavior means that birefringence, stress, and strain are linearly related at the instant of measurement. More of this subject later.
3. The C_σ is ordinarily determined from an experimental calibration test for each material used. It can be a function of wavelength, temperature, time, and other variables. This subject will also be discussed in more detail in another section.
4. There are still many research problems in the area of birefringence of materials.

From this discussion we conclude that birefringence in a material can be determined by measuring relative retardation. The birefringence information will yield information on directions and magnitudes of principal stresses.

4.7 Interferometric measurement of birefringence

Photoelastic stress analysis is based on the determination of birefringence in a body through the use of polarized light. The manner in which electromagnetic radiation is affected by a plate of birefringent material must be examined in detail. There are several ways to approach this problem. The one used here is consistent with all observed effects, and it is mathematically unsophisticated in that only trigonometric wave relationships are used. This treatment also preserves all absolute phase information, unlike more conventional approaches. These absolute phase data are important in certain measurement applications and should not be dismissed summarily. For the sake of simplicity, only the so-called linear polariscope is considered first. More complicated setups, which are more useful in practical experimentation, are treated as extensions of the linear setup.

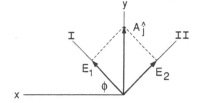

Figure 4.7. Resolution of
electric vector into
components in principal
directions.

Begin with light of wavelength λ, traveling in the z direction with velocity v, and polarized in, say, the yz-plane. Its electric vector will be

$$\mathbf{E} = A \cos\left[\frac{2\pi}{\lambda}(z - vt)\right]\mathbf{j} \qquad (4.20)$$

This ray impinges normally on a birefringent plate of thickness d whose principal refractive axes are inclined at angles ϕ and $\phi + 90°$ to the x-axis. The wave is split into two components that propagate through the plate at different velocities v_1 and v_2. Figure 4.7 shows the angular relationships. The two scalar component waves are seen to be

$$E_1 = E \sin\phi = A \cos\left[\frac{2\pi}{\lambda}(z - v_1 t)\right]\sin\phi$$
$$E_2 = E \cos\phi = A \cos\left[\frac{2\pi}{\lambda}(z - v_2 t)\right]\cos\phi \qquad (4.21)$$

Attenuation (reflection) and phase change at the interface have been ignored because they affect both components equally for most materials of practical interest.

As the two components leave the birefringent plate, they have been retarded by different amounts. These absolute retardations were calculated earlier (see eq. 4.9) and can be inserted into equation 4.21 to obtain expressions for the components leaving the plate and reentering air,

$$E_1 = A \cos\left\{\frac{2\pi}{\lambda}\left[z - vt - \left(\frac{n_1 - n_0}{n_0}\right)d\right]\right\}\sin\phi$$
$$E_2 = A \cos\left\{\frac{2\pi}{\lambda}\left[z - vt - \left(\frac{n_2 - n_0}{n_0}\right)d\right]\right\}\cos\phi \qquad (4.22)$$

The particular nature of the resultant of these components will be discussed later. Right now attention is focused on the phase difference. It cannot be observed directly because the eye, cameras, and other sensing devices are not sensitive to phase of the light. So, the fundamental idea of interferometry is brought to bear. A portion of each of the components E_1 and E_2 are brought

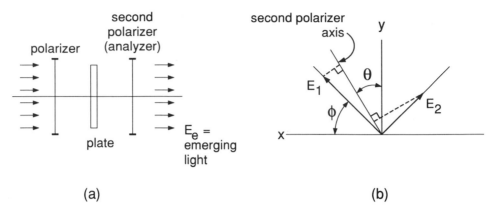

Figure 4.8. (a) Use of second polarizer to combine components of principal waves; (b) resolution of components of E_1 and E_2.

together in a mixing device where they interfere with one another. The phase difference is thereby converted to an intensity difference, which can be detected and measured.

To execute this process, a second polarizer is placed in the light path with its transmitting axis at inclination θ from the original axis of polarization (the y-axis), as shown in Figure 4.8. This polarizer will transmit only components that are parallel to its axis.

The light emerging from the second polarizer is

$$E_e = E_1 \sin(\theta + \phi) + E_2 \cos(\theta + \phi) \tag{4.23}$$

Substitute from 4.22 and use the appropriate identity to obtain

$$E_e = A \cos\left\{\frac{2\pi}{\lambda}\left[z - vt - \left(\frac{n_1 - n_0}{n_0}\right)d\right]\right\} \sin\phi \, (\sin\theta\cos\phi + \cos\theta\sin\phi)$$

$$+ A \cos\left\{\frac{2\pi}{\lambda}\left[z - vt - \left(\frac{n_2 - n_0}{n_0}\right)d\right]\right\} \cos\phi \, (\cos\theta\cos\phi - \sin\theta\sin\phi) \tag{4.24}$$

To simplify this result in the quickest way, let

$$P = A \cos\left\{\frac{2\pi}{\lambda}\left[z - vt - \left(\frac{n_1 - n_0}{n_0}\right)d\right]\right\}$$

and

$$Q = A \cos\left\{\frac{2\pi}{\lambda}\left[z - vt - \left(\frac{n_2 - n_0}{n_0}\right)d\right]\right\} \tag{4.25}$$

Then, we have

$$E_e = (P - Q) \sin\phi \sin\theta \cos\phi + \cos\theta(P \sin^2\phi + Q \cos^2\phi)$$

which becomes, after using the identity $\cos 2\phi = 2\cos^2 \phi - 1 = 1 - 2\sin^2 \phi$,

$$E_e = \frac{P-Q}{2}\sin\theta\sin 2\phi + \frac{P+Q}{2}\cos\theta - \frac{P-Q}{2}\cos\theta\cos 2\phi \quad (4.26)$$

Now, use the identity $\cos x - \cos y = -2\sin[(x+y)/2]\sin[(x-y)/2]$ to compute $P - Q$ and obtain, after some algebra and recognizing that $\sin x = -\sin(-x)$,

$$P - Q = 2A\sin\left[\frac{2\pi}{\lambda}\left(z - vt - \frac{n_1 + n_2 - 2n_0}{2n_0}d\right)\right]\sin\frac{\pi}{\lambda}\frac{n_1 - n_2}{n_0}d \quad (4.27)$$

Follow a similar path for $P + Q$, with the following result,

$$P + Q = 2A\cos\left[\frac{2\pi}{\lambda}\left(z - vt - \frac{n_1 + n_2 - 2n_0}{2n_0}d\right)\right]\cos\frac{\pi}{\lambda}\frac{n_1 - n_2}{n_0}d \quad (4.28)$$

Substitute equations 4.27 and 4.28 into 4.26 and simplify to obtain

$$\begin{aligned}
E_e &= A\sin\left[\frac{\pi}{\lambda}\left(\frac{n_1 - n_2}{n_0}d\right)\right](\sin\theta\sin 2\phi - \cos\theta\cos 2\phi) \\
&\quad \times \sin\left[\frac{2\pi}{\lambda}\left(z - vt - \frac{n_1 + n_2 - 2n_0}{2n_0}d\right)\right] \\
&\quad + A\cos\left[\frac{\pi}{\lambda}\left(\frac{n_1 - n_2}{n_0}d\right)\right]\cos\theta\cos\left[\frac{2\pi}{\lambda}\left(z - vt - \frac{n_1 + n_2 - 2n_0}{2n_0}d\right)\right]
\end{aligned}$$
$$(4.29)$$

This result is the final photoelastic equation for a linear polariscope with all absolute phase information preserved except for possible phase changes at interfaces. The meaning and utility of the general equation are established by considering some special cases, as follows.

4.8 Interpretation of equations

Equation 4.29 shows that the emerging wave is the scalar sum of two rays of equal wavelengths, out of phase by $\pi/2$, and with amplitudes depending on the retardations in the birefringent plate as well as the relative orientations of polarizers and plate. That is,

$$E_e = (\text{amplitude}) \times (\text{sin wave}) + (\text{amplitude}) \times (\text{cos wave})$$

Two special cases of equation 4.29 are particularly useful and instructive.

Special case A: Dark-field linear setup

Consider the case where the polarizers are crossed, so $\theta = 90°$. Equation 4.29 reduces to

$$E_e = A \sin\left[\frac{\pi}{\lambda}\left(\frac{n_1 - n_0}{n_0}\right)d\right] \sin 2\phi \sin\left[\frac{2\pi}{\lambda}\left(z - vt - \frac{n_1 + n_2 - 2n_0}{2n_0}\right)d\right]$$

(4.30)

This result describes a simple harmonic wave of wavelength λ and with amplitude dependent on the relative retardation as well as inclination of the principal stress axes. The irradiance (amplitude squared) of the emergent light is zero if *either*

1. $2\phi = m\pi$ (i.e., if the polarizers are aligned with the principal axes of refractive index, or stress) *or*
2. $[(n_1 - n_2)/(\lambda n_0)]d = 0, 1, 2, \ldots$ (i.e., if the relative retardation is an integer multiple of the wavelength).

This arrangement is very important in photoelastic interferometry.

Special case B: Light-field linear setup

In this case, the polarizers are parallel, so $\theta = 0°$. Equation 4.29 becomes

$$E_e = -A \sin\left[\frac{\pi}{\lambda}\left(\frac{n_1 - n_2}{n_0}\right)d\right] \cos 2\phi \sin\left[\frac{2\pi}{\lambda}\left(z - vt - \frac{n_1 + n_2 - 2n_0}{2n_0}\right)d\right]$$
$$+ A \cos\left[\frac{\pi}{\lambda}\left(\frac{n_1 - n_2}{n_0}\right)d\right] \cos\left[\frac{2\pi}{\lambda}\left(z - vt - \frac{n_1 + n_2 - 2n_0}{2n_0}\right)d\right] \quad (4.31)$$

This result contains two component waves that are out of phase by $\pi/4$ radians. The amplitude of one part depends on both relative retardation and direction, and the amplitude of the other part depends only on relative retardation. The emerging light is zero only if both waves are zero simultaneously. That is, zero light passes through the system only if both of the following are true at once:

1. $\phi = \pi/4$ (i.e., the polarizers are at 45° to the principal axes) *and*
2. $[(n_1 - n_2)/\lambda n_0]d = (2m + 1)/2$, where m is an integer.

This relationship is not commonly used in photoelasticity, but the information it gives can be important in some cases.

As a preliminary illustration of the meaning of the preceding special-case equations, Figure 4.9 shows two photographs of a plastic beam between polaroids. Light- and dark-field cases are shown. Do not worry about interpretation of these whole-field patterns at this point.

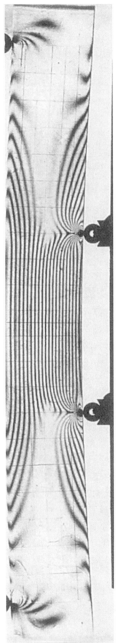

Figure 4.9 Sample photoelastic patterns, (a) light field and (b) dark field, obtained from beam in bending with linear polarization.

As mentioned, the analysis has preserved all absolute and relative phase information. Equation 4.29 shows clearly that the absolute retardation common to both components in the birefringent plate does not affect the results, although there are situations where it is important, as in materials studies. Only the relative retardation affects the light intensity. The common retardation,

$$\left(\frac{n_1 + n_2 - 2n_0}{2n_0}\right)d$$

appears only as a uniform phase shift, and it can be eliminated by a coordinate translation. Most authors ignore it completely. We shall eliminate it henceforth with few exceptions.

Again, notice that the equations show that two waves of radiation can combine to produce zero light. This interference phenomenon is something we can easily observe with our eyes. What has been accomplished is to measure phase difference by converting it to an easily seen intensity difference. This process is common to many measurement techniques. It is the unifying concept in such divergent areas as microwave interferometry, holography, photoelasticity, radar ranging, and laser Doppler velocimetry. Remember that interference can occur only when waves of the same wavelength and the same polarization are able to interfere with one another. This last condition means, for the amplitude-division common-path photoelastic interferometer, that the light waves come from the same source.

4.9 Whole-field analysis

Most of the theory so far has been for light passing through a specific single point in a birefringent plate. Let us now expand our thinking and consider light to be passing through all points in a plate, where the birefringence varies in magnitude and direction over the extent of the plate. If the plate is between crossed polarizers, then some areas of the system transmit the light and some do not. A lens can be used to construct a real image of the birefringent plate, and this image will show two families of dark bands. See Figures 4.10 and 4.11.

Two sets of dark bands or interference fringes can be identified, as follows:

Isoclinic fringes: loci of points of constant inclination of the principal axes of refraction

Isochromatic fringes: loci of points where relative retardation is an integral multiple of the wavelength of the radiation

In engineering and scientific work, the birefringence is usually produced by stress or flow. In such applications the isoclinics indicate points at which the principal stress axes have the inclination of polarizer and analyzer. Figure 4.12 illustrates this point.

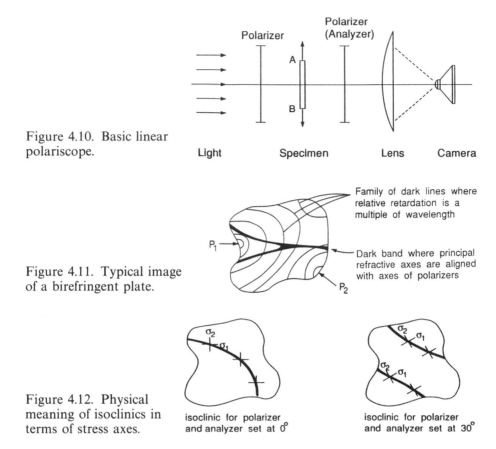

Figure 4.10. Basic linear polariscope.

Figure 4.11. Typical image of a birefringent plate.

Figure 4.12. Physical meaning of isoclinics in terms of stress axes.

As shall be seen, recording the isoclinics for many different polarizer settings enables one to construct a map of stress direction for the entire plate. The isochromatics yield information about stress magnitude. For the linear case, we had

$$R = C_\sigma(\sigma_1 - \sigma_2)d = \left(\frac{n_1 - n_2}{n_0}\right)d \qquad (4.32)$$

Along an isochromatic, $R = m\lambda$. Combining these equations gives the result that along a given isochromatic line,

$$C_\sigma(\sigma_1 - \sigma_2) = \frac{m\lambda}{d} \qquad (4.33)$$

Isochromatics are thus seen to be lines of constant principal stress difference or, equivalently, lines of constant maximum shear stress,

$$\sigma_1 - \sigma_2 = 2\tau_{max} = \frac{m\lambda}{dC_\sigma} \qquad (4.34)$$

where m is the isochromatic "order." Practice and study are needed before one can properly number isochromatics in a given field.

A basic difficulty with a system of the type discussed here is that both isoclinics and isochromatics appear in the same patterns. Among other problems, the isoclinics can mask the isochromatics. It is possible to modify the system to eliminate the isoclinics from the patterns. The technique involves the use of circularly polarized light, discussed in sections 4.10 and 4.11.

4.10 Circular polarization

Return to equations 4.22 for the light components emerging from the birefringent plate. Take the special case where $(n_1 - n_2)d/n_0 = R = \lambda/4 = 1/4$ wavelength and $\phi = 45°$. Such a birefringent plate is called a quarter-wave plate. If $\mathbf{p}$ and $\mathbf{q}$ are unit vectors along the principal axes of this plate, then the electric vector is

$$
\mathbf{E} = \frac{\sqrt{2}}{2} A \left\{ \cos \frac{2\pi}{\lambda} \left[z - vt - \left(\frac{n_1 - n_0}{n_0} \right) d \right] \mathbf{p} \right.
$$

$$
\left. + \cos \frac{2\pi}{\lambda} \left[z - vt - \left(\frac{n_1 - n_0}{n_0} \right) d + \frac{\lambda}{4} \right] \mathbf{q} \right\} \tag{4.35}
$$

Eliminate the retardation common to both waves and simplify to obtain

$$
\mathbf{E} = \frac{\sqrt{2}}{2} A \left[\cos \frac{2\pi}{\lambda} (z - vt)\mathbf{p} - \sin \frac{2\pi}{\lambda} (z - vt)\mathbf{q} \right] \tag{4.36}
$$

The two harmonic components are perpendicular and out of phase by $\pi/2$ radians.

By the Pythagorean theorem, the magnitude of the electric vector is found to be constant. The inclination of the vector with the $\mathbf{p}$ axis is

$$
\alpha(t) = -\left(\frac{2\pi}{\lambda} \right)(z - vt) \tag{4.37}
$$

At any instant of time, the orientation of the electric vector varies linearly with position along the z-axis. That is, the tip of the vector describes a circular helix. Alternatively, for a fixed z, the light vector rotates at constant angular velocity and describes a circle. Such radiation is said to be circularly polarized.

To summarize: A quarter-wave retarding plate is used to convert linearly polarized radiation to circularly polarized radiation.

As might be expected, states of polarization between linear and circular are possible and, in fact, are more common than the special cases discussed. These in-between polarization states are termed "elliptical" because the

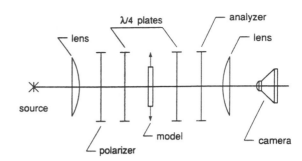

Figure 4.13. Polariscope
with quarter-wave plates.

electric vector describes an ellipse. Elliptically polarized light can be produced from linearly polarized light by means of a retarding plate whose relative retardation is different from $\frac{1}{4}\lambda$.

Circularly polarized light serves two useful functions in photoelastic interferometry:

1. It can be used to eliminate isoclinics from a photoelastic pattern, leaving only isochromatics.
2. It is used in compensation methods of accurately determining relative retardation at a point in a birefringent plate.

4.11 Circular polariscope

An especially useful form of the photoelastic interferometer uses circular polarization to eliminate isoclinics. To convert the linear instrument into a circular one, two $\lambda/4$ plates are inserted into the polariscope as shown in Figure 4.13. The axes of the $\lambda/4$ plates should be crossed and at 45° to the axes of polarizer and analyzer.

A good exercise is to derive the equation for the light vector emerging from such a system with various orientations of polarizer and analyzer relative to the crossed $\lambda/4$ plates. The development is similar to that leading to equation 4.29, except that more components are involved. Many of the terms drop out or combine, however, and the result is simpler than for the linear polarization case. The terms containing the directional dependence, those that contain ϕ, simply vanish. Only isochromatic fringes remain. All that has been said about the interpretation of the equations applies, except that nothing need be said about directional dependence.

A less-general, shorter, but still valuable exercise is to set up the equations with polarizer and analyzer crossed and the quarter-wave plates in at 45°. The common retardation can be dropped at the beginning. The result will be the same as outlined in the preceding paragraph. Neither of these derivations is reproduced here, although a portion of the general case appears in the section on goniometric compensation in Chapter 6.

Other methods for distinguishing isochromatic fringes from isoclinics will be outlined in the next chapter.

References

Born, M., and Wolf, E. (1975). *Principles of Optics*, 5th ed. New York: Pergamon Press.

Burger, C. P. (1987). Photoelasticity. *Handbook on Experimental Mechanics*, Ed. A. S. Kobayashi, Ch. 5. Englewood Cliffs: Prentice-Hall.

Coker, E. G., and Filon, L. N. G. (1957). *A Treatise on Photoelasticity*. Cambridge University Press.

Dally, J. W., and Riley, W. F. (1991). *Experimental Stress Analysis*, 3rd ed. New York: McGraw-Hill.

Frocht, M. M. (1941). *Photoelasticity*. New York: John Wiley and Sons.

Jessop, H. T., and Harris, F. C. (1949). *Photoelasticity, Principles and Methods*. London: Cleaver-Hume Press.

Wolf, H. (1961). *Spannungsoptik*. Berlin: Springer-Verlag.

5

Basic applied photoelasticity

Herein we describe how to set up a photoelasticity interferometer, calibrate it, manufacture models, obtain fringe patterns, and interpret them to obtain maps of stress directions and stress magnitudes. More can be said on all these topics; some are expanded in Chapter 6. Persons planning extensive experiments using photoelastic interferometry should also become familiar with the excellent treatments in the several available books and handbooks (e.g., Burger 1987; Dally and Riley 1991; Frocht 1941; Jessop and Harris 1949; Post 1989; Wolf 1961).

5.1 Polariscope optics

Many different choices of optical elements and systems are possible for conducting model analysis by photoelastic interferometry. The object here will be to describe a few practical basic arrangements for general use. Much confusion is avoided if a systematic approach is adopted. It is apparent that certain basic optical functions must be accomplished in a polariscope. As long as the basic functions are served, there is considerable latitude in the final choice of optical elements. These points are especially important when a polariscope is being built for a special research application.

The optical system can be represented in block diagram form as shown in Figure 5.1. The light source must be capable of providing fairly intense monochromatic radiation as well as white light. For most efficient operation, the radiation must be collimated. These requirements taken together mean that the lamp must be small, intense, and spectrally pure, although the last restriction may be eased if filters are used to separate monochromatic light from multicolor radiation. For most photoelastic investigations, it has proven best to use mercury vapor or sodium vapor discharge lamps. The mercury lamp radiation is easily filtered to produce a number of monochromatic wavelengths, and it can be used unfiltered for broad-band (white) light.

Figure 5.1. Polariscope system block diagram.

Table 5.1. *Monochromatic radiation sources and filters*

λ (nm)	Lamps	Filter
405	Mercury	Zeiss Jena (Schott) 267406
436	Mercury	Zeiss Jena (Schott) 013213
546	Mercury	Zeiss Jena (Schott) 050025 or Kodak 74
578	Mercury	Zeiss Jena (Schott) 373635 or Kodak 22
587	Helium	Zeiss Jena (Schott) 373635
589	Sodium	none needed
668	Helium	Kodak 92
707	Helium	Kodak 89-B
853	Incandescent or mercury	Kodak 87-C
1083	Helium	Kodak 87-C
1100	Arc	Zeiss Jena B8105/8
1145	Arc	Optics Technology 160
1244	Arc	Optics Technology 822

The sodium lamp is near enough to being spectrally pure at $\lambda = 589$ nm that it can be used unfiltered for monochromatic investigation. Another approach is to use broad-band radiation, such as from a tungsten arc or incandescent lamp, and filter it through a narrow band-pass interference filter. These filters are available only in small sizes, and they may be placed near the source or the camera lens. The variation of angle of incidence does cause the center frequency to change over the field in this case. A third useful type of light source is the laser. It is highly monochromatic and, usually, already polarized. Objectionable speckle effects will appear in the fringe photographs taken with laser illumination unless measures are taken to defocus the speckle. Table 5.1 lists several combinations of light sources and filters for producing monochromatic light at several wavelengths from deep violet through near infrared (Cloud and Pindera 1968).

The collimating device is usually a large-field lens with the light source at its focus. Because lenses are quite expensive, a practical (but less efficient) expedient is to replace the collimating lens with a diffuser that will scatter the light in all directions. With this system, care must be taken with the data recording apparatus to assure that only the light that travels parallel to the axis of the optical system is utilized, and the remaining light is rejected.

The polarizers are nowadays usually made of commercially available sheets of polaroid filter medium (Polaroid Corporation). These polarizers can be purchased in large sheets so that a large-field instrument is possible. Several types are available with various efficiencies and intended for use in various parts of the spectrum from ultraviolet through near-infrared. They are available laminated between glass plates if one wants to avoid the labor of mounting them.

The quarter-wave plates are also usually purchased in sheet or in mounted form from Polaroid Corporation or from suppliers of optical components. It is important that both $\lambda/4$ plates have the same relative retardation.

The field lens usually serves to collect the light in the large field and converge it to the camera lens, although it can act as an imaging lens when the photoelastic pattern is to be cast on a screen or a point sensor such as a photocell. To minimize distortion, this lens is placed as close to the model as practicable when a camera or other imaging lens is used. Because lenses sometimes contain residual birefringence, it is poor practice to have them inside the polarized light field. That is, the field lens should be between the analyzer and the camera.

As suggested in Chapter 4, the recording element may be a camera, the eye, a screen, or a photocell. In order that the camera will construct an image with only the light that travels normally through the model, the aperture of the camera is placed so that its distance from the field lens equals the focal length of that lens. This point is especially important if a diffuser is used at the light source. In addition, the camera aperture should be small (say, $f/11$ or smaller) so that the pencil of radiation received from each small area of the specimen has a small cone angle.

The chosen optical elements are usually held in retaining rings that are mounted on an optical bench, which maintains their alignment while allowing longitudinal adjustment. The model is usually held in a fixed loading frame, and the optical elements are adjusted relative to it as needed. Figure 5.2 is a schematic of a complete transmission polariscope of conventional type.

Figure 5.3 is a photograph of a simple and inexpensive general-purpose diffused light polariscope of a type designed by J. T. Pindera. In this instrument, a sodium lamp and a fluorescent fixture are both mounted in the large box so that either monochromatic or white light can be chosen at will. The polarizers and the field lens are held in mountings of wood that are bolted to adjustable pedestals. Although not entirely conventional, this polariscope is easy to build and is very flexible. It can be used with any loading frame or moved to a fixed testing machine. It is especially useful for educational purposes.

The exceedingly fundamental polariscope just shown can be modified and improved in several ways to improve convenience and precision and/or to control cost. The polarizers and $\lambda/4$ plates can be mounted in calibrated

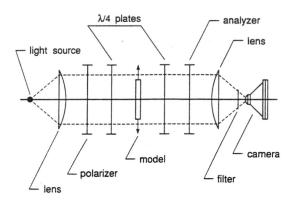

Figure 5.2. Schematic of conventional large-field photoelastic polariscope.

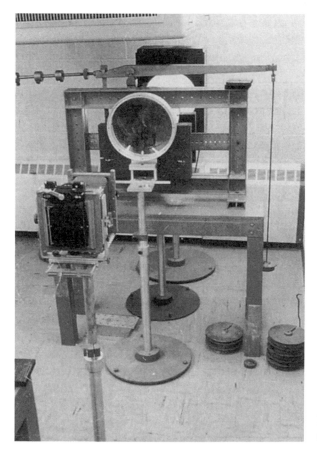

Figure 5.3 A simple diffused-light polariscope.

retaining rings. The field lens, always an expensive item, can be eliminated in favor of a telephoto lens on the camera, although this measure actually decreases precision. The long lens makes it possible to place the camera far

Figure 5.4. Polariscope typical of those available from commercial sources (courtesy of Measurements Group, Inc.).

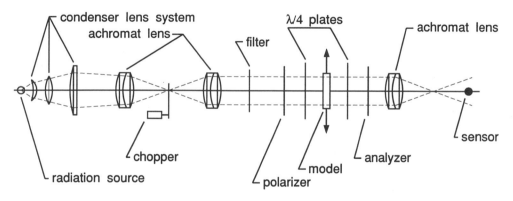

Figure 5.5. Schematic of small-field polariscope with point-sensing (Cloud and Pindera 1968).

enough from the model so that the light utilized in forming the image travels through the model at near-normal incidence. Similarly, the light source can be mounted in a projection housing that is placed at a distance from the model so as to eliminate a collimating lens or a diffuser. Finally, all the components can be mounted on precision benches.

Figure 5.4 shows a conventional polariscope that is marketed by a

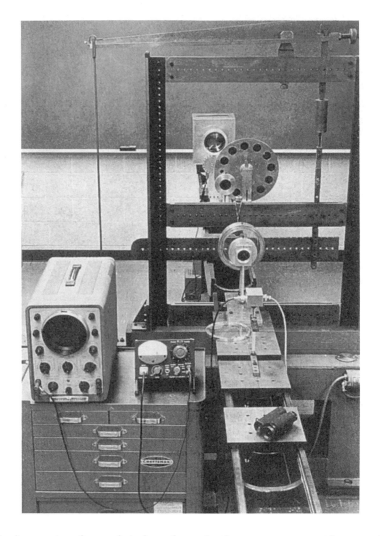

Figure 5.6. Apparatus for pointwise photoelastic measurements in near-infrared, visible, and ultraviolet (Cloud and Pindera 1968).

commercial firm. The optical elements are mounted on ways, and the $\lambda/4$ plates can be moved laterally to remove them from the optical system.

Figure 5.5 is a schematic of a small-field polariscope that has been used in point-by-point retardation measurements in infrared light (Cloud and Pindera 1968). Because point detection is used, there is no need for a large field. To measure fringe order at various locations on the specimen, the loading frame was moved relative to the optical axis. Amplification of the detected radiation signals was required, so a light-chopper was incorporated into the system. The photocell was connected to an amplifier that was tuned to the chopping frequency to give an improvement in the signal-to-noise figure. Light levels were read on a meter and an oscilloscope. Figure 5.6. is a photograph of this setup. Notice the heavy cast-iron optical bench with precision-ground ways, the component carriers, and the precision movable loading frame, which were designed by J. T. Pindera for photoelasticity and materials research and built at Michigan State University.

5.2 Calibration of polariscope

A problem to be faced after assembling the interferometer is how to establish the orientations of the axes of polarizers and $\lambda/4$ plates. This process is called "calibration of the polariscope." The following is a step-by-step procedure for getting the instrument properly set up for quantitative analysis.

1. Cross the transmission axes of the polarizer and analyzer by rotating them relative to one another until extinction is obtained.
2. Locate the directions of the transmission axes of the polarizer and analyzer. This can be done by examining a model with an axis of symmetry for both geometry and loading. A disc of Plexiglas or similar material works well as a specimen for calibrating the polariscope elements. Arrange the disc so it is loaded in diametral compression, with a plumb line in place to assure that the loading is along the vertical. Rotate the polarizer and analyzer, keeping their axes crossed, until the isoclinic line appears along the axes of symmetry of the model. For the disc it will look something like a Celtic cross. At this point, the transmission axes of the polarizer and analyzer are known to be lined up parallel and perpendicular to the principal stress axes on the axes of symmetry. Because the principal stress axes on the axes of symmetry are parallel and perpendicular to these axes, this procedure locates the directions of the transmission axes of the polarizer and analyzer with respect to the specimen. Because the specimen is plumb, the axes are known with respect to global coordinates. Mark the directions of the crossed axes on the holders, or mounts, of the polarizer and analyzer.
3. Locate the directions of the transmission axes in the quarter-wave plates as follows:
   (a) Place one of the quarter-wave plates between the crossed polarizer and analyzer, and rotate it until extinction is achieved. In the position for extinction, the transmission axes of the wave plate are lined up with the transmission axes of the polarizer and analyzer. Mark the directions of the two perpendicular transmission axes on the mount of the quarter-wave plate.

   The procedure just given does not tell us which of the two axes is the fast axis in the quarter-wave plate. This does not matter for the purpose of setting up the polariscope because there is a simple means of telling when the fast axes of the two quarter-wave plates are in opposition to each other, even if we don't know which of the two perpendicular directions corresponds to the fast or the slow axis.

   (b) After marking the positions of the transmission axes for the first quarter-wave plate, remove it from the polariscope and repeat the process with the second quarter-wave plate.

4. Now put both quarter-wave plates into the polariscope between the polarizer and analyzer, and check to see that the symmetrical isoclinic in the calibration specimen is reproduced. When the quarter-wave plates have been put into the polariscope, either a very small amount of light will be transmitted through the analyzer, or there will be a relatively large amount.

   If one uses monochromatic light whose wavelength corresponds to the design of the wave plate, then the small intensity of light should be essentially zero (i.e., one achieves extinction). However, if the wavelength of the light does not correspond to the quarter-wave plates, or if one uses white light, then a minimum amount of transmitted light (rather than extinction) will be observed.

   If the initial positioning of the wave plates results in minimum transmission of light, then the two fast axes of the quarter-wave plates are in opposition, which is the correct arrangement. On the other hand, if a large amount of light is transmitted, all one has to do to achieve the minimum transmission is to rotate either of the quarter-wave plates through 90°. These orientations should be marked so that the $\lambda/4$ plates can be removed and reinstalled with their fast axes crossed.

5. Now, to convert the linear polariscope to the circular configuration, the transmission axes of the $\lambda/4$ plates must be arranged so that they are at 45° to the axes of the polarizer and analyzer. To reach this state, simply start with the linear arrangement arrived at in the previous step, and rotate both $\lambda/4$ plates *in the same direction* 45°. Either clockwise or counterclockwise is fine.

   If one wants to change from dark field to light field, or vice versa, it is only necessary to rotate either the polarizer or the analyzer through 90°. This procedure places the polarizers parallel, with the two $\lambda/4$ plates still being crossed. In the dark-field arrangement, the whole-order isochromatic fringes will be dark. It is good practice to obtain additional data from both the light-field and dark-field arrangements. With the light-field arrangement, it is the half-order isochromatic fringes that appear dark, and they will fall between the fringes observed with dark field.

Examination of the photoelasticity equations indicates that there are several alternative arrangements of optical elements in the circular polariscope. The best optical setups are those that keep the axes of the $\lambda/4$ plates crossed; that is, the *fast* axis of one must coincide with the *slow* axis of the other. The reason is that it is difficult to make perfect $\lambda/4$ plates, but it is simple to make them in pairs that have similar errors. The "error" in a $\lambda/4$ plate is the amount by which the relative retardation differs from exactly one-quarter of the wavelength. If the plates are used in matched pairs and their axes are kept crossed, then the isochromatic errors will be minimized. In practice, it is not necessary to know which are the fast and slow axes of each of the quarter-wave plates. The orientations of their axes and the crossed configuration are easily established during the polariscope calibration with the disc, as discussed earlier.

5.3 Model materials and fabrication

The choosing of photoelastic model materials and the fabrication of models are somewhat arcane arts, and a great body of literature exists on these topics. The problem is not nearly as serious as it once was because sheets of materials developed specifically for photoelasticity are now available from suppliers at reasonable cost.

Some attention has already been given to the choice of material for isoclinic studies. The requirement is to have a material that has relatively low optical sensitivity and that is free of residual stress. The best choice is polymethylmethacrylate, sold under the trade names Plexiglas, Perspex, and Lucite.

For isochromatics, the material should be free of residual stress, and it should have a high stress–birefringence factor so that numerous fringes can be created with reasonable loads. In addition, it should not exhibit serious creep, and it should be easy to machine without chipping or acquiring edge stresses. The surfaces should be optically fine, and the material should be very transparent. Finally, it should have minimum susceptibility to time-edge effects. The phenomenon of time-edge effects has been a serious problem with many of the older materials. It appears as a pattern of fringes in the machined edges of a specimen when it has been stored for any length of time. With many materials, it has been necessary, for the best results, to machine the model no more than a few hours before testing. The effect is caused by migration of moisture into or out of the resin.

One of the classic model materials is Columbia Resin CR-39. Its optical quality and sensitivity are excellent, and it is quite easy to machine without creating edge stresses. Grades for photoelasticity are heat treated to reduce residual fringes. It is possible to do this heat treating oneself if an oven with good temperature control is at hand. The sheets are carefully supported in the oven, heated to just under the transition temperature, and then cooled for a day or two. The disadvantages of CR-39 are its relatively high notch sensitivity and, worse, its susceptibility to time-edge effect.

Paraplex resin P-43 and similar products are also very useful classic model materials. Its performance is similar to that of CR-39, but with higher notch sensitivity and lower edge effect.

Excellent model materials are those that are epoxy-based, such as are marketed for photoelasticity by Measurements Group. These are manufactured so as to be stress-free. The stress-birefringence coefficient is higher than most. Notch sensitivity is low, and they are relatively free of time-edge effects. They are more highly colored than are other materials, but the optical quality and surfaces are fine. A useful characteristic of these materials is that several formulations are available with different elastic moduli and optical sensitivities. One grade is very flexible and shows extremely high birefringence, so it can be used to study stresses in elastomers.

Finally, polycarbonate resin in the finer grades should be mentioned. This material tends to have high residual stress in the as-manufactured state, so it needs to be annealed as described earlier. Its other characteristics are advantageous. A useful characteristic of polycarbonate is that it is ductile, with a stress-strain curve that approximates in shape that of some steels and other alloys. It can be used for photoviscoelastic or photoplastic studies, although the inelastic deformation must be limited if fringes are to be seen.

Fabrication of all these materials generally follows normal machining practice with a few differences. The most important requirement is to avoid the induction of residual stresses. There are two approaches that work well, with many variations possible. One technique is to use relatively high cutting speeds with very small feeds. The other is to employ larger feeds and very low speeds. Sharp tooling is required. The best option is to use carbide tools because the resins shorten the life of high-speed steel cutters. Coolants or lubricants can be used, but they can cause compatibility problems. With careful technique, they do not seem to be necessary. Care should be taken with chip collection, and a protective breathing mask should be worn; remember that you are machining an organic compound.

One of the best approaches is to rough-saw the specimen on a bandsaw, leaving a few millimeters for trim to eliminate the saw marks and the stresses caused by sawing. The edges can then be finished with a router and a cross-cut carbide bit. Ordinary helical-tooth bits tend to cause edge stresses, and straight-cut bits tend to chatter. If several models are to be made, then it is worth making a master of thin steel plate. Then a flush-cutting trimmer bit can be used to finish the edge. The model and the master are simply held together with double-sided masking tape for the finish operation. A pantograph router (if available) is a good approach. The specimen can be cut from the sheet and finished in one operation if care is taken.

For a one-off engineering study, the specimen can be rough-sawn on a bandsaw, with a saber saw, or by hand with a coping saw. The important edges are then finished by hand with a good cross-cut mill file. Another option that works surprisingly well is to finish the edges with a small sanding drum in a hand grinder such as a Dremel tool or a die grinder. With light pressure and rapid tool movement, considerable stock can be removed with no heating or creation of residual stress in the edge. A master template or just some simple guide bars can be clamped to the specimen to provide a finish line. The sanding drum or file does not remove enough of the steel guide to worry about. Protective breathing gear, a vacuum system, and good ventilation are especially important if a sanding drum is used because the removed stock is in the form of a fine powder.

5.4 Calibration of model material, some basics

To convert isochromatic data to stress, the stress–optic coefficient C_σ for the model material must be determined through experiment. This requirement is, in general, not simple to meet because the materials are viscoelastic. The topic is discussed fully in the next chapter. The purpose here is to describe some simple methods that are adequate for determining the stress-optic coefficient with accuracy sufficient for basic engineering investigations of stress.

The first method is to fabricate a simple flat tension specimen of the material. The specimen is placed into the polariscope load frame, taking care to eliminate bending. As each increment of load is placed on the specimen, the flow of fringe orders is carefully watched. By watching the light and dark fringes form, one can develop the ability to estimate the fringe order to within one-fourth order or so at each load. Merely detecting whole- and half-orders is not good enough. The stresses are calculated for each load and plotted with the corresponding fringe orders. The maximum load is limited so as to fall within the linear range of response. The slope of a straight line drawn through the data points is the C_σ. The scatter of data points from the straight line is likely to be significant with this method because of the difficulty in estimating exact fringe order in the test section.

A better technique is to fabricate a beam specimen that is long enough to load in four-point bending, so that pure bending is produced in the center section. It is helpful to scratch some fiducial marks on the beam because the edges are not always visible in the photoelastic photographs. The maximum stress is calculated so as to be well within the linear range of response. The load is then applied, and the isochromatic pattern is photographed for a certain time after loading, in both light and dark field. The locations of the whole- and half-order fringes on the cross section are determined. The bending stresses at these locations are computed using simple beam theory. The pairs of fringe-order and stress data are then plotted. They should yield a good linear relationship with little scatter. The slope of the plot is the stress–birefringence coefficient for that time after load. If the isochromatics for the unknown shape are recorded for the same time after load, then viscoelastic effects are eliminated.

A similar method uses a disc specimen instead of a beam. The elasticity solution for a disc is used to calculate the principal stresses along the horizontal and vertical diameters; this solution is given in Chapter 6. The recording and data-reduction process is the same as that specified for the beam. An advantage of this approach is that a calibration disc can often be included within the load setup for the unknown specimen. The calibration data and the unknown data are then recorded at the same time on the same photographs.

5.5 Recording and interpreting isoclinic fringes

The family of isoclinic fringes is employed to determine the orientations of the principal stress axes for specimens of materials in which birefringence has been induced by stress. In engineering applications, the specimen is usually a plastic model of a structure or machine element. The similarity conditions, which will be discussed later, imply that, for elastic or linearly viscoelastic behavior, the principal axes are identical in the plastic model and prototype. The discussion here will be confined to two-dimensional analysis, although much of what is said applies to more general problems.

The first task is to record the isoclinic fringes in usable form. Although it seems rather crude, one of the easiest and best techniques is to trace the isoclinics by hand for successive settings of the polarizer and analyzer. Other approaches are to use a video camera and monitor, or else a computerized video image capture system. Remember that the polarizers must be kept crossed; the quarter-wave plates must be removed from the optical system or rendered ineffective by aligning their axes with the polarizer axes. Also recall that only one isoclinic fringe is seen for each orientation of the optical elements.

The photoelastic model is inserted into the optical system. If direct tracings are to be made, the image of the model is focused onto the ground-glass camera back or, after reflection, onto a horizontal table. A piece of transparent celluloid film (translucent paper, vellum, or mylar if a clear-glass tracing surface is used) is taped to the ground glass. When the model is loaded, a black isoclinic fringe, along with some isochromatics, will be seen on the image. The centerline of the isoclinic fringe is traced with a marking pen, and that line is labeled with the orientation of the polarizer axes with respect to some arbitrary reference, usually a model edge or the load axis. The polarizer and analyzer are then rotated through some increment in the same direction to keep them crossed, and the tracing is repeated for the new fringe. Increments of 10° are commonly used. The process is continued until the polaroids have rotated 90°, at which point the isoclinics will start repeating. The outline of the image of the model is also traced, and the sense of the rotation of the polarizers is clearly marked on the drawing. Because this procedure is carried out in the dark, it will seem quite difficult at first, and the results will appear rather shaky. It happens, however, that the human eye is quite good at locating the black center of a dark isoclinic fringe.

Another good technique for capturing isoclinics is to use an inexpensive monochrome television camera coupled directly with a monitor. A piece of tracing film can be taped to the monitor screen and the successive isoclinics traced with a marker. An advantage is that the brightness can be adjusted to give the best definition of the fringes. A disadvantage is that the tracing surface is well forward of the picture tube phosphor screen in the typical monitor, so the parallax will cause the tracing to shift with respect to the television image as the observing eye is moved. One way to solve this problem is to trace the outline of the specimen boundary first. Then, one keeps the traced outline coincident with the image boundary while tracing the isoclinics. Another disadvantage is that a television image is usually distorted unless one obtains top-quality electronics intended for metrological applications.

More elaborate techniques utilize a television camera with a computer digital image capture system. After being digitally recorded, the successive isoclinic images can be combined into one pattern, and further digital processing is possible. Problems are created by the widely variable breadth of the fringes; without rather extensive processing, the computer has trouble tracking fringe centers consistently.

Direct photography of the succession of isoclinics does not seem to work very well. The fringes are often quite broad in a photograph, and the subtle shades that the eye uses to find fringe center in the image are missing in the typical photograph.

In spite of its shaky appearance, the family of isoclinics developed by one of the tracing techniques will be quite accurate. The tracing will be easier if a bright light source is used. Often it is difficult to distinguish isoclinics from the isochromatics that appear in the image. For this reason, it is often advisable to use white light for isoclinic observation, because all but the isoclinic and the zero-order isochromatic will then be colored. Additional improvement is attained by using light loads and/or insensitive plastics for

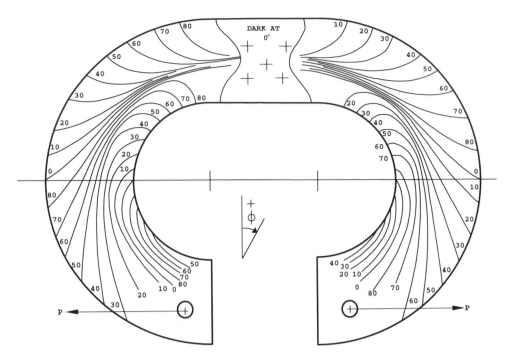

Figure 5.7. Isoclinic pattern tracing.

isoclinic work. Common practice is to make two photoelastic models, one of plexiglass (insensitive) for isoclinics and one of sensitive plastic for iso-chromatics.

For subsequent analysis, it is good practice to make an overlay tracing on paper of the original isoclinic family. Drafting instruments may be used to smooth the curves and obtain a neat rendering. A typical result from a student laboratory exercise is shown in Figure 5.7. The original pencil drawing was scanned into a computer graphics system for smoothing, labeling, and printing, but such refinements are not necessary.

The whole-field map of stress directions is not easily visualized from the isoclinic plot. When such a map is required, the isoclinic data are processed to obtain families of stress trajectories. These trajectories are lines that are everywhere tangent to the principal stress directions. There will be two such families, which together form an orthogonal network. One family will be for maximum principal stresses.

Plotting the stress trajectories is carried out by first making a duplicate of the isoclinic plot on a transparent overlay. Small crosses are marked along each isoclinic, with the axes of each cross parallel to the polarizer orientations that produced that isoclinic. Such a pattern of crosses is shown in Figure 5.8. The stress trajectories are then easily sketched by beginning a line at some

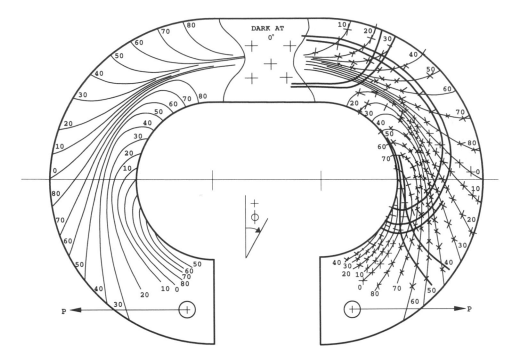

Figure 5.8. Constructing stress trajectories from isoclinics.

convenient point and carrying it across the drawing, making certain that wherever the line crosses an isoclinic it is parallel to the appropriate axis of the cross on that isoclinic, as shown in the figure. Again, computer graphics can be used to advantage. The result, after some computer processing, is shown in Figure 5.9.

The most serious requirement, which sometimes confounds even practiced photoelasticians, is keeping the two families of stress trajectories separate. The problem is minimized by careful and accurate use of the polarizer inclinations marked on the original isoclinic tracing. Familiarity with theory and solutions of elasticity and known related problems is also helpful. For example, stress trajectories near an unloaded boundary must always be either parallel or perpendicular to that boundary. Singular points, including points of zero stress or zero shear stress (hydrostatic tension or compression) also pose difficulties. One may find all the isoclinics passing through such a point, or else all of them going around such a point. Being aware of the existence of these singular points is half the battle. There is a considerable body of literature about interpreting isoclinic fringe patterns (e.g., Frocht 1941; Jessop and Harris 1949). Utilization of this knowledge is helpful in saving time and eliminating possible serious error.

Emphasis must be given to the necessity for establishing the polarizer and analyzer axes to begin with, in order to obtain a starting point. As suggested

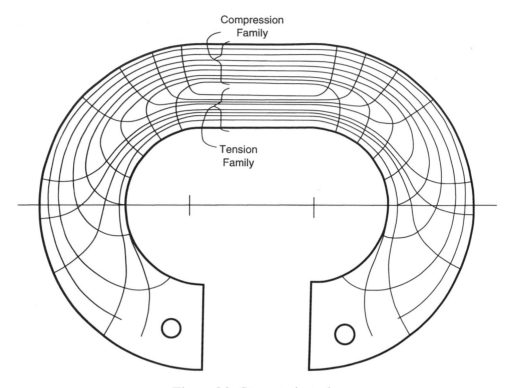

Figure 5.9. Stress trajectories.

in the section on polariscope calibration, the solution is to use a specimen for which the stress field is known and that exhibits isoclinics that are sensitive to error of setting polarizer axes. A good specimen for this purpose is a disc that is loaded in diametral compression. Symmetry demands that the zero-degree isoclinic be a cross whose axes correspond to the load axes. Minute angular deviations of either polarizer will force an obvious lack of symmetry in the cross. In practice, the polariscope base is made level, and the load axis for the disc made vertical with a plumb line. The polarizers are then adjusted to a perfect cross, and the final orientations are fixed and marked for the zero reference. This calibration need be done only once for any given instrument, although it is advisable to check the calibration before undertaking any critical photoelastic analysis.

5.6 Recording and interpreting isochromatics

The isochromatic fringes indicate lines of constant maximum shear stress (or principal stress difference, depending on your preference). With suitable calibration of the optically sensitive model material, the isochromatic observations will yield a quantitative map of the maximum shear stress over the extent of the model. Some basic techniques for material calibration have been presented, and more advanced methods will be discussed later, as will the procedure for transferring stresses from the photoelastic model to the prototype. For the moment, it is sufficient to observe that, for most problems, the stress distribution is independent of material properties. We need only multiply the stresses measured in the photoelastic model by a scaling factor that is a function of load magnitude and model size to obtain prototype stress.

To observe and record isochromatics without having them masked by isoclinics, one inserts the specimen into the polariscope as for the isoclinic observations, except that now the quarter-wave plates are in place with their axes at 45° from the axes of polarizer and analyzer, in order to create the so-called circular polariscope.

For highest contrast and sharpest fringes, as are needed for extremely precise photoelastic measurements such as those carried out in measuring dispersion of birefringence, it may be best to observe isochromatics with linear polarization. Quarter-wave plate factors, which are added unknown variables, are then completely eliminated.

As the specimen is loaded, the isochromatic pattern will probably be seen to change with time. This evidence of viscoelastic behavior is common to most photoelastic materials, and it can be a serious source of error unless properly accounted for. The section on material properties will discuss this point more fully. For the present, we only need to state that the time elapsing between the loading process and the isochromatic recording must be carefully noted. Photoelasticians, like primitive surgeons operating without anesthetic, used to take pride in the speed at which they could record isochromatics after loading, because high speed seemed to be the only way to minimize creep errors. Investigators now proceed at a leisurely rate, knowing that it is best to wait a fixed amount of time between loading and recording. The only requirement is that the time period be identical in the model study and in the material calibration. A 5-min wait is commonly employed, but 10–15 min is better for laboratory workers who tend to forget to load cameras and the like.

Isochromatics are best observed and recorded in nearly monochromatic light. A mercury or sodium lamp is most often employed with appropriate filtering to eliminate unwanted wavelengths. The next section will discuss the observation of isochromatics in white light; there is occasional need for this procedure in some applications.

The recording of the isochromatic fringes in light and dark field is best done photographically, although manual tracing also works well. Video imaging and tracing from the monitor screen, as described for isoclinics, works very well. Recording isochromatics by means of a frame grabber into a computer also serves nicely. This isochromatic recording step always seems

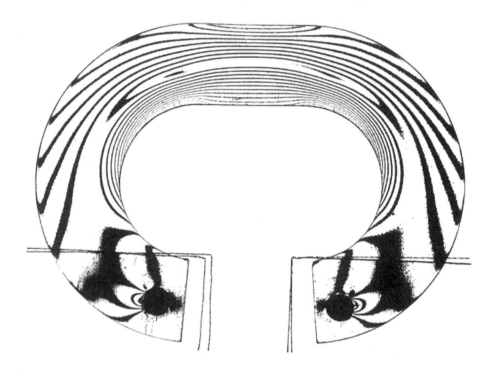

Figure 5.10. Light-field isochromatic pattern.

a bit anticlimactic after the intricacies of making the model, setting up the polariscope, and loading the model without breaking it.

A considerable body of literature has built up about the art of taking isochromatic photographs. Nowadays much time and expense can be saved by using Polaroid film. Polaroid high contrast in 4×5 format (type 55) works well with blue and green light, and 4×5 Polapan (type 52) works throughout the visible spectrum. In general, any technique that gets the isochromatics onto film or a tracing, or into a computer, is acceptable, and much can be done with even primitive cameras. The most critical point is that the optical setup be correct so that the picture is formed only with light that has passed through the model along a normal to its surface.

Figure 5.10 is a light-field isochromatic pattern for a c-shaped tension member. This result, which is a student's first effort, was obtained with a commercially available polariscope with 8-in. field using mercury green light ($\lambda = 546$ nm) and 8×10 Kodak film.

After recording the isochromatic patterns, the proper order (the parameter *m*) must be assigned to each isochromatic fringe. This task is often not easy because high-order fringes look exactly like the low-order ones in mono-chromatic light. Some study, practice, and tricks of the trade are needed before order can be accurately established in a complex stress field. Some useful techniques include the following:

1. Watching the isochromatic pattern form and spread as load is increased from zero, to establish directions of stress gradients.
2. Study and knowledge of related elasticity solutions.
3. Using white light in a qualitative observation. The colors of the high-order isochromatics tend to be less saturated.
4. Using an indentor method near a free edge of the model. The direction of motion of the isochromatics near the edge, resulting from addition of a small normal stress component, serves to establish the sign of the boundary stress and/or help fix the direction of the stress gradient. Some information about the stress field must be known beforehand.

Once isochromatic orders are established, a map of principal stress difference over the model can be obtained by simply multiplying the order by the material stress-optic coefficient, which was determined in the calibration experiment. Be sure that the result comes out dimensionally correct.

A boundary stress distribution is a direct product of this procedure. At an unstressed boundary, one of the principal stresses is zero. Thus, $\sigma_1 - \sigma_2 = \sigma_1 =$ normal stress along an edge. A visually convenient plot of boundary stress is made by erecting normals along the specimen edge, with the height of each normal proportional to the boundary stress magnitude at that point. The tips of the normals are connected by a continuous line. Such a plot (an example is shown in Figure 5.11) provides an immediate indication, even to laypeople, of the location and magnitudes of stress concentrations and stress gradients. For many engineering applications of experimental stress analysis, there is no need to explore a problem further. It is possible, however, to obtain the separate principal stresses if they are needed. Some procedures will be treated in the next chapter.

5.7 Use of non-monochromatic light

When a birefringent specimen is viewed with light of more than one wavelength, the isochromatic pattern is a system of colored lines. The relative retardation can be measured by carefully judging the color. This technique is not commonly used, but it is useful in some instances, such as with photoelastic coatings.

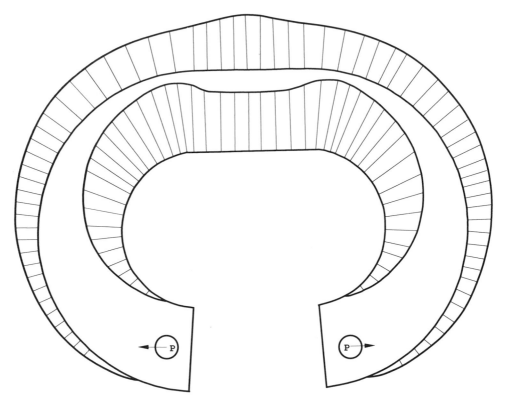

Figure 5.11. Boundary stress distribution.

To gain understanding of how the colored fringes are formed, consider again the basic equation for the light coming from a circular polariscope,

$$E = A \sin\left[\frac{\pi}{\lambda}\left(\frac{n_1 - n_2}{n_0}\right)d\right] \sin\left[\frac{2\pi}{\lambda}(z - vt)\right] \qquad (5.1)$$

The intensity of this wave is the square of the amplitude. It is convenient to normalize this intensity by the intensity I_0 of the incident light (i.e., the maximum available) and also to put it in terms of the relative retardation.

$$\frac{I}{I_0} = \frac{E^2}{A^2} = \sin^2\left(\frac{\pi R}{\lambda}\right) \qquad (5.2)$$

A plot of this equation, as shown in Figure 5.12, demonstrates that zero intensity results whenever $R = m\lambda$.

If two wavelengths are being used, then two separate curves can be plotted and the resultant light obtained by superposition, as shown by the following equations and Figure 5.13:

$$\frac{I_1}{I_0} = \sin^2\left(\frac{\pi R}{\lambda_1}\right), \qquad \frac{I_2}{I_0} = \sin^2\left(\frac{\pi R}{\lambda_2}\right) \qquad (5.3)$$

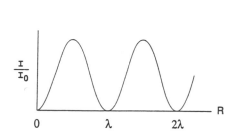

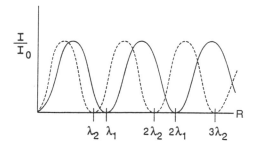

Figure 5.12. Typical normalized intensity as a function of relative retardation.

Figure 5.13. Typical normalized intensity as a function of relative retardation for two wavelengths.

Clearly, the two wavelengths are not extinguished at the same relative retardation. When $R = \lambda_2$, light of wavelength λ_2 is eliminated but some of λ_1 gets through. At $R = 2\lambda_1$, a great deal of λ_2 emerges, and so on. Sometimes, both colors are canceled at once but at different fringe orders; for example:

$$\text{Let } \lambda_1 = 660 \text{ nm (red)}$$
$$\lambda_2 = 440 \text{ nm (blue)}$$
$$\frac{(n_1 - n_2)d}{n_0} = R = 1320 \text{ nm}$$

$$\text{See that } 2 \times 660 = 1320 \text{ nm}$$
$$3 \times 440 = 1320 \text{ nm}$$

For this value of R, both wavelengths are eliminated. It is the second order of retardation for the red but the third order for the blue. The analysis is not precise because it has been simplified by ignoring dispersion of birefringence. Dispersion causes the second-order fringe for red not to exactly correspond to the third-order for blue.

As more wavelengths of light are added, a spectrum of colors is obtained. At any retardation, one or more wavelengths might be extinguished, and the eye sees the remainder (complementary color if white light is being used). This effect is useful for measuring small values of R. It is also beautiful and impressive to laypeople. Errors result from differences between individuals' eyes as well as from differences in lamps and other optical equipment. A chart of colors corresponding to various values of relative retardation appears in Table 5.2.

5.8 Separating isoclinics and isochromatics

In addition to using the circular polariscope to eliminate isoclinics from the photoelastic fringe pattern, other techniques can be used to distinguish the

Table 5.2. *Color sequence* (*complementary colors*) *in crossed linear polariscope, for sunlight as a function of the relative retardation R*

R (nm)	Extinguished	Observed
0	All colors	No color: darkness
100–300	None	Dark to light gray
440	Blue 1	Yellow 1
540	Green 1	Red 1
580		Tint of passage: purple
590	Yellow 1	Blue 1
660	Red 1	Green 1
880	Blue 2	Yellow 2
1080	Green 2	Red 2
1160		Tint of passage: purple
1180	Yellow 2	Blue 2
1320	Red 2 + blue 3	Green 2
1620	Green 3	Red 3
1760	Blue 4	[a]
1770	Yellow 3	[a]
1980	Red 3	Green 3
2160	Green 4	Red 4
2200	Blue 5	[a]
2360	Yellow 4	[a]

[a] As the retardation is increased, the overlapping of the extinguished colors becomes more and more pronounced: the color sequence above the third order gets paler until about the seventh order, when the transmitted light is practically white.
Source: G. Mesmer (1939).

types of fringes from one another. The most satisfactory methods are summarized briefly here:

1. As mentioned previously, to maintain isoclinics and reduce the number of isochromatics in the field, a model can be made of a relatively insensitive plastic. A disadvantage is that two models must be made, one for isoclinics and one for isochromatics. The gain in quality of results is usually worth the effort if good stress orientation data are required.
2. Use light loads on the model; only light loads are required to adequately develop the isoclinics.
3. Use polychromatic (white) light. Only the isoclinic and the zero-order isochromatic will be black, and the high-order isochromatics will be faint. Isoclinic patterns are commonly recorded in white light for this reason, and also to take advantage of the increased illumination level.

4. Examine the fringe pattern while varying the load. All the isochromatics except that of zero-order will move as the load is changed.
5. To separate the zero-order isochromatic from the isoclinics, watch the fringes as the crossed polarizers are rotated with respect to the specimen. The isoclinics will move, but the isochromatics will remain fixed.

5.9 Transfer from model to prototype

Engineers and laypeople alike, when they learn about photoelastic inter-ferometry, ask the obvious question: "What good does it do to determine the stresses in a model that is made of plastic when we need the stresses in a mechanical part that is made of metal?" In fact, the quantities desired in many experimental studies, such as stress, deformation, velocity, or strain, are determined by experiments on models rather than on a prototype, a production model, or a feature in the natural environment. Photoelasticity is a good example of a technique in which models are convenient and usually necessary. The information gained from a photoelasticity experiment, most often stress or strain in the model, must be related to the stress or strain in the prototype, which is likely of different size and different material and is subjected to different loads.

Attention must be given to two separate aspects of the problem. The first is the dependence of the measured stress and strain on material properties. The second has to do with the effects of model and load scaling. Both of these questions are treated in detail in Chapter 6. Only the important conclusions are given in this chapter on basic photoelasticity.

If the region being studied is "simply connected," meaning it has no holes, and if there are no displacement boundary conditions, and if the body forces are zero or constant, then the *stress distribution is completely independent of material properties.* If the body forces do not equal zero or a constant, then the Poisson ratio is an important parameter, but other material coefficients are not. If displacement boundary conditions are given, then, obviously, the stress magnitudes are simply proportional to the modulus of elasticity. These last two cases are relatively rare in applied photoelasticity. The conclusion is that in the great majority of practical applications of photoelasticity, the differences in properties of model and prototype are not relevant as long as both are behaving in a momentarily linearly elastic way. This result tends to startle inexperienced stress analysts. Even if the body force terms do not meet the conditions specified, the error in ignoring the difference between Poisson ratios for model and prototype is usually negligible.

For multiply connected regions (ones with holes) that are free of un-equilibrated traction on any boundary, the stress is independent of properties. If any boundary carries an unequilibrated load – such as usually happens with fasteners, for example – then the stress distribution depends on Poisson's ratio. The error resulting from ignoring this dependence will be less than 7% for ordinary photoelastic materials. Of course, if displacements or strains are specified, then the elastic modulus is important; but it is easily accounted for.

The laws of dimensional similarity turn out to be almost intuitively obvious. Geometric similarity requires that the model be linearly scaled up or down from the prototype. Its basic shape cannot be changed. A similar result holds for loads. That is, the ratios of like loads between model and prototype must be constant. A similar law holds for load direction. These conditions are sufficient for cases where no displacement boundary conditions are specified; otherwise the absolute load magnitudes must be scaled also.

If the conditions just specified are met for a two-dimensional problem, then the stresses in model and prototype satisfy the following equation,

$$\sigma_p = \sigma_m \left(\frac{P_p}{P_m}\right)\left(\frac{a_m}{a_p}\right)\left(\frac{d_m}{d_p}\right) \tag{5.4}$$

where   subscript m means model and p means prototype

P_p/P_m is the load scale factor

a_m/a_p is the dimensional scale factor

d_m/d_p is the thickness scale factor

Note that this result allows the thickness scale factor to be different from the scale factor for lateral dimensions. Under certain types of loading, such as tension or pure bending, the similarity conditions may be relaxed further. In other cases, such as for composite structures and laminates, the similarity conditions are more stringent because material properties enter the formulation in a more complex way.

References

Burger, C. P. (1987). Photoelasticity. In *Handbook on Experimental Mechanics*, Ed. A. S. Kobayashi, Ch. 5. Englewood Cliffs: Prentice-Hall.

Cloud, G. L., and Pindera, J. T. (1968). Techniques in infrared photoelasticity. *Experimental Mechanics*, 8, 5: 193–210.

Dally, J. W., and Riley, W. F. (1991). *Experimental Stress Analysis*, 3rd ed. New York: McGraw-Hill.

Frocht, M. M. (1941). *Photoelasticity*. New York: John Wiley and Sons.

Jessop, H. T., and Harris, F. C. (1949). *Photoelasticity, Principles and Methods*. London: Cleaver-Hume Press Ltd.

Mesmer, G. (1939). *Spannungsoptik*. Berlin: Springer-Verlag.

Post, D. (1989). Photoelasticity. In *Manual on Experimental Stress Analysis*, 5th ed., Eds. J. F. Doyle and J. W. Phillips, pp. 80–106. Bethel, CT: Society for Experimental Mechanics.

Wolf, H. (1961). *Spannungsoptik*. Berlin: Springer-Verlag.

6

Photoelasticity methods and applications

This chapter treats several topics that are important not only in photo-elasticity but also in other areas of interferometry. For example, many engineering experiments require good knowledge of the viscoelastic properties of materials and how to measure them. The transfer of experimental results from model to prototype is important to anyone contemplating experiments that might involve models, and such experiments are very common. Measurement of exact fractional fringe orders is useful in many applications of interferometry, and the methods used in photoelasticity provide a basis for similar methods in other areas, such as phase shifting in electronic speckle pattern interferometry.

Even so, inclusion of all the material that should appear in a chapter on applied photoelastic interferometry has been problematical because of space limitations and possible perceived distortion of emphasis. A factor is that many optical methods courses still spend a good deal of time on applied photoelasticity, and rightly so because it is an important engineering tool. Industrial users of the book probably need more on the subject than is given in Chapter 5. The coverage, in the end, represents a compromise among utility, educational value, length limitations, and the overall balance of the book; and it should be viewed in that light.

Finally, many of the references cited in the preceding two chapters are relevant to this one, even though they might not be mentioned specifically. The literature on techniques of photoelasticity is very large, and practitioners of the art will need to examine a good deal of it. Burger (1987) provides a particularly extensive and valuable list of references along with a wealth of technical detail. The treatment of many of these details by Post (1989) is compact, lucid, and useful. Dally and Riley (1991) provide a work that has become a standard in the field.

6.1 Birefringence in materials

The stress–strain–birefringence properties of materials are of interest for two main reasons. First, the relationship between stress and birefringence must be established before quantitative results can be obtained from a photoelastic model analysis. Second, improved understanding of the atomic and molecular structure and the behavior of materials will result from careful study of their mechanical-optical properties. Here material calibration is discussed, but observations about microstructural aspects are not pursued. The approaches used are drawn largely from the work of J. T. Pindera (1959a, 1959b, 1960, 1962), who was a pioneer in the characterization of time-dependent materials.

Experimental mechanicians sometimes are tempted to expend less than adequate effort in characterizing their model materials. Remember that the precision of quantitative results can never be better than the precision with which the material properties are determined.

6.1.1 General behavior

In approaching the problem of material calibration, it is important to establish the parameters that might be important. Certainly, time and temperature strongly affect the properties of plastics. Also, we would soon learn that the wavelength of the light used is important. In connection with the time effects, keep in mind the following two factors:

1. The constitutive relationships are themselves time-dependent.
2. The range of linearity of the basic constitutive relationship is time-dependent.

Because time is a very important parameter, the concepts of viscoelasticity can be used to advantage. There are two approaches. One is to try to model the material mathematically by viewing it as a system of spring and dashpot elements. For real materials, this is difficult to do and is of questionable validity if models are simple enough to be practical. Simple models are, however, useful in establishing concepts and guiding one's thinking. A more rewarding approach is to develop graphical and numerical representations of material behavior from experimental testing, which can involve a broad range of parameters. Such results can be used readily in various experimental and analytical situations. Where appropriate, but only where appropriate, the measured properties can be used in developing values of the compliance coefficients that are used in viscoelasticity. The point is, one should not assume linear viscoelasticity before the fact is demonstrated.

Ordinarily, the mechanical-optical constitutive relationships for the isotropic materials used in standard photoelasticity are determined from simple one-dimensional tests. If the behavior is linear, the results can be applied to multidimensional states.

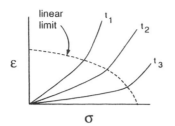
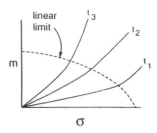

Figure 6.1. Mechanical-optical properties typical of some photoelastic model materials.

An important step is to decide how many measurements or parameters must be determined to completely characterize the material. This question has been answered as follows (Theocaris 1965):

A simultaneous measurement of the mechanical and optical quantities of a linear viscoelastic material in a *simple tension test* supplemented with one other characteristic function suffices for the complete characterization of the material.

These "mechanical and optical quantities" may be, for example, a stress–strain curve and a stress–birefringence curve from some given time after load or at a given strain rate. The "one other characteristic function" may well be the "lateral contraction ratio," which has been called a "time-dependent Poisson ratio" (Daniel 1965). From such a test, all other functions characteristic of a linear viscoelastic material can be derived. Alfrey (1948) shows that there are (at least) seven equivalent and related ways of defining the viscoelastic behavior of materials. For most plastics in the glassy state, the lateral contraction ratio can be taken as constant, and it is easily determined.

It is unwise to assume that photoelastic materials are linearly viscoelastic, although they might be so over a certain range of stress and time. The range would have to be established as part of the test. Guided by the previous comments about viscoelastic materials, one can summarize the minimum requirements for a calibration test of a material. For any single given wavelength, the plots of strain versus stress (ϵ–σ) and fringe order versus stress (m–σ) for constant times, (t), as identified in Figure 6.1, plus the lateral contraction ratio, are needed.

The implications of the nonlinear portions of Figure 6.1 are important. For the time being, discussion will be confined to the linear portion for which the experimental creep results are found to be related to the parameters used in linear viscoelasticity. Observe that the linear portions of the curves shown above can be described mathematically at any instant as follows:

$$\epsilon(t_i) = D(t_i)\sigma \qquad \text{(for given } \lambda) \tag{6.1}$$

(a)
specimen

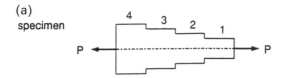

(b)
load
diagram

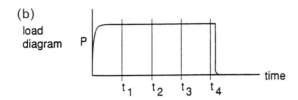

(c)
plot

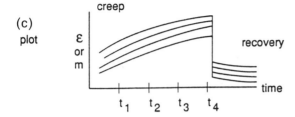

Figure 6.2. Creep test using a stepped specimen. (a) specimen, (b) loading curve, and (c) typical response curves.

where D is the instantaneous value of the "extension compliance." Its reciprocal is the "relaxation modulus," E.

$$\epsilon(t_i) = \frac{\sigma}{E(t_i)} \tag{6.2}$$

Clearly, the preceding equation is a time-dependent Hooke's law. Similar expressions hold for the optical compliance.

6.1.2 Creep and relaxation tests

The important parameters discussed previously can be established directly for broad spectra of stress, strain, birefringence, wavelength, and time from a complete creep or relaxation test of the material. For a given engineering model analysis, a complete test is not often necessary; appropriate simplifications will be discussed later. Here, general aspects of creep tests and their interpretations are offered. Then, particular tests that are of practical use will be described.

In a creep test, the load is applied and held constant over a period of time. By using several specimens at different loads or a single specimen with a spectrum of stress levels, the strain or optical response at several stresses and times is determined. Figure 6.2 pictures a specimen, a history of load P, and typical response curves for such a creep test.

The problem now is to develop meaningful measures of material response from the test results.

6.1.3 Linear viscoelastic response

Some of the concepts and quantities in the theory of linear viscoelasticity will now be summarized. Keep in mind that one cannot assume a material to be linearly viscoelastic (or momentarily linearly elastic). The concepts of visco-elasticity are helpful mainly in deciding what types of calibration tests to conduct, what data to collect, and how to present it.

The viscoelastic response of a material can be written in terms of the "creep compliance" by Boltzmann's superposition principle,

$$\epsilon_{ij}(t) = \frac{1}{2} \int \left[J(t - \tau) \frac{dS_{ij}(\tau)}{d\tau} \right] d\tau \qquad (6.3)$$

where J is the creep compliance (strain response to unit sustained shear stress as determined from experiment or with theoretical models);
S_{ij} is the time-dependent deviatoric stress;
ϵ_{ij} is the time-dependent deviatoric strain;
τ is the integration variable for the time domain

A similar equation can be written for the bulk compliance.

For the uniaxial case,

$$\epsilon(t) = \int \left[D(t - \tau) \frac{d\sigma(t)}{d\tau} \right] d\tau \qquad (6.4)$$

where $D(t)$ = extension compliance (see eq. 6.1). For a creep test, the stress is constant,

$$\sigma(t) = \text{const} = \sigma_0 \qquad (6.5)$$

So, for the creep test,

$$\epsilon(t) = \sigma_0 D(t) \qquad (6.6)$$

Equation 6.6 tells us that a creep test can be used to determine the extension compliance, which is a special case of the creep compliance J.

The preceding equations can be put in terms of the extension modulus $E(t)$, with the result, for a relaxation test where $\epsilon = \epsilon_0 = \text{const}$,

$$\sigma(t) = \epsilon_0 E(t) \qquad (6.7)$$

The two compliances are related by the integral

$$\int E(t - \tau)D(\tau) \, d\tau = t \qquad (6.8)$$

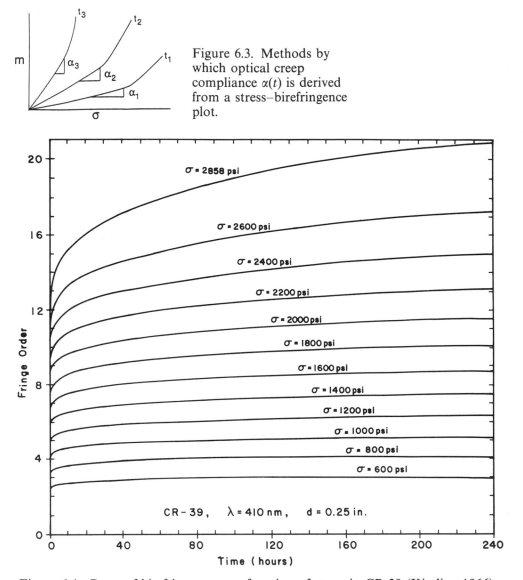

Figure 6.3. Methods by which optical creep compliance $\alpha(t)$ is derived from a stress–birefringence plot.

Figure 6.4. Creep of birefringence as a function of stress in CR-39 (Kiesling 1966).

An approximation is

$$E(t) \approx \frac{1}{D(t)}$$

If the Poisson ratio is taken as constant, then a creep test completely characterizes the material as far as its mechanical response is concerned.

The optical response is determined in a similar way except that the isoclinic information complicates matters. A typical equation is

$$\sigma_{xy} = \frac{1}{2} \int \beta_\sigma(t - \tau) \frac{d}{d\tau} \left[\frac{n_1 - n_2}{n_0} \sin 2\phi_n(\tau) \right] d\tau \qquad (6.9)$$

where ϕ_n are stress directions;
          $\beta_\sigma(t)$ is the stress difference per unit birefringence

A creep test gives the birefringence per unit stress difference, called the optical creep compliance $\alpha(t)$, which is just the inverse of the β_σ and which can be defined as the slope of the m–σ curve expressed as a function of time. For the uniaxial creep test,

$$(\sigma_1 - \sigma_2) = \sigma_0$$

$$m(t) = \alpha(t)(\sigma_1 - \sigma_2) \qquad (6.10)$$

$$\phi_n = 0$$

Figure 6.3 clarifies this concept.

The $\beta_\sigma(t)$ comes from an optical relaxation test. The two optical parameters are related as follows,

$$\int \alpha(t - \tau)\beta_\sigma(\tau)\,d\tau = t \qquad (6.11)$$

The approximate relationship is

$$\alpha(t) \simeq \frac{1}{\beta_\sigma(t)} \qquad (6.12)$$

In the past sections we have used a stress-optical coefficient C_σ. Clearly, it must be a function of time and wavelength. How is it related to the $\alpha(t)$ that results from the creep calibration? We would find that, for thickness d and wavelength λ,

$$\alpha(t) = C_\sigma(t)\frac{d}{\lambda} \qquad (6.13)$$

6.1.4 A typical result

Figures 6.4 and 6.5 are examples of the stress-birefringence characterization curves that were obtained for the popular photoelastic material CR-39 (Kiesling 1966; Kiesling and Pindera 1969) using techniques developed by Pindera (1959a, 1959b, 1960, 1962). Figure 6.6 shows some of the viscoelastic optical and mechanical characterization parameters that can be developed from such test results (Kiesling 1966). The details of the material tests differ from what has been suggested here so far, but the basic ideas are the same. Many interesting

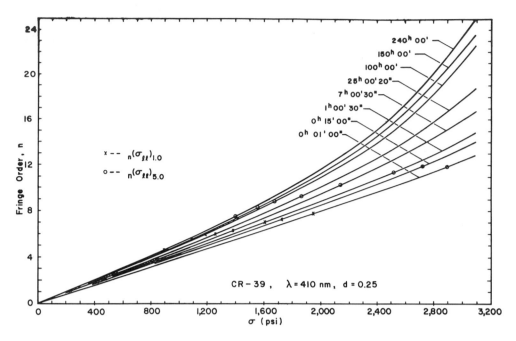

Figure 6.5. Birefringence as function of stress for constant times in CR-39 (Kiesling and Pindera 1969).

research possibilities remain in this field of viscoelastic materials characterization.

6.1.5 Specimens and procedures for material characterization

So far, only the tension test has been suggested for obtaining the stress–strain–optic properties of a material. The tension test is certainly the best one to use, especially where nonlinear behavior must be considered. The outstanding advantage of this test is that the stress field does not depend on material properties; that is, it does not require a calculation that presumes ordinary elastic behavior. The stress state is a function of only the cross-sectional area, symmetry, and load. The main disadvantage of the simple versions of the tension test is that several specimens must be used at several loads in order to obtain a complete picture of the response over a spectrum of stress–time–wavelength. This difficulty can be eliminated by utilizing a more complicated tensile model, such as one having several different cross sections, as was pictured in Figure 6.2. The following are summaries of various models and testing procedures that can be employed to investigate properly the stress–strain–birefringence properties of materials. Attention is focused on

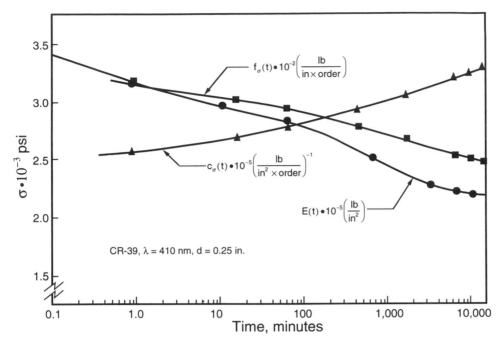

Figure 6.6. Time-dependent stress-optic coefficient (C_σ), material fringe value (f_σ), and extension modulus (E) as functions of stress (σ) and time (t) for CR-39 (Kiesling 1966).

the stress–birefringence determination for the sake of brevity. Identical principles apply to establishing strain–related properties.

Simple tension test

The simplest test in principle, although not so in practice, involves the typical tensile "dog-bone" specimen having constant cross section through the test area and enlarged ends for load application. To obtain a spectrum of stress–time response, the load must be varied. The loading and recording must be done in one of two ways to properly characterize the viscoelastic response. These techniques are as follows:

1. Increase load stepwise, taking isochromatic or strain data at constant times after changing the load. The loading and recording history are shown graphically in Figure 6.7. Note that the load increments need not be equal. The restriction that the intervals between changing load and taking data be constant must be adhered to for valid results. The results of this test are calibration curves for the chosen times after loading. See Figure 6.8.

2. Rather than a stepwise increasing load, the specimen can be loaded and unloaded. Results are recorded at the chosen times after applying each load. The specimen is allowed to recover to near-original state between load periods. Figure 6.9 shows

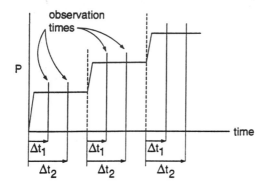

Figure 6.7. Load and observation history for tensile test with increasing load.

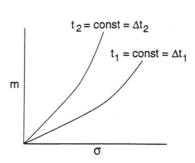

Figure 6.8. Stress-optic properties as derived directly from a tension test.

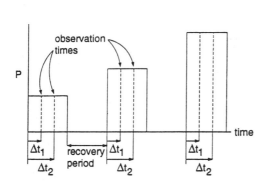

Figure 6.9. Alternate loading and recording history for simple tension test.

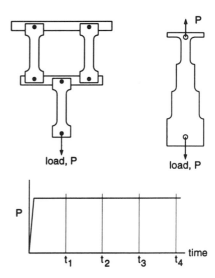

Figure 6.10. Specimens, load, and observation histories for the tensile creep test.

the loading and recording histories. The problem with this test is that the time for a reasonable degree of recovery might be quite long. As nonlinear response is approached, this period increases. If plastic behavior ensues, recovery can never be attained, and the test results are not valid beyond this point.

Tensile creep test

An alternative form of the tension test utilizes several specimens loaded simultaneously in a special load-dividing jig, or else a single, stepped tension

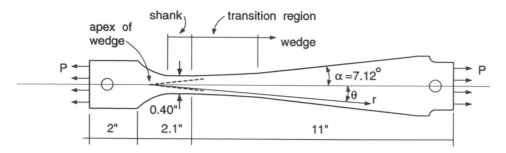

Figure 6.11. Typical tapered tension specimen.

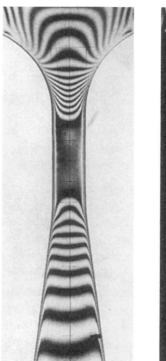
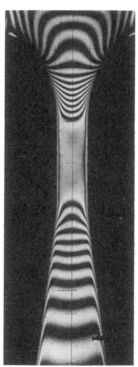

Figure 6.12. (a) Light-field and (b) dark-field isochromatic patterns for a portion of tapered specimen (Cloud 1966).

specimen. This approach is, of course, equivalent to conducting several different creep tests on fresh specimens. Figure 6.10 illustrates two types of specimen arrangements with the load and resultant stress histories for this type of test. Observations are taken at chosen times (e.g., t_1, t_2, t_3). From these data a plot similar to that of Figure 6.8 can be constructed.

Creep test with tapered specimen

The idea of using a tapered tensile model was developed and validated by
Pindera (1959a, 1959b, 1960). Such a specimen affords a continuous distribution of stress between certain limits. Various investigators (e.g., Cloud
1969; Kiesling 1966; Pindera 1966; Pindera and Cloud 1966) have employed
the tapered specimen with success in precise measurement of properties of
photoelastic materials, and the results presented in Figures 6.4, 6.5, and 6.6
were obtained in this way. The wedge angle should be kept smaller than
about 14° in order to hold uniaxiality of the stress state to within 2%.
Figure 6.11 shows the version of this specimen used by the authors just
mentioned. Figure 6.12 shows typical isochromatic patterns for this specimen.

Because of the particular stress distribution in the tapered specimen, it is
convenient to add an intermediate stage to the data processing. The stress in
the wedge portion can be calculated from the equations (Pindera 1959a),

$$\sigma_{radial} = K\sigma_{nom}$$

$$\sigma_{nom} = \frac{P}{2rd\cos\theta\tan\alpha} \tag{6.14}$$

$$K = \frac{4\cos^2\theta\tan\alpha}{2\alpha + \sin 2\alpha}$$

where α is the wedge angle
             θ is the angle off the centerline at which the stress is desired

Usually one takes data only along the centerline. The stress in the uniform
part is found by the usual P/A method. Because of stress–concentration
effects, the stresses in the transition region are not accurately known. The
known stresses are conveniently plotted on large-size graph paper or
tabulated on computer along with isochromatic orders obtained from
photographs taken at the specified times after load. Strain plots can also be
included. All data are plotted or tabulated versus position on the specimen.
There will, of course, be different isochromatic plots for each time (or
wavelength) used. The final calibration curves are obtained by plotting the
pairs of stress and fringe orders that occur at chosen specimen cross sections.
The result will be a comprehensive graph such as was presented in Figure 6.5.

There are several other valid approaches to processing the data from a
creep test of a tapered specimen. The one just described is reasonably efficient
and gives good accuracy while automatically giving some necessary smoothing to the data. The plotting must, of course, be carefully done. Certain
computer plotting routines can be used to advantage. Care must also be
exercised in reading the isochromatic photographs. Considerable time and
effort will be saved if one studies the literature on the subject before
undertaking an extensive study of mechanical birefringence in materials.

Beam bending test

The beam in pure bending has a spectrum of uniaxial stress, which makes it a useful calibration model. Unfortunately, the calculated stress distribution depends on the assumption of linear material and geometric behavior. If any of the beam in the region studied evidences nonlinear response, then *none* of the computed stresses are any good. The *strain* distribution, however, is calculated on the basis of symmetry alone, so such a model can be used to determine nonlinear strain-optic parameters. The strain in the beam can be determined directly by measuring radius of curvature or by using strain gages on the cross section. Handling of observations and data is similar to the methods used for the tapered model. For reasons not well understood, the beam calibration test often gives trouble in practice. It is a good quick way to get C_σ to within about 5% when linearity is maintained.

Compression of disc

A disc in diametral compression is another model that is useful for calibration (recall that a disc is also used in calibrating the photoelasticity instrument). Data analysis techniques parallel those used for the tapered tension model, so little more need be said about it except to comment again that the calculated stresses depend on the assumption of linear behavior. The elasticity solution for the disc with diametral load is given below (Frocht 1941). Refer to Figure 6.13 for the specimen geometry and coordinate system.

The stress state at any point (x, y) is

$$\sigma_x = \frac{-2P}{\pi d}\left[\frac{(R-y)x^2}{r_1^4} + \frac{(R+y)x^2}{r_2^4} - \frac{1}{D}\right]$$

$$\sigma_y = \frac{-2P}{\pi d}\left[\frac{(R-y)^3}{r_1^4} + \frac{(R+y)^3}{r_2^4} - \frac{1}{D}\right]$$

$$\tau_{xy} = \frac{2P}{\pi d}\left[\frac{(R-y)^2 x}{r_1^4} - \frac{(R+y)^2 x}{r_2^4}\right]$$

On the horizontal diameter,

$$y = 0, \qquad r_1 = r_2 = (x^2 + R^2)$$

$$\sigma_x = \frac{2P}{\pi dD}\left[\frac{D^2 - 4x^2}{D^2 + 4x^2}\right]^2$$

$$\sigma_y = \frac{-2P}{\pi dD}\left[\frac{4D^4}{(D^2 + 4x^2)^2} - 1\right]$$

$$\tau_{xy} = 0$$

(6.15)

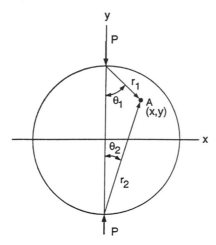

Figure 6.13. Coordinate system for a disc in diametral compression.

On the vertical diameter,

$$x = 0$$

$$\sigma_x = \frac{2P}{\pi dD}$$

$$\sigma_y = \frac{-2P}{\pi d}\left[\frac{2}{D-2y}+\frac{2}{D+2y}-\frac{1}{D}\right]$$

$$\tau_{xy} = 0$$

(6.16)

Equations 6.15 and 6.16 are the ones most used in photoelasticity studies of materials. The reference cited earlier (Frocht 1941) also gives solutions for the disc with loads on any chord, as well as other results.

Comments and tips on laboratory procedures for calibration studies and other model tests

1. Care must be used in keeping the loading truly uniaxial in the tensile tests. The isochromatic pattern in white light is a good detector of poor symmetry. Grips with adjustable links such as those pictured in Figure 6.14 should be employed if possible (Pindera 1959a). Apply a small load to the specimen and adjust the grips to get a uniform isochromatic order (color), then proceed with the test.
2. A plumb line and lines lightly scratched into the specimen surface are helpful in assuring proper alignment of load in calibration and model tests.
3. Establish exposures, waiting periods, load magnitudes, and so on before beginning the test.
4. Be aware at all times of the drastic effects of temperature on photoelastic plastics. A calibration at one temperature is not valid for any other temperature. Humidity seems to be not so important, but it may have an effect on results.

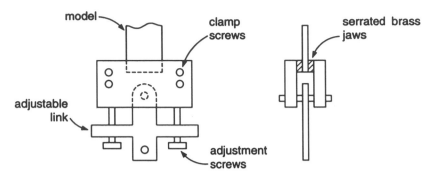

Figure 6.14. Adjustable grips for the tension test.

5. Don't forget to determine and record the wavelength of light used.
6. Record all time data immediately.

6.1.6 Optical and mechanical creep coefficients

Constructing and using the complete calibration curves, such as shown in Figures 6.5 and 6.6, can be laborious. The job would be simpler if data obtained from one batch of material could be used for another lot. This cannot be done directly, because both the mechanical and the optical responses on different lots of the same material can vary greatly. The author has observed, however (Cloud 1969), that the creep, meaning the change in $C_\sigma(t)$ or $E(t)$, seems to be relatively independent of lot number for some materials. In such situations one can determine the creep response of the material once and use it for this material. A base value of the C_σ must still be determined for each batch at one time only. To accomplish all this, an optical creep coefficient is used for a given wavelength λ_0:

$$r_t = \frac{m(t)}{m(t_0)} = \text{optical creep coefficient} \qquad (6.17)$$

where t_0 = a "base time" established arbitrarily. As mentioned, r_t is established once and for all from a complete calibration test on a given lot of material. A typical result is given in Figure 6.15 (Cloud 1969).

To obtain $C_\sigma(t)$ for another lot of the material, it is necessary to test it for only *one* time t_0 (and wavelength). The complete $C_\sigma(t)$ can then be calculated as follows:

$$C_\sigma(t) = C_\sigma(t_0)r_t \qquad (6.18)$$

The r_t is related to the viscoelasticity parameters mentioned previously. Mechanical creep can be handled in a similar way.

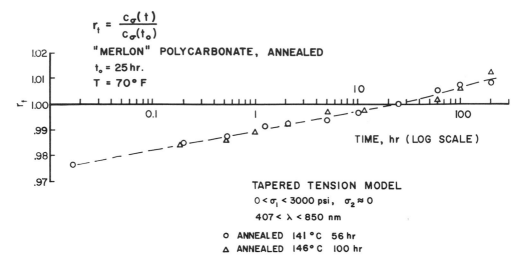

$$r_t = \frac{c_\sigma(t)}{c_\sigma(t_o)}$$

"MERLON" POLYCARBONATE, ANNEALED

t_o = 25 hr.

T = 70° F

TIME, hr (LOG SCALE)

TAPERED TENSION MODEL

$0 < \sigma_1 < 3000$ psi, $\sigma_2 \approx 0$

$407 < \lambda < 850$ nm

○ ANNEALED 141 °C 56 hr

△ ANNEALED 146° C 100 hr

Figure 6.15. Optical creep coefficient (r_t) for polycarbonate resin (Cloud 1969).

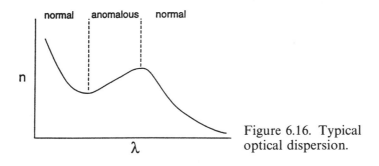

Figure 6.16. Typical optical dispersion.

6.1.7 Dispersion of birefringence

Basic photoelasticity theory says that fringe order is inversely proportional to wavelength. That is, the relative retardation $R = m\lambda$ should be constant for a given stress state regardless of the wavelength of light used to observe it. Experimental observations show that the basic theory is simplistic, and the photoelastic coefficient is, in fact, a function of wavelengths as well as time. In studying this phenomenon, the concepts of ordinary optical dispersion are useful.

The ordinary index of refraction n of a material is dependent on wavelength. If it were not, then a simple prism would not separate colors. This dependence of n on wavelength is called *optical dispersion*. Normally, n decreases with increasing λ, and such behavior is called normal dispersion. In some instances, the opposite is true; and this behavior is called anomalous dispersion (Coker and Filon 1957). Figure 6.16 shows the typical behavior in a qualitative way.

A theoretical model for optical dispersion can be found in optical physics

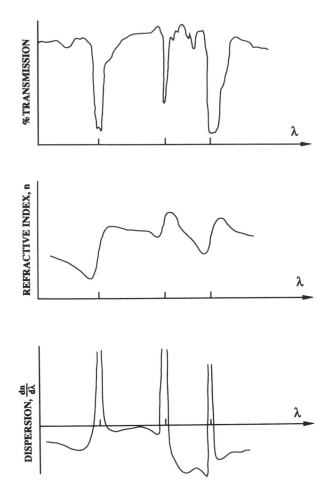

Figure 6.17. Typical relationship between transmission spectrum, refractive index, and optical dispersion.

books (Garbuny 1965). The anomalous dispersion regions are explained as regions where an electron in the harmonic oscillator model is in resonance with the radiation (at optical frequency). Figure 6.17 summarizes the essential details of experimental observations.

Similar remarks can be made about the dependence of relative retardation $R = m\lambda = (n_1 - n_2)d/n_0$ on wavelength (Pindera 1959b). It is called *dispersion of birefringence*, for obvious reasons. Regions of anomalous and normal dispersion can be observed experimentally. Such observations are quite difficult, however, because of the nature of the measurements required.

Two parameters have been used to describe the dispersion of birefringence of a material (Pindera and Cloud 1966). The most useful is the "normalized retardation,"

$$r_\lambda = \frac{m\lambda}{m_0\lambda_0} = \frac{R_\lambda}{R_{\lambda 0}}$$ (6.19)

where λ_0 is some reference wavelength.

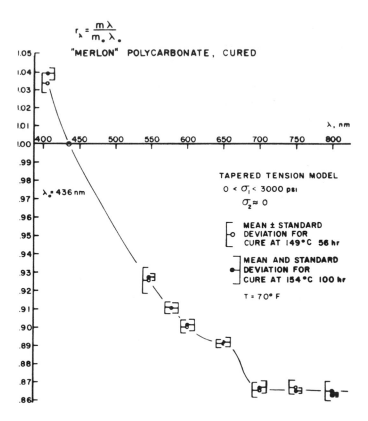

Figure 6.18. Normalized retardation as a function of wavelength for polycarbonate (Cloud 1969).

The "normalized dispersion of birefringence" is the derivative of the preceding quantity:

$$D_\lambda = \frac{dr_\lambda}{d\lambda} \tag{6.20}$$

Note that, in a linear range,

$$C_\sigma(\lambda) = C_\lambda(\lambda_0) r_\lambda \tag{6.21}$$

Equation 6.21 allows one to determine the stress-optic coefficient at one wavelength from a calibration at another wavelength. Compare with the use of the creep coefficient in equation 6.18.

As an example of the magnitude of dispersion of birefringence (Cloud 1969, 1970), Figure 6.18 shows the normalized retardation as a function of wavelength in visible and near-infrared light for polycarbonate resin. The dispersion amounts to about 18% through the visible range, and there is a region of anomalous dispersion. Such results are quite typical of photoelastic materials, and this nonlinear aspect of photoelastic response should not be ignored in experiments where more than one wavelength is used.

It is instructive to compare the normalized dispersion of birefringence data with the transmission spectra (Cloud 1970), as it suggests the nature of the relationships between birefringence and material structure. The change of stress-optic coefficient with wavelength is marked by three general characteristics:

1. There is a region of anomalous dispersion of birefringence on the short-wavelength side of each absorption band.
2. The dispersion of birefringence is large and normal on the long-wavelength side of each absorption band.
3. The dispersion of birefringence remains uniformly small in the portions of the spectrum where the transmittance remains uniformly high.

The molecular and atomic aspects of birefringence will not be discussed further, owing to space limitations and the general nature of this treatise. The development of physical models of birefringence that are based on experiment is an interesting area of research in which much remains to be done.

6.2 Similarity and scaling in model analysis

In many engineering investigations, the desired quantities, such as stress, deformation, velocity, or strain, are determined by experiments on physical models rather than on a prototype, a production model, or a feature in the natural environment. Often this use of a model is dictated by necessity, as when the physical prototype is large (consider the investigation of flow in a harbor) and/or expensive (as in determining stresses in a dam); and sometimes investigating a model is simply more informative or convenient. In all these applications, the findings must be properly transferred from the model to the prototype. Or the problem might be inverted. Given a prototype (perhaps a proposed design), how should a model be constructed to give useful results within the limitations of the technique at hand?

Photoelasticity is a good example of a technique in which physical models are convenient and usually necessary. The information gained from photoelasticity experiments, most often stress or strain in the model, must be related to the stress or strain in the prototype, which is likely of different size, different material, and subjected to different loads. The dependence of stress or strain on material properties, model size, and model load must be established.

Many of the problems faced in photoelastic model similarity are common to other areas of model analysis. The same concepts of similarity and dimensional analysis apply for wind tunnel testing, tank testing of ship models, pilot plant testing, and stress analysis of structures or mechanical components.

Here, attention is focused on two separate aspects of the problem. The first is the dependence of the measured stress and strain on material properties; the conditions of material similarity are established. Then, the factors of scale in load and model size are developed.

6.2.1 Accounting for material properties

Discussion will be confined to the two-dimensional case for simplicity and maximum usefulness. Similar reasoning is valid for three dimensions.

The sum of principal stresses in a two-dimensional stress field (which includes plane stress, plane strain, and generalized plane stress) must satisfy the compatibility equations of classical elasticity. These equations are

$$\nabla^2(\sigma_{xx} + \sigma_{yy}) = -(v + 1)\left(\frac{\partial F_x}{\partial x} + \frac{\partial F_y}{\partial y}\right)$$

for plane stress, and

$$\nabla^2(\sigma_{xx} + \sigma_{yy}) = -\left(\frac{1 - 2v}{1 - v}\right)\left(\frac{\partial F_x}{\partial x} + \frac{\partial F_y}{\partial y}\right) \tag{6.22}$$

for plane strain, where F_x and F_y are components of the body force field.

If the region being studied is simply connected, meaning it has no holes, and where there are no displacement boundary conditions, then equations 6.22 are sufficient for defining the stress distribution. If the body force terms on the right-hand side are zero (e.g., body forces are zero or constant), then the stress *distribution* is completely *independent of material properties*. If the body force terms do not equal zero, then the Poisson ratio is an important parameter but other material coefficients are not. If displacement boundary conditions are given, then, obviously, the modulus of elasticity governs the stress magnitudes. These last two cases are relatively rare in applied photoelasticity. The conclusion is that in the great majority of practical applications of photoelasticity, the differences in properties of model and prototype are not relevant as long as both are behaving in a momentarily linearly elastic way. This result tends to startle inexperienced stress analysts. Even when the body force terms in the preceding equations are not zero, the error in ignoring the difference between the Poisson ratios for model and prototype is usually negligible.

In a multiply connected region (one with holes), the compatibility conditions are necessary but not sufficient. The requirement that displacements be single-valued must be imposed also. Mathematically, this condition is

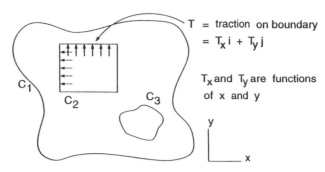

Figure 6.19. Multiply connected region with a loaded inner boundary.

expressed by stating that the resultant displacement integrated around each and every boundary is zero. After considerable algebra, this statement leads to "Mitchell's conditions," which are given in equations 6.23 with reference to Figure 6.19.

$$\int_{C_i} \frac{\partial}{\partial n} (\nabla^2 \phi) \, ds$$

$$\int_{C_i} \left[y \frac{\partial}{\partial n} (\nabla^2 \phi) - x \frac{\partial}{\partial s} (\nabla^2 \phi) \right] ds = - \frac{1}{1-v} \oint_{C_i} T_y \, ds \qquad (6.23)$$

$$\int_{C_i} \left[y \frac{\partial}{\partial s} (\nabla^2 \phi) - x \frac{\partial}{\partial n} (\nabla^2 \phi) \right] ds = - \frac{1}{1-v} \oint_{C_i} T_x \, ds$$

where C_i is the ith boundary
T_x, T_y are the components of traction on the ith boundary
ϕ is the stress function

Each of these integral expressions is extended over each region enclosed by a boundary.

Only the expressions on the right-hand sides of the preceding equations contain a material property: Poisson's ratio, to be specific. If these integral terms on the right are all zero, then, clearly, the stress distribution is again independent of material constitutive relations. The conclusion is that in a multiply connected region free of unequilibrated traction on any boundary, the stress is independent of properties. Otherwise, the distribution depends on Poisson's ratio. Of course, if displacements or strains are specified, then the elastic modulus is important. Clutterbuck (1958) has shown that in cases where the stress distribution depends on v, the error resulting from ignoring this dependence will be less than 7% for ordinary photoelastic materials.

A classic example of a practical case where there is a resultant force on an inner boundary is a machine component that has holes for fasteners.

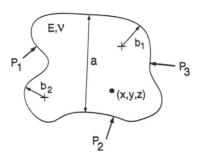

Figure 6.20 Hypothetical two-dimensional elasticity problem.

6.2.2 Accounting for model and load dimensions

The transfer of photoelastic data to a prototype can be handled by application of the principles of dimensional analysis. The results are called the laws of model similarity.

The principal weapon of dimensional analysis is Buckingham's theorem. For a linearly elastic solid, this theorem states that the stress can be expressed as a function of a number of dimensionless products of the variables involved. Figure 6.20 shows these variables.

For stress σ at (x, y, z), Buckingham's pi theorem gives the following functional relationships:

$$\frac{\sigma a^2}{P} = f\left[\frac{x}{a}, \frac{y}{a}, \frac{z}{a}, v, \frac{P}{Ea^2}, r_1, r_2, \ldots, r_n, p_1, p_2, \ldots, p_n, d_1, d_2, \ldots, d_n\right]$$

(6.24)

where
    x, y, z locate a point in the body
    a is a representative dimension
    b_i are other dimensions
    r_i are size ratios describing geometry of body, $r_1 = b_1/a$,
        $r_2 = b_2/a, \cdots$
    P is one of the loads
    p_1 are load ratios, $p_1 = P_1/P$, $p_2 = P_2/P, \cdots$
    d_1 are direction ratios for loads
    E is the elastic modulus
    v is the Poisson ratio

A key point is that this functional equation for stress must apply to model and prototype alike. If the groups on the right-hand side are made the same for model and prototype, the result on the left must also be the same for both. These conditions lead to the basic similarity laws, which are summarized here:

1. Geometric similarity requires corresponding dimensions to be in the same ratio,

$$r_{1m} = r_{1p}, \qquad r_{2m} = r_{2p}, \qquad \cdots$$

where subscripts m and p identify model and prototype. Using the definitions of the r_i and doing a little algebra yields

$$\frac{a_m}{a_p} = \frac{b_{1m}}{b_{1p}} = \frac{b_{2m}}{b_{2p}} = \cdots = \frac{b_{im}}{b_{ip}} \qquad (6.25)$$

This result says that like linear dimensions of model and prototype must occur in constant ratio: the model can be scaled up or down as long as its shape is not changed. A special case, which we will not study, shows that for a two-dimensional problem, the condition of geometric similarity need not apply to the thickness.
2. Applying similar reasoning to the load ratios establishes load distribution similarity.

$$P_{1m} = P_{1p}, \qquad \cdots \qquad (6.26)$$

or, after some algebra,

$$\frac{P_m}{P_p} = \frac{P_{1m}}{P_{1p}} = \frac{P_{2m}}{P_{2p}} = \cdots = \frac{P_{im}}{P_{ip}} \qquad (6.27)$$

That is, the ratios of like loads between model and prototype must be constant. A similar law holds for load direction.
3. To establish load magnitude similarity, observe that the load magnitude appears in only one ratio of equation 6.24. In cases where E is important, such as when displacement boundary conditions are specified, the load scaling factor is

$$\left(\frac{P}{Ea^2} \right)_m = \left(\frac{P}{Ea^2} \right)_p \qquad (6.28)$$

4. Poisson's ratio must be the same if it enters the formulation, as in multiply connected regions with a resultant force on a boundary. This condition has been analyzed in another way in the section on material similarity.

If the preceding model laws are satisfied, then the left-hand side of equation 6.24 is the same for model and prototype.

$$\left(\frac{\sigma a^2}{P} \right)_m = \left(\frac{\sigma a^2}{P} \right)_p$$

or

$$\sigma_p = \sigma_m \left(\frac{P_p}{P_m} \right) \left(\frac{a_m}{a_p} \right)^2 \qquad (6.29)$$

If the photoelasticity study is strictly two-dimensional, then equation 6.29 can be relaxed slightly. The a^2 can be replaced by $a \times d$, where d is the thickness. Equation 6.29 is changed to

$$\sigma_p = \sigma_m \left(\frac{p_p}{p_m}\right) \left(\frac{a_m}{a_p}\right) \left(\frac{d_m}{d_p}\right) \qquad (6.30)$$

Under certain types of loading, such as tension or pure bending, the similarity conditions can be relaxed further. In other cases, such as for composite structures and laminates, the similarity conditions are more stringent because material properties enter the formulation in a more complex way.

6.3 Obtaining separate principal stresses

We have seen how photoelastic analysis yields directly the following information:

stress directions in whole region
difference between principal stresses (maximum shear stress) in whole region
boundary stress

For many problems in analysis and machine design, the study need be carried no further. Certain problems, however, require knowledge of individual principal stresses. There are several methods – analytical, experimental, and combined – for obtaining this information.

6.3.1 Laplace equation

Clearly, if the sum of the principal stresses in the region could be determined, then, because the stress difference is known from photoelasticity, then, because the problem of obtaining separate principal stresses is solved. It is easy to show that the sum of principal stresses must satisfy the Laplace equation with given boundary values. Begin with the equilibrium equations of two-dimensional elasticity,

$$\frac{\partial \sigma_{xx}}{\partial x} + \frac{\partial \tau_{xy}}{\partial y} + F_x = 0$$

$$\frac{\partial \sigma_{yy}}{\partial y} + \frac{\partial \tau_{xy}}{\partial x} + F_y = 0 \qquad (6.31)$$

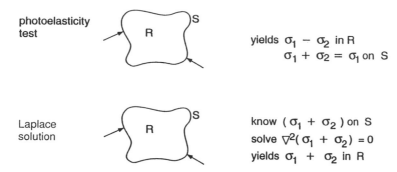

Figure 6.21. Stress separation through use of the Laplace equation.

Then, by using the compatibility equations, it is easily demonstrated that

$$\nabla^2\left(\sigma_{xx} + \sigma_{yy} - \frac{\Omega}{1-v}\right) = 0 \qquad (6.32)$$

where $\qquad \Omega$ is the body force potential;
$\nabla^2 = \partial^2/\partial x^2 + \partial^2/\partial y^2$ is the Laplace operator

A special case having broad application is when $\nabla^2\Omega = 0$. For this case,

$$\nabla^2(\sigma_{xx} + \sigma_{yy}) = 0$$

The principal stresses σ_1 and σ_2 must satisfy this equation. Now, on the boundary of the region, $\sigma_1 + \sigma_2 = \sigma_1 = \sigma_1 - \sigma_2$; so $(\sigma_1 + \sigma_2)$ on the boundary has already been determined from the photoelasticity study.

The problem just formulated is a classic in mathematical physics, and it is called the *Dirichlet problem*. Simply stated, one seeks to find a function $(\sigma_1 + \sigma_2)$ where the boundary values $(\sigma_1 + \sigma_2)_s$ are known and where $\nabla^2(\sigma_1 + \sigma_2) = 0$. Figure 6.21 summarizes the problem. Many solutions, analogies, and techniques are available for solving such a problem. A few of the techniques will be described.

6.3.2 Numerical iteration scheme

A simple numerical scheme for solving the Laplace equation with known boundary values can be developed from finite differences. Consider first a simple function $z = z(x)$, such as that shown in Figure 6.22.

The function is divided into finite segments, and approximations to the first derivatives, written as partials because that is what will be needed, can be devised for each segment as

$$\left.\frac{\partial z}{\partial x}\right|_{wo} \approx \frac{z_0 - z_w}{h} \qquad \text{and} \qquad \left.\frac{\partial z}{\partial x}\right|_{oe} \approx \frac{z_e - z_0}{h}$$

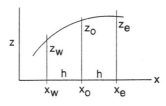

Figure 6.22. Development of finite difference approximations for a simple function.

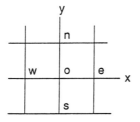

Figure 6.23. Node points for a finite difference approximation in two dimensions.

These approximations are then used to develop an estimate of the second derivative,

$$\frac{\partial^2 z}{\partial x^2} = \frac{\left.\frac{\partial z}{\partial x}\right|_{oe} - \left.\frac{\partial z}{\partial x}\right|_{wo}}{h}$$

$$= \frac{z_e + z_w - 2z_0}{h^2} \tag{6.33}$$

Now, think of the function as a surface over a two-dimensional space, $z = a(x, y)$. Refer to Figure 6.23 for identification of node points.

The approximate partial derivatives for the y-direction are developed as they were before,

$$\frac{\partial^2 z}{\partial y^2} = \frac{z_n + z_s - 2z_0}{h^2} \tag{6.34}$$

When the above expressions are put together, an approximation to the Laplacian operator is obtained,

$$\frac{\partial^2 z}{\partial x^2} + \frac{\partial^2 z}{\partial y^2} = \frac{z_e + z_w + z_n + z_s - 4z_0}{h^2} \tag{6.35}$$

To use this in solving the stress separation problem, recall that

$$\nabla^2 \Phi = 0, \qquad \text{where } \Phi = \sigma_1 + \sigma_2$$

In finite difference form,

$$\frac{\Phi_e + \Phi_w + \Phi_n + \Phi_s - 4\Phi_0}{h^2} = 0 \tag{6.36}$$

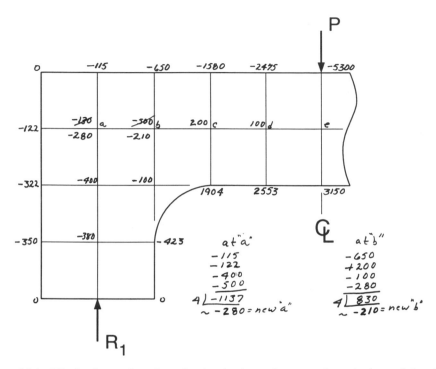

Figure 6.24. Work sheet showing the beginning of a sample solution of Laplace equation by iteration.

or

$$\Phi_0 = \frac{\Phi_e + \Phi_w + \Phi_n + \Phi_s}{4} \tag{6.37}$$

In simple language, equation 6.37 states that the function Φ at any point is the average of the function at the neighboring points. This, of course, is what the Laplace equation is all about, whether used in the context of fluid mechanics, heat transfer, or stress analysis.

The procedure for separating stresses is to divide the region studied by a mesh. At boundary nodes, the known $\sigma_1 + \sigma_2$ is shown. An "educated guess" is made for $\sigma_1 + \sigma_2$ at each interior node. Then, the values are improved by averaging adjacent points, starting at an edge and working through the region. This process is continued until the values of $\sigma_1 + \sigma_2$ converge. Note that the process is self-healing: a mistake merely causes the convergence to be slower. Figure 6.24 shows a sample solution in the early stages of iteration. Convergence is rapid, but some drudgery is involved. A computer helps and programs are available. After values converge, the grid elements can be subdivided and the process repeated to obtain better accuracy. Near boundaries some unequal node spacings may have to be used.

Note that the methods involving solution to the Laplace equation do not involve isoclinic data. This fact is important because isoclinic information is usually less accurate than information derived from isochromatic fringe patterns.

6.3.3 Series solution

Other numerical schemes for solving the Laplace equation have been developed. One such method (Dally and Erisman 1966) uses a series of harmonic functions to represent the stress sum in the interior and minimizes the error at the boundaries in order to establish the coefficients in the series. A brief outline of the technique is as follows:

given a Dirichlet problem
$\nabla^2 H = 0$
H known on boundary
$H = H(x, y)$ is the first invariant of stress

Separation of variables in the Laplace equation yields a sequence of harmonic functions, which can be combined to give a series representation for H. As an example in Cartesian coordinates, the harmonic functions involved include:

$$1, x, y, xy$$
$$\sinh kx \sin ky \qquad \text{All or part may be used} \qquad (6.38)$$
$$\sinh kx \cos ky$$
$$\text{etc.}$$

The coefficients in the series representation can be found by Fourier analysis if the region conforms to a regular coordinate system. Otherwise, a least squares procedure can be used. Choose a finite number of harmonic functions to form a series solution. Then establish the coefficients so that the mean-square difference between the given boundary values and the boundary of the series is minimized.

In mathematical terms, begin by choosing

$$H = \sum C_n F_n \qquad (6.39)$$

where   the C_n are undetermined coefficients
             the summation is over N unknowns

Let S be distance along boundary of total length L, and $I(s)$ the boundary values of H. Then, we want

$$\int [I(s) - \sum C_n F_n]^2 \, ds = \text{minimum} \tag{6.40}$$

The N coefficients C_n are determined to make the integral of Equation 6.40 a minimum. This task is accomplished by taking the derivative with respect to each coefficient and equating it to zero,

$$\frac{\partial}{\partial C_k} \int [I(s) - \sum C_n F_n]^2 \, ds = 0 \tag{6.41}$$

which gives N equations for N unknowns C_n. Equation 6.41 can be reduced to the following equation, which is in useable form,

$$\sum C_n \int F_n F_k \, ds = \int I(s) F_k \, ds \tag{6.42}$$

The integrals are in each case extended over the whole boundary.

It can be shown that the accuracy of the series in the interior is always better than the accuracy achieved at the boundaries. An upper bound on the accuracy of our solution for the whole region is, therefore, obtained automatically.

Note that this technique displays some similarities to variational methods of analytical mechanics, in particular, the Rayleigh–Ritz method.

6.3.4 Using thickness change

If z is taken as the coordinate direction normal to the model used for photoelastic observation, then the strain component in this direction is

$$\epsilon_{zz} = \frac{\sigma_{zz} - \nu(\sigma_{xx} + \sigma_{yy})}{E} \tag{6.43}$$

But $\sigma_{zz} = 0$, so,

$$\sigma_{xx} + \sigma_{yy} = \frac{-E\epsilon_{zz}}{\nu} = \frac{-E(\Delta d/d)}{\nu} \tag{6.44}$$

where $\Delta d =$ change in thickness d.

This result implies that the stress sum can be determined by measurement of the thickness change with loading of the photoelastic model. The experimental details are not quite so simple. The basic problem is one of sensitivity. Much work has been done in this area. Mention of some techniques used to facilitate this difficult measurement is made here:

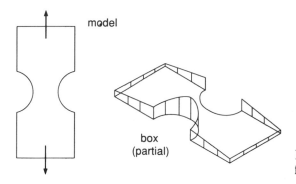

Figure 6.25. Model used
for soap-film analogy.

1. Frozen models and rubber models increase the Δd while making measurement easier.
2. Mechanical devices for measuring Δd tend to be cumbersome, but they can be obtained with sensitivities of better than 2 μm.
3. Electrical methods, including strain gages and LVDTs can give the required precision.
4. Optical techniques are now the most used; in general terms, techniques include interferometry, moire, holointerferometry, and schlieren.

6.3.5 Experimental analogs

As mentioned in section 6.3.1, several natural phenomena are described by the Laplace equation. These problems can be used as analogs for solving the stress-separation problem (or vice versa). The same procedure is used, for example, when solving a torsion problem by the soap-film or sand-hill analogy. With the advent of computers in the laboratory, these analog methods have given way to totally numerical or "hybrid" approaches. The analogs are surprisingly simple to use, and they can give good quick answers. They should not be forgotten. The following summarizes a few of the useful techniques. The list is not exhaustive.

Membrane or soap-film analogy

This procedure is the same as that used for the torsion problem. A box is constructed in the shape of the model being studied, as suggested by the sketch of Figure 6.25. The height of the edge of the box is proportional to the boundary stress at the point. A membrane of soap film is formed over the box. The height of the membrane at any point is proportional to the stress sum at the point. The constant of proportionality is the same as that used for establishing the height of the edge of the box.

Conduction-paper analogy (or conducting-tank analogy)

A conducting sheet is cut in the shape of the region being studied. To the boundary, a voltage E is applied that is proportional to the boundary stress at each point. In the interior, $\nabla^2 E = 0$, so measured values of E will be proportional to the stress sum at each point. Note that the boundary values are a discretization of the actual stresses; that is, the boundary voltage distribution cannot easily be made continuously varying around the boundary. With enough voltage dividers at hand, this method gives very good results, and it is easier and quicker to do than one might think.

Electro-optic analogy

An electro-optical fluid, such as an aqueous suspension of the organic dye milling yellow, is placed in a tank that is in the shape of the region being studied. A voltage is applied to the boundary, with the voltage being proportional to the boundary stress. The potential causes the fluid to become birefringent. Isoclinics indicate the direction of voltage gradient, and isochromatics are lines where $|\text{grad } E| = \text{const.}$ For some problems, the magnitude and direction of voltage gradient is enough. For a stress analysis problem, the gradient must be integrated to obtain E for stress separation. This technique is satisfying in that it uses the polariscope to obtain the stress sum directly. It is, however, somewhat messy and unwieldy.

6.3.6 Oblique incidence

By observing isochromatics in a model that is tilted with respect to the optical axes of a polariscope, one is able to determine the individual principal stresses with an accuracy that is reasonable for many practical applications. Here, discussion will be confined to the simplest case; that is, when point-by-point oblique incidence measurements are made with the principal stress directions known from previous isoclinic observation. For this case, the model is tilted on angle θ about the axis of a principal stress at the point, as illustrated in Figure 6.26.

Now, the concept of "secondary principal stress" or "effective principal stress" is introduced. This stress is what affects the light; it is the principal stress in the plane normal to the light path. From equilibrium of a triangular element, this secondary principal stress σ_2' can be related to the real principal σ_2 in the plane of the specimen, as follows,

$$\sigma_2' d' = \sigma_2 (d' \cos \theta) \cos \theta$$

and so,

$$\sigma_2' = \sigma_2 \cos^2 \theta \tag{6.45}$$

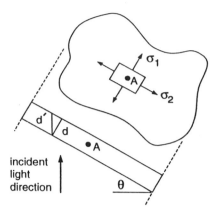

Figure 6.26. Geometry of oblique incidence stress separation.

Now, relate the isochromatic orders m to the stresses. In the following, subscript $\perp$ means normal incidence, and subscript θ means oblique incidence.

For normal incidence,

$$(\sigma_1 - \sigma_2) = \frac{f_\sigma m_\perp}{d} \tag{6.46}$$

where $f_\sigma = \lambda/C_\sigma$ is introduced to simplify the equations. From equation 6.46,

$$m_\perp = (\sigma_1 - \sigma_2)\frac{d}{f_\sigma} = \frac{\sigma_1 d}{f_\sigma} - \frac{\sigma_2 d}{f_\sigma} \tag{6.47}$$

Introduce the definitions,

$$\frac{\sigma_1 d}{f_\sigma} = m_{\sigma 1} \quad \text{and} \quad \frac{\sigma_2 d}{f_\sigma} = m_{\sigma 2} \tag{6.48}$$

and equation 6.47 becomes

$$m_\perp = m_{\sigma 1} - m_{\sigma 2} \tag{6.49}$$

Similarly, for oblique incidence, after recalling the definition of secondary principal stress,

$$m_\theta = (\sigma_1 - \sigma_2')\frac{d'}{f_\sigma} \tag{6.50}$$

Use equation 6.45 and $d' = d/\cos\theta$ in equation 6.50,

$$m_\theta = (\sigma_1 - \sigma_2\cos^2\theta)\frac{d/\cos\theta}{f_\sigma}$$

$$= \frac{\sigma_1 d}{f_\sigma \cos\theta} - \frac{\sigma_2 d\cos\theta}{f_\sigma}$$

Use the definitions of equation 6.48 to get

$$m_\theta = \frac{m_{\sigma 1}}{\cos\theta} - m_{\sigma 2}\cos\theta \tag{6.51}$$

Then use equation 6.49 in 6.51 to get

$$m_\theta = \frac{m_{\sigma 1}}{\cos\theta} - m_{\sigma 1}\cos\theta + m_\perp\cos\theta$$

$$m_{\sigma 1}\left(\frac{1}{\cos\theta} - \cos\theta\right) = m_\theta - m_\perp\cos\theta$$

$$m_{\sigma 1} = \frac{m_\theta - m_\perp\cos\theta}{\dfrac{1}{\cos\theta} - \dfrac{\cos^2\theta}{\cos\theta}} = \frac{m_\theta - m_\perp\cos\theta}{1 - \cos^2\theta}\cos\theta$$

So,

$$\frac{\sigma_1 d}{f_\sigma} = m_{\sigma 1} = (m_\theta - m_\perp\cos\theta)\frac{\cos\theta}{\sin^2\theta} \tag{6.52}$$

giving for the first principal stress

$$\sigma_1 = \left[(m_\theta - m_\perp\cos\theta)\frac{\cos\theta}{\sin^2\theta}\right]\frac{f_\sigma}{d} \tag{6.53}$$

A similar procedure yields for the remaining principal stress,

$$\sigma_2 = \left[\frac{(m_\theta\cos\theta - m_\perp)}{\sin^2\theta}\right]\frac{f_\sigma}{d} \tag{6.54}$$

If the following data are known for the point being studied:

> stress direction
> tilt of model
> m_θ and $m_\perp$
> material fringe coefficient f_σ $(= \lambda/C_\sigma)$

then the individual principal stresses at the point can be determined by equations 6.53 and 6.54.

It is possible to use a different procedure that does not require isoclinic data. There are three major advantages. First, isoclinic data are usually not as accurate as isochromatic data. Second, rotation can be about an arbitrary axis, so less equipment is needed. Third, all needed data can be obtained from three photographs, which is better than making observations point-by-point

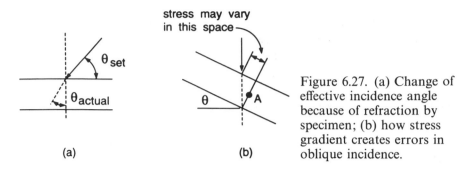

Figure 6.27. (a) Change of effective incidence angle because of refraction by specimen; (b) how stress gradient creates errors in oblique incidence.

over the model. Experimental data needed include:

> normal incidence photograph to obtain $m_\perp$
> oblique incidence photograph no. 1 to obtain $m_{\theta 1}$
> oblique incidence photograph no. 2 to obtain $m_{\theta 2}$
> material fringe coefficient

Usually, $m_{\theta 1}$ and $m_{\theta 2}$ are obtained for rotation about orthogonal axes. The price is greater complexity in the equations that replace 6.56 and 6.57. This more general procedure will not be developed further.

In principle, the oblique incidence techniques are very simple and useful, but there are some important limitations:

1. Initial residual stress effects.
2. Refractive index of plastic changes θ from that set by the equipment, as shown in Figure 6.27a. This error can be eliminated by using an immersion tank or by calculation of the correct θ.
3. Errors in establishing principal stress directions.
4. Errors resulting from stress gradient normal to rotational axis. Figure 6.27b illustrates this difficulty. To minimize these errors, it is desirable to use small θ; but then m_θ and $m_\perp$ do not differ appreciably, and errors of measurement are magnified when the two fringe orders are subtracted.

The oblique incidence technique is used in photoelasticity applications where some accuracy can be sacrificed. Its greatest area of application is in reflection photoelasticity.

6.3.7 Shear difference

One of the oldest methods for obtaining the separate stresses involves integration of the equilibrium equations along selected axes in the body. The integrals are approximated by their finite difference equivalents, which are of the form

$$\sigma_x = (\sigma_x)_0 - \sum \left(\frac{\Delta \tau_{xy}}{\Delta x} \right) \Delta y$$

The $\tau_{xy} = [(\sigma_1 - \sigma_2)/2] \sin 2\theta$ is determined from the photoelasticity model. The θ is the inclination of the principal axis as obtained from the isoclinics, and $\sigma_1 - \sigma_2$ comes from the isochromatic patterns. One starts the finite difference integration at a point where the stress component is known, such as at an edge point.

A large body of literature about this method exists, and it will not be pursued further here. It seems to have fallen from favor, but it offers good results with careful work. The main disadvantage seems to be the potential for accumulated error as one integrates in steps along an axis. Isoclinic data usually are not as precise as isochromatic data. Because the principal angles enter each finite difference calculation, the precision of the stress magnitude results is degraded to the precision level of the stress direction information. Another difficulty has to do with the problem of attaching the correct sign to the τ_{xy} at each integration. The method is also tedious.

6.4 Determining exact fringe order

6.4.1 Summary of approaches

Often, it is necessary to measure with precision the birefringence at a particular point in a plate. For example, research on the birefringent properties of relatively insensitive materials requires measurements of relative retardation to within 1/100 of an isochromatic order. There are six basic approaches to this problem of measurement, although the approaches are not entirely independent. For orientational purposes, the concepts behind these techniques are summarized in the following list, after which some specific methods are described in more detail.

1. Interpolation and extrapolation from light- and dark-field photographs. Often, graphs or numerical procedures are used to improve the interpolation. This method will give measurements to about 1/10 order. The approach is intuitively logical, and it has been discussed indirectly in the sections on material calibration and fringe interpretation, so no more is said.
2. Interpretation of colors from isochromatic patterns in white light. This method is tricky. See the section on the use of nonmonochromatic light for the concepts behind this technique. This idea is not developed further here.
3. Introduce into the optical path a device whose birefringence is known or can be controlled so that it (usually) cancels the birefringence in the plate. Such a device is called a compensator. They are available commercially, or a simple compensator can be built. If such an instrument is carefully used and calibrated, it will easily give the desired accuracy.
4. One of the elements in the polariscope, usually the analyzer, is rotated to produce predetermined (usually zero) light intensity at the point in the image of the model.

otation is directly related to the fraction of isochromatic order.
ires no auxiliary equipment and, with care, gives the desired
ral term for this measurement process is goniometric com-
common procedures are the Tardy method and the Senarmont
ote that all these compensation methods are forerunners of what are
ow called phase-shifting interferometry, which is discussed in the context of video
holography or electronic speckle pattern interferometry in Chapter 22.

5. Multiply the fringe orders by causing the light to traverse the model several times. The relative retardation occurring in each traverse is accumulated, and the total retardation will cause the fringe orders to be greatly increased in the fringe photographs (e.g., Post 1970, 1989). Tilted partial mirrors are used to force the light to pass back and forth through the specimen. This technique is used as an adjunct that expedites and improves the process of interpolation as previously mentioned. It is especially valuable in examining the slices in frozen-stress three-dimensional photoelasticity. Given space limitations, this interesting technique is not discussed further.

6. C. P. Burger and colleagues have introduced a technique called "half-fringe photoelasticity" that exploits the data processing capability of computer-based image analysis systems (Burger 1987). Basically, the computer system measures precisely the light irradiance at every point of the model and converts it to fringe order through use of the basic retardation-intensity relation discussed in Chapter 5. The irradiance is normalized for a given specimen through use of goniometric compensation. This technique functions only when the maximum fringe order is less than one-half, hence the name. Methods such as this offer much promise in facilitating photoelasticity studies of complicated problems, such as those involving composite materials.

6.4.2 Tardy goniometric compensation

The Tardy method is simplest and easiest to remember, but it is not always the best method of fractional fringe measurement. It can be developed intuitively without reference to any theory, although the photoelastic equations must be utilized to justify intuition.

To use the method to measure the fractional isochromatic order at a point P in a birefringent plate, first establish the stress directions at the point by rotating polarizer and analyzer to make the isoclinic cover the point. The polarizers are thus aligned with the principal axes. Then insert the $\lambda/4$ plates at $45°$ to the polarizer axes to obtain the usual circular polariscope dark-field setup. An ordinary isochromatic pattern with the point P somewhere between two fringes will be seen, as in Figure 6.28. At this point, it is necessary to figure out the adjacent isochromatic orders.

Now, one rotates the polarizer alone through an angle θ to cause

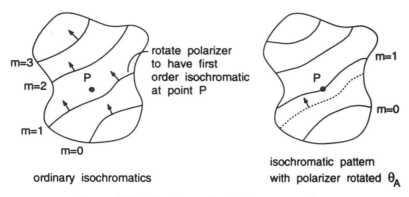

Figure 6.28. Tardy method of compensation.

an adjacent isochromatic to move to point P. In the example pictured, either the first order will move up to P or the second order will move back to P. Recall that 90° of rotation will change from dark to light field, that is, $\frac{1}{2}$ order. The fraction of an order at P can be found from a simple proportion,

$$m_f = \pm \frac{\theta}{180} \qquad (6.55)$$

Suppose, for example, we rotate the polarizer θ_A to move order 1 to the point P. The order at P is

$$m_p = 1 + \left(\frac{\theta_A}{90}\right) \times \frac{1}{2} = 1 + \frac{\theta_A}{180}$$

On the other hand, we might have rotated the polarizer through θ_B to move order 2 back to the point. In this event,

$$m_p = 2 - \frac{\theta_B}{180}$$

The major problem here is deciding whether to add or subtract the fraction and then measuring the angle rather than its complement. Experience and practice help. A dry run on a simple tension specimen will also clarify matters. The sign of the fraction cannot, in general, be correlated with direction of polarizer rotation because it depends on the "handedness" of the quarter-wave plates.

The Tardy method is used in the reflection photoelasticity apparatus to measure fractional orders. It is especially important there because isochromatic orders rarely go beyond 3 in reflection photoelasticity.

The theoretical foundation of Tardy compensation will not be developed further. The equations are somewhat similar to those of the Senarmont method, which are presented in the next section.

6.4.3 Senarmont method

This technique is similar to the Tardy method except that only one $\lambda/4$ plate is used. The polarizer axis is set at $45°$ from a principal stress direction at the point of interest. The first $\lambda/4$ plate is eliminated. The second $\lambda/4$ plate is made parallel to the polarizer. The starting (zero) position of the analyzer is $90°$ from the polarizer. The rotation θ_s of the analyzer from the starting position that causes a dark fringe to cover the point P is related to the fractional fringe order at P by the same equation utilized with the Tardy technique. The nearest whole orders are established from an isochromatic fringe record taken with the standard polariscope. A problem with the technique is that the point being studied is not neatly bracketed by two isochromatics that are then shifted by analyzer rotation. With the Senarmont approach, the pattern is a rather meaningless jumble of partial isochromatics. The analyzer rotation causes a dark fringe to suddenly appear at the point of observation.

As with the Tardy method, the sign of the fractional order depends on the handedness of the $\lambda/4$ plates, the photoelastic material properties, and the direction of analyzer rotation. It is wise to establish the signs, for each combination of polariscope and material, by a preliminary test on a simple tension specimen. The procedure is to increase the load on the tensile bar slightly and see in which direction the analyzer must be rotated to bring the intensity of light back to zero.

The equations describing the Senarnot method of fractional fringe measurement are not readily available in simple form in the literature on photoelasticity. They are reproduced here for those wanting to gain a full understanding of goniometric compensation.

Begin with linearly polarized light that is incident on a birefringent plate of thickness d and whose principal stress axes at the point are inclined at $45°$ to the axis of polarization (the y axis). The components of the ray emerging from the plate are

$$
\begin{aligned}
E_1 &= \frac{\sqrt{2}}{2}\, A \cos\left\{ \frac{2\pi}{\lambda}\, [z - vt] \right\} \\[2mm]
E_2 &= \frac{\sqrt{2}}{2}\, A \cos\left\{ \frac{2\pi}{\lambda} \left[z - vt - \frac{(n_1 - n_2)}{n_0} \right] \right\} d
\end{aligned}
\tag{6.56}
$$

The common absolute retardation has been neglected. Next allow the rays to pass through a quarter-wave plate whose axes are at $45°$ to the principal axes, that is, parallel and perpendicular to the polarizer axis. The components as they enter the $\lambda/4$ plate with axes x and y are

$$E_x = \frac{\sqrt{2}}{2} E_1 - \frac{\sqrt{2}}{2} E_2$$

$$E_y = \frac{\sqrt{2}}{2} E_1 + \frac{\sqrt{2}}{2} E_2 \tag{6.57}$$

While traversing the $\lambda/4$ plate, the rays will be retarded absolutely by a common amount and with respect to one another by $\lambda/4$. The components are

$$E_x = \frac{1}{2} A \cos\left\{\frac{2\pi}{\lambda} [z - vt]\right\} - \cos\left\{\frac{2\pi}{\lambda}\left[z - vt - (n_1 - n_2)\frac{d}{n_0}\right]\right\}$$

$$E_y = \frac{1}{2} A \cos\left\{\frac{2\pi}{\lambda}\left[z - vt - \frac{\lambda}{4}\right]\right\} - \cos\left\{\frac{2\pi}{\lambda}\left[z - vt - (n_1 - n_2)\frac{d}{n_0} - \frac{\lambda}{4}\right]\right\} \tag{6.58}$$

These components reduce to

$$E_x = -A \sin\left\{\frac{2\pi}{\lambda}\left[z - vt - \frac{n_1 - n_2}{2n_0}d\right]\right\} \sin\left\{\frac{2\pi}{\lambda}\left[\frac{n_1 - n_2}{2n_0}d\right]\right\}$$

$$E_y = A \sin\left\{\frac{2\pi}{\lambda}\left[z - vt - \frac{n_1 - n_2}{2n_0}d\right]\right\} \cos\left\{\frac{2\pi}{\lambda}\left[\frac{n_1 - n_2}{2n_0}d\right]\right\} \tag{6.59}$$

The rays are then passed through the second polarizer, which makes an angle of $(\pi/2 - \theta)$ with the first. Angle θ is measured from the position that would cause dark field. The ray leaving the analyzer will be

$$E = E_x \cos\theta + E_y \sin\theta$$

$$E = -A \sin\left[\frac{2\pi}{\lambda}\frac{n_1 - n_2}{2n_0}d - \theta\right] \sin\left\{\frac{2\pi}{\lambda}\left[z - vt - \frac{n_1 - n_2}{2n_0}d\right]\right\} \tag{6.60}$$

Equation 6.60 describes a traveling wave of amplitude

$$-A \sin\left[\frac{2\pi}{\lambda}\frac{n_1 - n_2}{2n_0}d - \theta\right] \tag{6.61}$$

whose irradiance or intensity is

$$I = A^2 \sin^2\left[\frac{\pi(n_1 - n_2)d}{\lambda n_0} - \theta\right] \tag{6.62}$$

The intensity is zero when the argument is zero or $m\pi$. So $I = 0$ when

$$\frac{\pi(n_1 - n_2)d}{\lambda n_0} - \theta = 0 \qquad \text{or} \quad \pm m\pi \tag{6.63}$$

The angle to produce zero intensity at the point is related to the birefringence,

$$\theta = \frac{\pi(n_1 - n_2)d}{\lambda n_0} \pm m\pi \tag{6.64}$$

The fraction m_f of a whole order is,

$$m_f = \frac{(n_1 - n_2)d}{\lambda n_0} = \frac{\theta(\text{rad})}{\pi} = \frac{\theta^\circ}{180^\circ} \tag{6.65}$$

The $m\pi$ represents the whole order next lowest or highest. Designate this whole order m_w, and the total fringe order at the point is

$$m = m_w \pm \frac{\theta^\circ}{180^\circ} \tag{6.66}$$

As mentioned earlier, the sign of the fractional portion of the order is best correlated with the direction of rotation through simple experiments on tension or compression specimens. Once determined for a given setup, they will not change.

6.4.4 Birefringent compensators

A logical method for ascertaining the exact fringe order at a given point in a photoelastic model is to superimpose over the point a device for which the birefringence is adjustable and accurately known. If its birefringence is equal in magnitude and opposite in sign to that of the unknown, then the isochromatic order of the combination will be reduced to zero, which is easy to detect accurately. The isochromatic order in the unknown is thereby established. Devices of this type are called compensators.

Many varieties of these instruments have been invented, and a few types are marketed by optical instrument makers. The techniques for using them are basically of a pattern. The principal axes are determined for the point in question by rotating the polarizer and analyzer so that the isoclinic covers the point. The polariscope is then converted to circular. The compensator is introduced into the field and rotated in its mounting so that its principal axes are parallel to the model axes at the point. The instrument is then adjusted to bring the resultant fringe order at the point to null. The fringe order is then read from the calibrated scale. Some of these instruments have a range of only one isochromatic order, so the nearest whole order must be established by other means. Much literature is available about the design and construction of compensators, as well as the relative merits of each type. Only a few basic types are described in brief here.

One of the simplest and oldest of birefringent compensators is the Coker type (Jessop and Harris 1949). This instrument is merely a calibrated tension

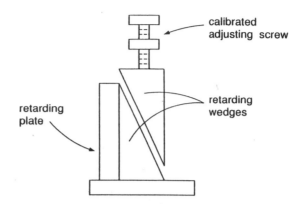

Figure 6.29. Schematic of Babinet–Soleil birefringent compensator.

model mounted in a small screw-operated load frame, which is itself carried in a mounting that permits rotation. The instrument must be carefully calibrated to relate fringe order accurately to rotation of the load screw.

An approach that is similar replaces the tension model and the entire load frame by a slice of material that carries a gradient of permanent birefringence. The compensator element is again mounted in a holder that permits rotation. The birefringent slice is made by the techniques described for the "frozen-stress" method of three-dimensional photoelasticity, as described near the end of this chapter. The birefringence may be locked into a beam specimen, for example. A cross-sectional slice will then exhibit the desired gradient of fringe orders. Calibration is accomplished by using another compensator or by a goniometric method.

Several compensator designs utilize wedges of naturally birefringent quartz. The Babinet–Soleil type is one that is sophisticated and precise. Figure 6.29 illustrates the construction of this device, which is mounted in a ring that allows calibrated rotation.

The heart of the instrument is a quartz plate of constant thickness and two overlapping quartz wedges. The total birefringence in the optical path is proportional to the thickness of the quartz traversed by the light. It is controlled by a calibrated micrometer screw that causes one of the wedges to slide over the other. These devices are carefully calibrated for specific wavelengths of light and are capable of precisions on the order of 0.01 wavelength or better.

6.5 Reflection photoelasticity

A characteristic of two-dimensional transmission photoelasticity is that the analysis must usually be made with a model of the structure being studied. In some cases, particularly with various industrial problems, it is advantageous to analyze the machine part itself; strain gages are often used when

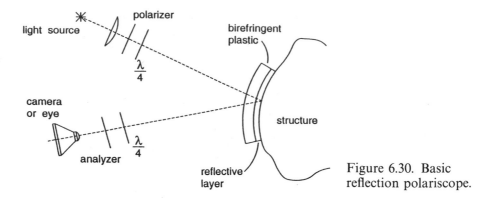

Figure 6.30. Basic reflection polariscope.

only point-by-point data are desired. Reflection photoelasticity is another technique that has the outstanding advantage of giving relatively accurate strain measurement over the whole surface of the structure being studied.

The basis of the reflection photoelasticity method is that a birefringent material is bonded to the structure so that it undergoes the same strains as are imposed on the structure. Observation of the birefringence in the coating can then be directly related to the strain. A good bit of special technology is involved, but the basic apparatus is just a normal polariscope which is "folded" in the middle. A reflective light-scattering layer is bonded between the plastic and the surface of the structure being investigated, or else the shiny surface of the structure is used to reflect the light. Figure 6.30 illustrates these basic ideas.

Reflection photoelasticity is quite easy to use and is a quick way to obtain reasonably accurate readings of surface strains in a structure of any size or geometry. It has been widely accepted and used (or misused in some cases) in industry. The apparatus and materials are readily available commercially, as is manufacturers' literature.

The theoretical aspects of reflection photoelasticity are similar to those presented for the transmission variety, except that (a) the light travels twice through the coating and (b) the birefringence is related to the surface strain in the test piece. Stress separation is done by oblique incidence with a system of mirrors. The fringe orders are usually very low, and fractional fringe measurement through Tardy compensation is necessary for precise work.

With careful work, the method is capable in practical applications of 10–20% accuracy. For many design situations, such accuracy is more than acceptable; and in these cases, reflection photoelasticity is a fine time-saving tool. Incidentally, many designers do not utilize quantitative reflection data. They are interested only in locating and minimizing stress concentrations while improving strength-to-weight ratios. Qualitative reflection observations are very helpful in such work, and they are easily and quickly obtained.

In view of space limitations and the purpose of this text, details of the reflection photoelasticity method are not offered. Interested readers should refer to manufacturers' literature and to the several texts on the subject (e.g., Dally and Riley 1991).

6.6 Three-dimensional photoelasticity

So far, discussion has been limited strictly to two-dimensional photoelasticity. These methods, including transmission and reflection photoelasticity, suffice for any problem that can be reduced to a plane or where only surface strains are required. Most practical photoelasticity applications are approached in this way through careful thought and technique.

Photoelasticity can be applied in three dimensions. The theory and procedures are complex and time-consuming in comparison with the two-dimensional method. Before three-dimensional photoelasticity studies are undertaken, the analyst should be certain that his problem cannot be reduced to two dimensions and that the needs justify the undertaking.

Because the medium of observation is light, the analysis of a three-dimensional problem must, with limited exceptions, be reduced to an assembly of two-dimensional cases. There are three classic approaches to this problem, including:

> slicing after stress freezing
> layered models
> scattered light

A new and different approach to three-dimensional photoelasticity is offered by holographic-photoelasticity or stress-holo-interferometry. These holographic methods do not require physical or optical slicing of the specimen.

The art of three-dimensional photoelasticity is more complicated and critical than are theory and general procedure, which are extensions of basic transmission work. Before undertaking a three-dimensional project, the photoelastician should read related texts and technical papers as well as consult, if possible, with persons experienced in this line of experimentation.

Again, for reasons of space and emphasis, the subject of three-dimensional potoelasticity is not pursued. As with reflection photoelasticity, the giving of scant attention to this interesting material is cause for regret on the part of the author.

References

Alfrey, T., Jr. (1948). *Mechanical Behavior of High Polymers*, App. II. New York: Interscience Publishers.

Burger, C. P. (1987). Photoelasticity. In *Handbook on Experimental Mechanics*, Ed. A. S. Kobayashi, Chap. 5. Englewood Cliffs: Prentice-Hall.

Cloud, G. L. (1966). *Infrared Photoelasticity: Principles, Methods, and Applications*. Ph.D. dissertation, Michigan State University, East Lansing, MI.

Cloud, G. L. (1969). Mechanical-optical properties of polycarbonate resin and some relations with material structure. *Experimental Mechanics*, 9, 11: 489–500.

Cloud, G. L. (1970). Correlations between dispersion of birefringence and transmittance of three polymers. *Journal Optical Society of America*, 60, 8: 1032–45.

Clutterbuck, M. (1958). The dependence of stress distribution on elastic constants. *British Journal of Applied Physics*, 9, 8: 323–9.

Coker, E. G., and Filon, L. N. G. (1957). *A Treatise on Photoelasticity*. Cambridge University Press.

Dally, J. W., and Erisman, E. R. (1966). An analytic separation method for photoelasticity. *Experimental Mechanics*, 6, 10: 493–9.

Dally, J. W., and Riley, W. F. (1991). *Experimental Stress Analysis*, 3rd ed. New York: McGraw-Hill.

Daniel, I. M. (1965). Quasi-static properties of a photoviscoelastic material. *Experimental Mechanics*, 5, 3: 83–9.

Frocht, M. M. (1941). *Photoelasticity: Vol. I*. New York: John Wiley and Sons.

Garbuny, M. (1965). *Optical Physics*. New York: Academic Press.

Jessop, H. T., and Harris, F. C. (1949). *Photoelasticity*. London: Cleaver-Hume Press and New York: Dover Publications.

Kiesling, E. W. (1966). *Nonlinear Quasi-Static Behavior of Some Photoelastic and Mechanical Model Materials*. Ph.D. dissertation, Michigan State University, East Lansing, MI.

Kiesling, E. W., and Pindera, J. T. (1969). Linear limit stresses of some photoelastic and mechanical model materials. *Experimental Mechanics*, 9, 8: 337–47.

Pindera, J T. (1959a). Investigations of some rheological photoelastic properties of some polyester resins, Part 1: Methods and techniques of investigations (in Polish). *Rozprawy Inzynierskie (Engineering Transactions)* 7, 3: 363–411. Warsaw: Polish Academy of Sciences.

Pindera, J. T. (1959b). Investigations of some rheological photoelastic properties of some polyester resins, Part 2: Experiments, evaluated data, interpretation of results (in Polish). *Rozprawy Inzynierskie (Engineering Transactions)* 7, 4: 483–520. Warsaw: Polish Academy of Sciences.

Pindera, J. T. (1960). Some research in the field of photoelasticity performed in the Polish Academy of Sciences (in Russian). *Transactions of the Conference of 13–21 February 1958*, pp. 32–44. St Petersburg: The Leningrad University.

Pindera, J. T. (1962). Einige rheologische probleme bei spannungsoptischen untersuchungen. *Internationales spannungsoptisches symposium, Berlin, 10–15 April 1962*, pp. 155–72. Berlin: Academie-Verlag.

Pindera, J. T. (1966). Remarks on properties of photoviscoelastic materials. *Experimental Mechanics*, 7, 6: 375–80.

Pindera, J. T., and Cloud, G. (1966). On dispersion of birefringence of photoelastic materials. *Experimental Mechanics*, 6, 9: 470–80.

Post, D. (1970). Photoelastic-fringe multiplication – for tenfold increase in sensitivity. *Experimental Mechanics*, 10, 8: 305–12.

Post, D. (1989). Photoelasticity. In *Manual on Experimental Stress Analysis*, 5th ed., Eds. J. F. Doyle and J. W. Phillips, pp. 80–106. Bethel, CT.: Society for Experimental Mechanics.

Theocaris, P. S. (1965). A review of rheo-optical properties of linear high polymers. *Experimental Mechanics*, 5, 4: 105–14.

Part III
Geometrical moire

7

Geometrical moire theory

This part of the book deals with geometrical moire, an optical effect that is useful, interesting, and, to many minds, esthetically pleasing. It is also the only optical approach discussed here that does not rely on optical wave interference and diffraction. Rather, the geometrical moire fringe patterns are created entirely by mechanical occlusion of light by superimposed gratings. There are other moire techniques, to be examined in Parts IV and V of this text, that do utilize interference and diffraction. The fundamental concepts supporting those more exotic methods are to be found in the basic theory of geometrical moire, to be discussed here.

7.1 The moire effect

The moire effect is the mechanical interference of light by superimposed networks of lines. The pattern of broad dark lines that is observed is called a moire (or Moiré) pattern. Such a pattern is formed whenever a repetitive structure, such as a mesh, is overlaid with another such structure. The two structures need not be identical. The effect was evidently noted in ancient times. Modern examples easily observed include the effect when two layers of coarse textile are brought together, the bars observed on television when the scene includes a striped shirt or a building with regular joinings at the proper distance, and the pattern seen through two rows of mesh or picket fence from a distance.

Only a little study of the moire effect uncovers a very striking and useful characteristic: a very large shift in moire pattern is obtained from only a small relative motion between the superimposed networks. A logical conclusion is that the moire pattern is a sort of "motion magnifier," which might be used to give a highly sensitive measurement of relative motion. This idea was described by D. Tollenar in 1945, and it was immediately employed in the study of deformation, motion, and strain by various researchers including Weller and Shepard in 1948 and Dantu in 1954 (Dally and Riley 1991).

147

Subsequently, the method was refined and applied in a host of circumstances (e.g., Durelli and Parks 1970; Morse, Durelli, and Sciammarella 1961; Post 1965; Riley and Durelli 1962; Theocaris 1969). The moire method can be employed in the precise determination of translational and rotational movements, and the output is easily incorporated into measuring and automatic control circuits. The effect is also useful in measuring deformation and strain in elastic bodies. A quite different application is to use the moire effect to generate contour maps for all sorts of complex shapes.

7.2 Occlusion by superimposed gratings: parametric description

The first step in general moire analysis is to establish the relationship between the superimposed network configurations and the resulting moire pattern. A parametric approach will be utilized to develop the needed relationships in the simplest and most intuitive manner (Oster, Wasserman, and Zwerling 1964; Parks 1987). A common alternate approach will be used later. With either method, the task is not simple when the structures to be superimposed are complex: say, for example, a set of concentric ellipses. There is little need for the experimental stress analyst to become involved with complicated moire geometries. It will be sufficient to discuss a simple specific case to illustrate the general procedures; this case also happens to be the one that is most vital in strain and motion measurement.

Superimpose two sets of equidistant straight lines of slightly different spacings (pitch) p and q, the superimposition taking place so that the lines of one set intersect those of the other set at a "small" angle θ. Figure 7.1 shows the two gratings and how the moire fringes are created.

The moire pattern is formed by the intersecting lines. The eye integrates these areas of maximum light blockage into a family of heavy dark moire fringes. The effect is more pronounced as the grating pitch becomes small enough so that the eye has trouble resolving individual grating lines.

Mathematical relationships between grating parameters and moire fringe geometry are established through application of elementary analytical geometry. The lines of the vertical grating can be described by the equation

$$x = lp \tag{7.1}$$

where l is an integer and p is the pitch, or spacing, of the grating. The second family of lines is expressed by the equation

$$x \cos \theta = mq - y \sin \theta$$

or

$$x = \frac{mq}{\cos \theta} - y \tan \theta$$

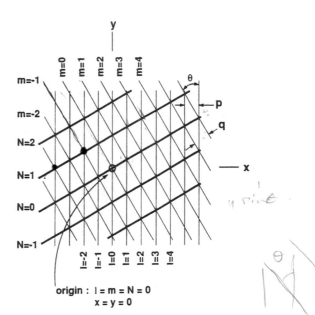

Figure 7.1. Formation of
moire fringes by
superimposed line gratings.

which converts to

$$m = \frac{x \cos \theta + y \sin \theta}{q} \tag{7.2}$$

where m is an integer
         q is the pitch of this family

The origin of the coordinate system is, of course, where $l = m = 0$.

The moire fringe or line of intersection that passes through the origin is taken as the zero-order fringe. Inspection of Figure 7.1 shows that along this zero-order moire fringe, the intersecting grating lines are of the same order, $m = l$. Let the adjacent fringe be numbered one. Along this fringe, all the intersections correspond to $m - l = 1$. Continue with this process to find that a general equation for a moire fringe order in this simple case is

$$m - l = N \tag{7.3}$$

where N is an integer designating the moire fringe number. The signs in this equation are changed if one of the three line arrays is arbitrarily numbered in the sense opposite to that shown in the figure. Such a sign change would be carried through subsequent equations, but there are no serious consequences.

The analysis of the moire patterns resulting from superimposing more complex arrays of lines or dots involves similar procedures.

Expressions for the spacing and inclinations of moire fringes in terms of differences of spacing or orientation of the two line gratings are developed by pressing the analysis a bit further. Substitution of equations 7.1 and 7.2 into 7.3 yields

$$\frac{x \cos \theta + y \sin \theta}{q} - \frac{x}{p} = N$$

or

$$\frac{(p \cos \theta - q)x + py \sin \theta}{pq} = N \tag{7.4}$$

Along any given moire fringe order, N is constant. This realization leads to the definition of a moire fringe as the locus of points for which the left-hand side of equation 7.4 is a constant. Although it is not necessary, we chose to let N be an integer for a fringe that is dark.

A linear relationship results if relative rotation is kept small,

$$(p - q)x + p\theta y = Npq \tag{7.5}$$

At this point, strain and relative rotation quantities begin to appear independently in the equation. Division of equation 7.5 by p yields,

$$\left(\frac{p - q}{p}\right)x + \theta y = Nq \tag{7.6}$$

Suppose that one of the gratings had been applied to a deformable solid and that the two gratings were initially identical with pitch p. In this situation, the first term in equation 7.6 is just the change of pitch divided by original pitch, or, obviously, normal strain along the original x axis; that is,

$$\frac{p - q}{p} = \frac{\text{length change}}{\text{old length}} = \epsilon_x \tag{7.7}$$

With this result, equation 7.6 can be expressed as

$$\epsilon_x x + \theta y = Nq \tag{7.8}$$

The preceding equations imply that moire fringe order depends on the initial pitches of the gratings and their initial relative position and orientation. If either the initial pitch, the relative orientation, or the relative position is changed, then the moire fringe pattern will change. This fringe shift can be used as a device to measure the change of pitch (strain), change of relative position (translation), or change of relative orientation (rotation). All moire measurement applications, whether interferometric or not, utilize this basic idea.

In this derivation, rigid-body translation of one of the gratings relative to the other has not been allowed. Inclusion of this possibility is not complicated and is left as an exercise.

7.3 Moire fringe, displacement, strain relations

Let us examine more closely how the moire effect is used in the analysis of strain in deformable bodies. Equation 7.8 contains the strain term; but to clarify the situation when one grill is deformed, a common alternative derivation of the fringe-order–strain relationship is worth pursuing (Dally and Riley 1991; Theocaris 1969).

The sketch in Figure 7.2 shows, in cross section, light passing through a deformed specimen grill and an undeformed master. In certain areas, the light is blocked, causing a moire "fringe." Remember that the eye smoothes out or averages light intensities over a small area, so alternate bands of light and dark are seen. The eye is a low-pass filter in this situation.

Examination of the sketch will show that one dark band appears every time six grill lines on the specimen have been stretched to fill the space of seven on the undistorted master array. Number the moire fringes consecutively, starting anywhere. Then, beginning from the corresponding point on the specimen, the relative displacement between specimen and master is easily calculated to be

$$u = Np \qquad (7.9)$$

where N is the moire fringe order
  p is the pitch of master
  u is the x-component of displacement

Recall the strain–displacement relations from strength of materials,

$$\epsilon_x = \frac{\partial u}{\partial x}$$

Combine the preceding two equations to get

$$\epsilon_x = \frac{\partial(Np)}{\partial x}$$

But p is constant, so

$$\epsilon_x = p \frac{\partial N}{\partial x} \qquad (7.10)$$

Equations 7.8 and 7.10 are clearly different statements of the same thing. When equation 7.7 was written, the pitches were implicitly assumed constant, which implies that $\partial n/\partial x = n/x$. In other words, equation 7.8 strictly applies only to a uniform strain field, although it can be easily extended to nonuniform fields. Also, relative rotation was taken as zero in developing equation 7.10.

The equations just developed allow us to make a clear statement about what a moire fringe is in the context of strain analysis. Ignore for the moment

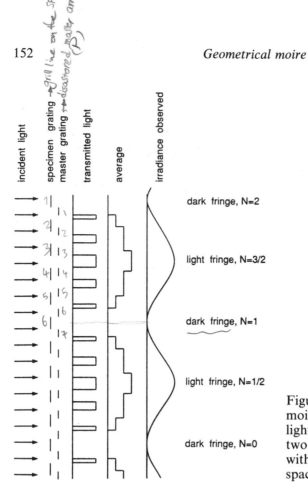

incident light
specimen grating
master grating
transmitted light
average
irradiance observed

dark fringe, N=2

light fringe, N=3/2

dark fringe, N=1

light fringe, N=1/2

dark fringe, N=0

Figure 7.2. Formation of
moire-fringe pattern as
light is transmitted through
two parallel line arrays
with slightly different line
spacings.

any contribution from rotation; that this is valid will be shown later. A moire
fringe is a *locus of points of constant displacement component*. In general,
displacement components normal to the grid orientation are the ones that
create moire fringes.

Furthermore, the analysis shows that the strain in the vicinity of a point
is proportional to the *spacing* of the moire fringes near the point. To find
strain, the derivative with respect to position must be found. This derivative
is taken along an axis that is normal to the moire grill lines. Accurate
determination of the derivative of experimental data is not trivial.

Some specific techniques for finding rigid-body motions and strain by the
geometrical moire technique, along with some examples, are offered in the
next chapter.

7.4 Whole-field analysis

The theoretical developments just presented tend to imply that moire analysis
applies only to large fields in which the displacement fields are uniform or
slowly varying. That this implication is not valid can be demonstrated
mathematically, but a reasoned approach is adequate. Each of the derivations

applies to every small piece of a large field. For a given small piece, the displacements in the neighboring portions of the deformed grating combine to create a shift of the origin for the piece being examined. This translational shift only adds to the absolute moire fringe order at the origin for the piece. This addition will be about the same for closely adjacent small pieces.

Now, think of the whole field as rotations and displacements of a large number of the small portions. The rotations and displacements are continuous but not necessarily uniform. The moire fringes will vary in a complicated but continuous way over the extent of the field. At any point, the fringe order will indicate the appropriate mixture of displacement and rotation.

Further, for strain determination in the piece, it is not essential that the fringes be numbered beginning from zero because only the spacing of the fringes (the partial derivative with respect to position) is of any consequence.

7.5 Using moire to obtain full-field derivatives

Various space and time derivatives of physical parameters can be obtained by the moire method for full-field visual examination and/or photographic recording. This subject has been explored extensively (e.g., Parks and Durelli 1966; Post 1965), and it should be studied by anyone contemplating measurements in field mechanics that might involve the finding of derivatives. The remainder of this section is devoted to a few examples to illustrate some possibilities of the moire technique.

Suppose, as a first example, that two isochromatic photographs are taken with slightly different loads on the photoelastic model. When the two photographic negatives are superimposed, and if the isochromatic density is high enough, a moire fringe pattern will be seen. Recalling that isochromatics are lines of constant maximum shear stress τ_{max}, it is apparent that a moire fringe caused by superimposing two isochromatic patterns will be a locus of constant *change* of maximum shear stress with load.

For the whole field, therefore, $\Delta\tau_{max}$ resulting from load increment ΔP is obtained. If the rate of loading is $\Delta P/\Delta t$, then the moire fringes yield $d\tau_{max}/dt$, which is the rate of increase of maximum shear stress.

Another possibility is to superimpose two moire fringe patterns for two different loads. The resulting "moire pattern of moire patterns" will give the change of displacement for the change of load. If the loading rate is known, it is possible in this way to obtain directly the *velocity* distribution for the full field. Such a technique would be extremely useful in plasticity research and the like.

The two procedures just outlined give time derivatives of stress or displacement. Space derivatives can be obtained by a shift of one moire or isochromatic pattern relative to another one for the same load. The resulting "moire of moire" or "moire of isochromatics" gives the whole-field partial space derivative of displacement or of maximum shear stress in the direction of the relative shift. The application of this procedure in moire analysis of strain fields is obvious because it is always necessary to evaluate the partial space derivative of moire fringe order.

The most serious difficulty is to obtain sets of moire or isochromatic fringes that are dense enough so that, when superimposed, a moire fringe pattern is visible. Fringe multiplication techniques can be used to advantage.

Higher-order derivatives can, in principle, be obtained by successive moire superpositions. In practice, the requirement for dense patterns tends to discourage the practice. Means to surmount such difficulties can now be found in electronic imaging and digital processing.

In any case, the methods described are probably the only known ways to directly obtain full-field representation of various partial derivatives in field mechanics.

References

Dally, J. W., and Riley, W. F. (1991). *Experimental Stress Analysis*, 3rd ed. New York: McGraw-Hill.

Durelli, A. J., and Parks, V. J. (1970). *Moire Analysis of Strain.* Englewood Cliffs: Prentice-Hall.

Morse, S., Durelli, A. J., and Sciammarella, C. A. (1961). Geometry of moire fringes in strain analysis. *Transactions of American Society of Civil Engineers*, 126-I.

Oster, G., Wasserman, M., and Zwerling, C. (1964). Theoretical interpretation of moire patterns. *Journal Optical Society of America*, 54, 2: 169–75.

Parks, V. J. (1987). Geometric Moire. *Handbook on Experimental Mechanics*, Ed. A. S. Kobayashi, Ch. 6. Englewood Cliffs: Prentice-Hall.

Parks, V. J., and Durelli, A. J. (1966). Moire patterns of partial derivatives of displacement components. *Journal Applied Mechanics*, E33, 4: 901–6.

Post, D. (1965). The moire grid-analyzer method for strain analysis. *Experimental Mechanics*, 5, 11: 368–77.

Riley, W. F., and Durelli, A. J. (1962). Application of moire methods to the determination of transient stress and strain distributions. *Journal of Applied Mechanics*, 29, 1.

Theocaris, P. S. (1969). *Moire Fringes in Strain Analysis.* New York: Pergamon Press.

8
In-plane motion and strain measurement

This chapter describes some practical applications of geometric moire analysis in the measurement of displacement, deformation, and strain. Questions of sensitivity, some extensions of the method, and laboratory details are also discussed.

8.1 Determination of rigid-body motion

The moire phenomenon has been used, but not extensively, to sense the rigid-body rotations and translations of objects relative to a fixed coordinate system or to another object. The devices can give visual readout or serve as a source of an electronic signal that can be used in feedback control. As an illustrative application, consider the problem of developing a simple method for precise angular positioning of a shaft over a small range of motion. The apparatus must have low mass, be portable, and give visual readout. Application possibilities are automatic or manual control of a steering mechanism, directional control of an antenna, measurement of wind direction, and so on.

One way to accomplish the desired result is to utilize two rigid moire grills. One of the grills is attached to the end of the shaft. The other is placed in close proximity to the first, but rigidly fixed. Figure 8.1 illustrates the general scheme. The two grills are initially aligned so that no moire fringes can be seen. Then, by simply counting the fringes as they form, one can determine N at a given distance y from the shaft center; θ is calculated by equation 7.8. In practical terms, the goal is to establish θ as a function of fringe spacing or the number of fringes that move through a chosen point on the plate. Keep in mind that equation 7.8 is restricted to small θ. For larger rotations, one would need to use the nonlinear equation 7.4.

Typical sensitivities are calculated as follows:

$$1/p = 500 \text{ lines/in.} = 500/\text{in.; so } p = 1/500$$
$$y/N = 0.5 \text{ in.} = \text{fringe spacing; so } N/y = 1/0.5$$
$$\theta = 1/500 \times (0.5) = 1/250 \text{ radian}$$

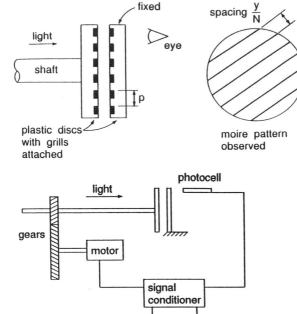

Figure 8.1. Optical measurement of shaft rotation using two moire gratings.

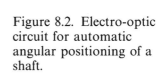

Figure 8.2. Electro-optic circuit for automatic angular positioning of a shaft.

Suppose y/N can be observed to within only 1/10 inch. What is the sensitivity or least reading of the method? It will be 1/5,000 radian. This result emphasizes the statement that moire fringes undergo large movements with small relative motions of the gratings. It is a motion magnifier. The point is that precise measurements can be made even with relatively coarse gratings.

The measurement process just described is easily interfaced with electronics for automatic shaft positioning or speed control. For an example, see Figure 8.2. With even simple optics, very precise control can be obtained with minimum equipment.

Similar procedures can be used for observations and control of translatory motion. A combination technique will give observation of rotation and translation separately when motion of the body being studied is complex.

8.2 Strain analysis

Figure 8.3 illustrates the steps to measure normal strain by the moire technique. A deformable specimen grating is superimposed with a master grating to create moire fringes, as shown in Figure 8.3a. The derivative of the

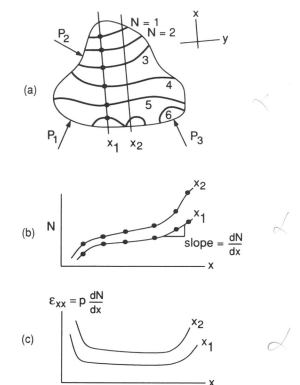

Figure 8.3. Conceptual procedure of moire strain analysis. (a) picture of loaded model with moire fringes; (b) plot of moire fringe order along x_1 and x_2 axis on model; (c) plot of strain ϵ_{xx} along x_1 and x_2 axis.

moire fringe order with respect to position coordinate must be found. A host of techniques can be employed to perform these basic tasks, and many of these are described in the extensive literature on moire analysis of deformation and strain (e.g., Chiang 1989; Dally and Riley 1991; Parks 1987; Post, Han, and Ifju 1994; Sciammarella 1982).

The final steps in the process of determining strain can be troublesome from an experimental viewpoint because of the difficulty of differentiating experimental data. Some practical procedures will be discussed later. To establish concepts, observe that moire fringe order N can be plotted as a function of position coordinate along a specific set of axes that run normal to the grating direction, as shown in Figure 8.3b. The slopes of this curve can be evaluated point by point and the results plotted. When multiplied by the pitch of the grid used, these last plots become a plot of the strain along the given axes (Fig. 8.3c). With enough axes covered, a map of the distribution of the strain component over the whole field results.

To fix these ideas, turn now to two examples that illustrate the application of the most basic geometrical moire techniques to strain analysis. Figures 8.4 through 8.7 show the steps in an actual application of basic geometric moire methods in analyzing the longitudinal strain as a function of time in a tapered

Figure 8.4. Typical apparatus and optical system for one- or two-dimensional moire strain analysis: a, light source; b, collimating lens; M, model; D, master grating; d, condenser lens; and C, camera (Kiesling 1966).

Figure 8.5. Photograph of geometric moire fringe pattern for axial strain measurement in tapered tension specimen (Kiesling 1966).

tension model as part of a useful materials characterization study (Kiesling 1966; Kiesling and Pindera 1969). In this example, the grills were reproduced on photographic stripping film and cemented to the specimen and master plate with contact adhesive. The master grill was likewise mounted on a plastic plate, which was clamped to one end of the specimen with a shim between the two to provide a small clearance. Photography of the fringe patterns was by a good 35-mm SLR camera with panchromatic medium-speed film.

Figure 8.8 is an example of moire fringes obtained by direct superimposition for a plastic ring and beam specimen that is loaded in bending. Grating lines are perpendicular to the load axis.

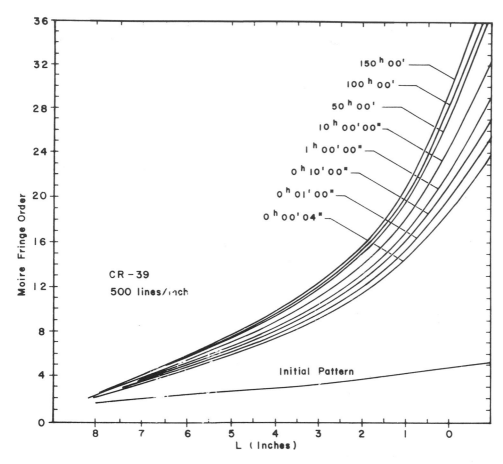

Figure 8.6. Moire fringe order *m* vs. distance along the centerline of tapered tension specimens, $t = t_1, t_2, \ldots, t_n$, for CR-39 (Kiesling 1966).

8.3 Observing techniques for geometric moire

Direct superimposing

The moire fringes are formed by superimposing two grid or grating arrays. The superimposition can be done directly (as suggested by the previous examples), optically, or photographically. An important requirement is that it be done without optical distortions; otherwise, a false strain will be indicated. The following is a summary description of various successful techniques, along with some merits and demerits of each.

Direct superimposition approach. Direct superimposition of model and master grating is intuitively the simplest. This method automatically eliminates distortional effects; and resolution requirements for the fringe recording medium are not severe because there is no need to resolve the grating lines.

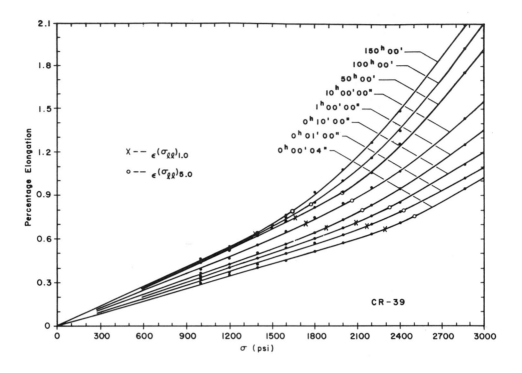

Figure 8.7. Percentage elongation vs. stress (σ) for CR-39 at constant times as determined by geometric moire (Kiesling and Pindera 1969).

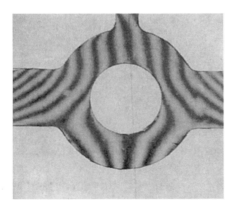

Figure 8.8. Geometric moire fringe pattern for plastic ring and beam model.

The approach works best for transparent materials. Good fringe visibility is difficult to attain when the specimen is opaque, so the superimposition is done in reflection mode.

Maintaining good contact between model and master grills is important and often difficult. If separation between the grills is allowed, then fringe visibility suffers. If the gratings are reproduced photographically, then the emulsion sides should face one another. One useful trick is to merely clamp the plate carrying the master grating to the specimen at a corner, using a light spring clip or even a clothespin. The two gratings move as one at the clamp point, and the displacement fringe field is referenced to that point. This absolute reference is not important in strain analysis anyway because it is the fringe gradient that is sought. Another good laboratory trick is to use a few drops of immersion fluid, such as oil or xylene, between the two gratings.

With this approach, resolution requirements for camera and recording medium are not stringent; it is only necessary to resolve the fringes. Instant recording films serve nicely, as does any ordinary television system if video display or recording is desired.

Finally, note that the optical system for direct superimposition of the gratings (e.g., Figure 8.4) is similar to the optics, minus the polarizers, of the transmission photoelasticity apparatus illustrated in, for example, Figures 5.2 and 5.3. Some practitioners use the same optics for both moire and photoelasticity.

Double-exposure technique. If the recording system (camera) is capable of resolving individual grating lines, the double-exposure technique is a good one. The superimposition occurs at the film plate, with the grating that is on the specimen serving as both specimen and master arrays. The specimen is photographed with no load (or while in some other initial state), the load is applied, and the specimen is photographed again on the same film plate. Optical distortions and film shrinkage, which affect both initial and final exposures equally, create no problems with the double-exposure approach. Real-time viewing of fringes as the model is loaded is not possible with this technique. Some degree of signal-to-noise improvement (fringe contrast) as well as some sensitivity multiplication can be had by optically processing (spatial filtering) the doubly exposed photographic negative (see Chapter 11).

The greatest problem with the direct photography approaches is a result of photographic limitations and their effects on measurement sensitivity, which is directly proportional to the inverse of the grating pitch (its spatial frequency as lines/millimeter). There is a direct tradeoff between field size and resolution capability in imaging by a lens. For example, a good view camera lens at 1:1 magnification and a 100-mm field is able to reproduce gratings that have spatial frequencies on the order of 20 lines/mm (500 lines/in.). If an excellent lens is used and steps are taken to stabilize the camera and specimen, and if a magnifier is used to focus the real image in the plane of the emulsion, then resolution can be increased to as much as 80 lines/mm (2,000 lines/in.). Resolution and contrast can be increased further by use of custom apertures, and it is even possible to use slotted apertures to multiply the grating frequency (Cloud 1976). Slotted apertures have also been used to photograph gratings from more-or-less random surface patterns that have

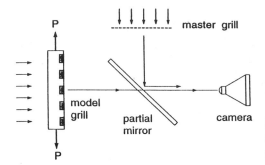

Figure 8.9. Moire superimposition with a partial mirror.

been stamped on large components (Burch and Forno 1975). That technique is closely related to white-light speckle photography. Of course, much higher photographic resolution is attainable; witness what is done in semiconductor chip technology. The catch is in the field size requirement. For moire strain analysis of typical mechanical components, formats are limited to something between 35-mm and 8 × 10. Simple calculations show that changing formats will not solve the problem. The smaller format, for example, will have at its film plane a better resolution than specified above. But its resolution at the specimen will be about the same as for the 100-mm format at 1:1 conjugates. Also keep in mind that the problem does not appear strictly in black and white. With any lens, as grating frequency increases, contrast diminishes so that the high-contrast grating lines are reproduced in shades of gray. Finally, there seems no possibility with current or near-future technology of using video-based systems for direct imaging of gratings that are fine enough to be of use in most strain analysis.

Double-exposure technique with two film plates.　A variation of the double-exposure technique is to make the first and second exposures on separate film plates. The negatives are developed normally and are superimposed directly for observation of the moire fringes. Some flexibility is gained in that the negatives can be shifted slightly to reduce the effects of rigid-body motions. The data are stored in raw form, which is convenient for experiments involving several nonreversible steps. The second plate can be an exposure of the specimen, or it can be a submaster grating. A high degree of data improvement and sensitivity multiplication can be achieved with the use of this approach along with optical processing (Chapter 11). Naturally, this direct photography method suffers the same limitations as does the double-exposure technique discussed previously.

Optical superimposition approach.　A lens system can be used to create a real image of the model grating in a plane containing the master array. A camera can be used for this purpose, and the master grating simply replaces the film holder. The master can be a photographic negative of the model grating in its unloaded state. The method is similar to the photographic superposition techniques described earlier, except that real-time viewing of fringes as they form is possible. A variation is to utilize a separate grating as the master, which allows the use of pitch mismatch (see section 8.9). This approach is the one that is used in the moire grid-analyzer apparatus, which is based on the work of Zandmann (1967) and which has been commercially available.

Optical distortions are serious with this method, but pitch adjustment can be attained by shifting a lens. Because direct imaging of the specimen grating is involved, maximum sensitivity is again limited by the field-size versus resolution problem.

Method of superposition with a partial mirror. The superposition can be performed in the plane of a partial mirror with an optical setup such as that shown in Figure 8.9. Considerable care is needed in the setup of this method, but some flexibility is gained. It gives direct superimposition without the necessity of bringing the master into close proximity with the model grill. There is light loss at the partial mirror, so contrast is reduced. Little use seems to have been made of this approach.

8.4 Creating gratings, grills, and grids

The grating and grid arrays that are suitable for moire strain analysis are not easy to obtain, make, or handle. In addition to being fine, with a spatial frequency of several hundred to several thousand lines/in., they must also be sharp and of high contrast. They should be delicate so that they will not change the mechanical properties of the object on which they are mounted. In fact, it is good practice to work with a dot array (negative of grid) instead of a grid, because the lack of mechanical interconnection in the dot array means that it will disturb the strain field less.

The techniques and art of moire strain analysis have developed substantially in the last decade or so. One is well advised to investigate newly developed techniques before undertaking a moire experiment. In fact, there is a large body of literature about the making and attaching of moire master and specimen gratings (e.g., Holister and Luxmoore 1968).

Some specific examples of procedures for reproducing moire gratings and grids appear in section 12.2 within the context of intermediate sensitivity moire. Those techniques are largely applicable to simple geometric moire, and they should be reviewed by any novice who anticipates involvement in grating replication.

Because quality gratings in large sizes are quite expensive, most practitioners purchase a few in needed line densities and sizes and use them only as masters. Submaster gratings are usually photoreproduced from the masters, and specimen gratings are third-generation photoreproductions.

Master gratings are usually purchased from suppliers. Ronchi rulings from scientific and optics companies such as Edmund Scientific Co. (101 Gloucester Ave., Barrington, NJ) are fine for spatial frequencies up to a few hundred lines/in. Superb gratings intended specifically for moire use are sold by Graticules Ltd. (Sidcup, Kent, U.K.). Two types are available, one being a photographic replica on glass substrate and the other being a mylar-based photoreproduction. These have spatial frequencies up to 3,000 lines/in., and some are provided in sets of three gratings so that plus and minus pitch mismatch can be introduced without much effort. Another good source of grids (two-way gratings) is often overlooked. The fine metallic screens used in the printing industry come in spatial frequencies up to about 1,000 lines/in.; they are accurate and not very expensive. One good source is Buckbee-Mears Corp. (St. Paul, MN). These high-quality screens are a fine, nickel mesh, and they are economical enough so that they can be cut up and used for specimen gratings (Cloud and Bayer 1988).

Submaster gratings can be made by contact printing the master grating. Care must be taken to bring the grating into close contact with the photoemulsion. Usually monochromatic light is best, and glycerine or xylene can be used between the layers to reduce lenticulation and diffraction effects. Loss of quality owing to diffraction will occur, especially with very fine gratings.

Given the extensive history of the subject, it would be possible to write an entire chapter on techniques for reproducing gratings on specimens. The following list summarizes only various basic techniques for joining moire grating with model or prototype. The important characteristics of each approach are described, and some pertinent references are given in order to direct the user to useful literature. The list is certainly not exhaustive, and anyone who contemplates extensive moire work, geometric or other, is well advised to study some of the literature and to collect practical information from experienced practitioners.

Ruled gratings. The most fundamental approach is to rule the grills directly onto model or master with straight edge and pen. A modern variation is to use a computer graphics system. Densities are limited, of course, but in some applications, the specimen can be made large enough to counteract the resulting sensitivity problem. The ruling technique has been used in studies of deflections of plates, where sensitivity requirements are not so severe as in strain analysis.

Machine tools. A related approach is to use a lathe or milling machine to rule the grating into the specimen. There are many variations of this method.

Remember that diffraction gratings have been made in this way for many years.

Direct photographic printing.   The grating or grill may be photographically printed onto the specimen surface (Austin and Stone 1976; Luxmoore and Hermann 1970). This technique is derived from that used in printed electronic circuit work. The specimen is coated with a light-sensitive emulsion such as Kodak or Shipley Photo-Resist which, after drying, is exposed to the grill pattern in a contact printer. The exposed emulsion is then developed and washed. The resulting photographic reproduction of the grill is very thin and closely bonded to the surface. The photoresist grating works very well by itself, or it can be dyed to improve contrast. If the specimen is of metal, the emulsion can serve as a mask while the grill is etched into the surface by acid. Or, a metallic film can be vacuum-deposited over the grating, after which the photoresist and its metallic coating are removed with a solvent. A fine metal-dot pattern that is quite robust remains on the specimen. The photoresist technique in its many variations has been one of the most used, for the good reason that it is easy, after some practice, to obtain fine results for a variety of testing environments including high temperatures (Cloud, Radke, and Peiffer 1979). If something goes wrong, the poor grating is simply washed off and the process repeated, with little loss of time or investment in specimens.

Photographic film.   The grills can be reproduced on photographic film and then glued to the specimen. Ordinary photo negative backing tends to be overly strong for this purpose, so it is wise to use stripping film of the type used by lithographers for composite photographs. One procedure is to fasten the grill reproduction to the specimen with the emulsion side in contact. Eastman 910 contact adhesive seems ideal for this purpose if care is taken to align the grill axes properly. Upon completion of the bonding, the emulsion backing is stripped away, leaving the thin emulsion on the specimen.

Metal transfer grids.   A scheme has been developed for reproducing grills and grids in a bilayer metallic form that is handled like the stripping film (Zandman 1967). The grills are reproduced and etched by the photoresist technique through a foil that is glued to a stainless steel backing. The composite is glued to the specimen with an epoxy, and the backing is pulled away after the adhesive has set. These grills can be used in transmission work if a transparent cement is employed. Cementing is done with a black epoxy to obtain extremely good contrast for reflection moire studies. These moire grids have been rather expensive and available only in small sizes. They were sold commercially, but their availability now is doubtful.

Stencil method. A stencil method that is very quick and effective has been developed (Luxmoore and Hermann 1971). In one version, a fine mesh, such as the photolithography screen mentioned earlier, is temporarily adhered to the surface with a soap film. The excess soap is blotted from the holes, and then a metal layer is vacuum-deposited over the mesh. When the mesh is stripped away, a metallic-dot pattern remains. Rather than using vacuum deposition, ink or dye can be applied using the "pouncing" technique of stencil artists.

Phase gratings. Methods to create and reproduce phase gratings of very high spatial frequency for moire interferometry have been developed. These approaches are summarized in Chapter 15. There seems no reason that they could not be employed to reproduce coarse gratings for geometrical moire.

Metallic mesh. As suggested earlier, the metallic mesh grating intended for screen printing can be cut into pieces for bonding directly to the specimen. Various cements can be used for testing at moderate temperatures. For high temperatures, the bonding can be accomplished with ceramic paint (Cloud and Bayer 1988). Care must be taken or the fragile mesh will be distorted by handling. Specimen grids produced in this way are of excellent contrast and regularity.

8.5 Complete analysis of two-dimensional strain field

The moire technique for measuring normal strain along specified axes in a plane body has been described. This basic idea will now be extended to determine completely the state of strain throughout a general two-dimensional strain field. Recall that there are three unknown strain components $(\epsilon_{xx}, \epsilon_{yy}, \epsilon_{xy})$ at every point in the plane elasticity problem. It will not be necessary at this stage to distinguish between "plane stress," "plane strain," and "generalized plane stress." A complete analysis of one or more of these problems may eventually require measurement of the third normal strain (ϵ_{zz}).

The theory presented in Chapter 7 demonstrates that the displacement component in a given direction can be measured by orienting the moire gratings (master and model) so that their lines run perpendicular to the displacement direction. That is, to measure u_x, the grills must be parallel with the y-axis. Clearly, both displacement components, u_x and u_y, can be found from two separate moire experiments. One fringe pattern is obtained with the grills in the x-direction, and the other results from a pattern obtained with the grills in the y-direction. It will be shown presently that the model and procedure do not have to be duplicated totally in order to obtain the two fringe patterns.

With u_x and u_y known everywhere, all three strain components can, in principle, be established through use of the strain–displacement equations,

$$\epsilon_x = \frac{\partial u_x}{\partial x} = p\,\frac{\partial N_x}{\partial x}$$

$$\epsilon_y = \frac{\partial u_y}{\partial y} = p\,\frac{\partial N_y}{\partial y} \tag{8.1}$$

$$\gamma_{xy} = 2\epsilon_{xy} = \frac{\partial u_y}{\partial x} + \frac{\partial u_x}{\partial y} = p\left(\frac{\partial N_y}{\partial x} + \frac{\partial N_x}{\partial y}\right)$$

where p is the pitch of the grills

N_x is the moire fringe order obtained with grills running in the y-direction

N_y is the moire fringe order obtained with grills running in the x-direction

The derivatives of moire fringe order with respect to the space coordinates can be evaluated in various ways, including the graphical procedure already discussed for the uniaxial case. It will be shown in the next section that, owing to rotation errors in the cross-derivatives, the technique of using two moire fringe photographs and equations 8.1 for measuring shear strain is subject to large errors unless special precautions are taken.

A very practical approach for two-dimensional moire strain analysis is to obtain three normal strain components directly with three different grill orientations (Dantu 1964). The potential errors in shear strain that are caused by relative rotations between master and specimen grills are automatically eliminated. The procedure is analogous to using a foil strain-gage rosette. The three normal strain readings completely describe the state of strain at a point. The strain transformation relations are used to determine principal values, maximum shear strain, or the normal and shear strains in some arbitrary coordinate system.

As an example, consider a case in which three moire fringe patterns are obtained with the moire grills oriented at 0°, 90°, and 135°. The displacement components, which can be found from the fringe patterns, are u_0, u_{90}, and u_{45}. Differentiation yields ϵ_0, ϵ_{90}, and ϵ_{45}. At this point, it is convenient to recall the equation commonly used for finding strain components from normal strain measurements with the rectangular strain-gage rosette.

$$\epsilon_x = \epsilon_0$$

$$\epsilon_y = \epsilon_{90} \tag{8.2}$$

$$\gamma_x = 2\epsilon_x = 2\epsilon_{45} - \epsilon_0 - \epsilon_{90}$$

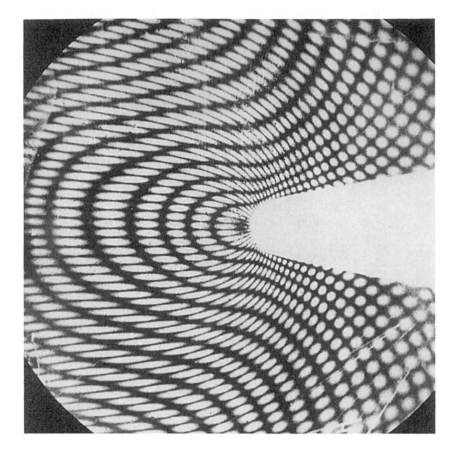

Figure 8.10. Moire pattern obtained using a two-way grid; orthogonal displacements can be determined from one photograph (courtesy of Daniel Post).

The transformation equations or Mohr's circle can then be used to find the principal values and directions.

Rather than use several different arrays of parallel lines (grills) to obtain the various needed displacement components, arrays of dots or crossed lines are used for model and master. Then, the moire fringe pattern consists of two families that form an orthogonal network. Two displacement components can be found from the single photograph. An example of such a result is shown in Figure 8.10, which is similar to those published by Post (1965). A problem with these patterns, which is serious near stress concentrations and the like, is to separate the two families correctly in subsequent data reduction.

Post (1965) has demonstrated a very useful advantage of procedures that involve the use of orthogonal master gratings to obtain the displacement components, whether on one fringe pattern or as separate patterns. If the master gratings are exactly perpendicular when the two fields are determined, and if the derivatives are precisely determined, then the effects of rotations on the cross-derivative terms in the shear strain–displacement equation are canceled. In this instance, the shear strains can be accurately established without the need for a third normal strain measurement.

A technique that eliminates the potential confusion of the two orthogonal families of fringes is to utilize a dot or grid array on the model and a line array (grating or grill) as the master. Only one model grid is required, but two displacement components can be found separately. One fringe pattern is photographed with the master grill at 0° and another with the master at 90°. In the primitive versions of the technique, a third specimen carrying a grating at 45° must be used to get the remaining normal strain for the rosette reduction equations. This limitation is eliminated by incorporating a third grating with the orthogonal set, as explained in section 8.7.

8.6 Errors in moire strain measurement from rotation

In Chapter 7, it was shown that the rotation of one grill array relative to another creates a moire fringe pattern. Because deformation of an elastic body is usually accompanied by local or gross rotations, there seems to be a possibility that the moire measurement of strains will be adversely affected. As a result, the accuracy and range of application of the moire technique are subject to question.

These questions are answered easily after study of moire patterns resulting from rotation and strain, as shown in Figure 8.11. These figures suggest that the fringes created by small relative rotations are nearly perpendicular to the grating lines, whereas those caused by normal strain are parallel to the gratings. The rotation of a moire fringe from a perpendicular to the grating lines is, in fact, one-half the relative rotation of the gratings; this fact can be useful in measuring angular displacements by the moire technique. The results summarized here can be proven by extension of the development leading to equation 7.8.

Suppose, for example, that ϵ_x is to be measured. Gratings are oriented in the y-direction for this purpose. As the final result, the gradient or spacing of the moire fringes in the x-direction is sought. The rotation-induced fringes run roughly parallel to the x-axis and, therefore, have negligible gradient in that direction. The conclusion is that small rotation does not affect the measurement of normal strain. For subsequent use, it is worthwhile to express this reasoning in algebraic form. The total fringe order is expressed as the sum of a part resulting from strain and another part caused by rotation,

$$N_x = (N_x)_\epsilon + (N_x)_\theta \tag{8.3}$$

$$\epsilon_x = p\,\frac{\partial N_x}{\partial x} = p\left[\frac{\partial (N_x)_\epsilon}{\partial x} + \frac{\partial (N_x)_\theta}{\partial x}\right]$$

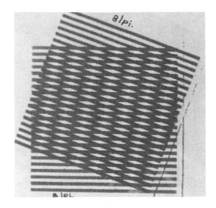

(a)

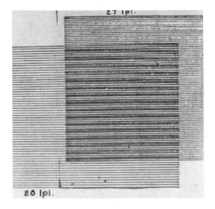

(b)

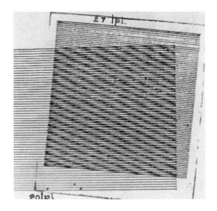

(c)

Figure 8.11. Moire fringes caused by (a) purely relative rotation of two identical gratings, (b) pure extension of one of the gratings, and (c) combined rotation and extension.

but (8.4)

$$\frac{\partial(N_x)_\epsilon}{\partial x} >>> \frac{\partial(N_x)_\theta}{\partial x}$$

In practice, the moire fringe order is not neatly separated into the two parts. The rotational element merely causes the moire strain fringes to deviate from their usual orientations parallel to the grill. This effect is seen clearly in Figure 8.5 near the large end of the tapered tension model.

The error analysis gives quite a different result if the strain–displacement relations, equation 8.1, are used to determine shear strain. The partial derivatives in the shear equation are gradients in the direction parallel to the grating lines. The gradient of rotational fringes in that direction can be at least as large as the gradient of strain fringes. In algebraic notation,

$$\gamma_{xy} = 2\epsilon_{xy} = p\left(\frac{\partial N_x}{\partial y} + \frac{\partial N_y}{\partial x}\right)$$

$$= p\left[\frac{\partial(N_x)_\epsilon}{\partial y} + \frac{\partial(N_x)_\theta}{\partial y} + \frac{\partial(N_y)_\epsilon}{\partial x} + \frac{\partial(N_y)_\theta}{\partial x}\right] \qquad (8.5)$$

but

$$\frac{\partial(N_x)_\theta}{\partial y} \geq \frac{\partial(N_x)_\epsilon}{\partial y}$$

$$\frac{\partial(N_y)_\theta}{\partial x} \geq \frac{\partial(N_y)_\epsilon}{\partial x}$$

(8.6)

So, the rotational contribution may seriously contaminate the shear strain, and large errors may result. The situation is worsened by the lack of any direct way of estimation or of eliminating the error.

The conclusion is that it is unwise to try to determine the shear strain with two measured displacement components and the strain–displacement equations unless care is taken to obtain N_x and N_y with master grills that are exactly orthogonal, as mentioned in the previous section. It is often best to obtain shear strains through the moire-rosette technique involving three measurements of normal strain and the transformation equations. An alternative approach is to utilize the rotation–displacement equation,

$$\theta = \frac{\partial u_x}{\partial y} - \frac{\partial u_y}{\partial x} \qquad (8.7)$$

Write the displacements in terms of the moire fringe orders to estimate and eliminate the rotational error in the shear strain given by the strain–displacement equation. Although reasonable in theory, this approach seems difficult to apply.

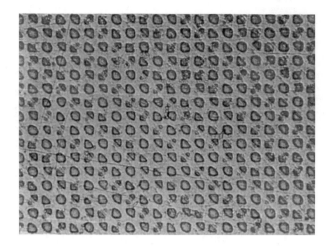

Figure 8.12.
Photomicrograph of 39
lines/mm three-way grating
in photoresist on
aluminum specimen.

8.7 Moire strain rosette

As suggested, a good way to completely determine the strain at a point is to obtain a third measurement of normal strain, and perhaps a fourth for checking purposes, by repeating the experiment with another two-dimensional grid that is inclined with respect to the first. The required three strain components (or principal strains plus principal angles) are found by application of the strain rosette equations commonly used for reducing resistance strain-gage data. At first glance, this procedure seems to require that two identical specimens be fabricated and that these be treated identically. Error propagation tends to be unfortunately large because of the experimental scatter between two separate tests and because the strain rosette equations are not very forgiving of errors in determining the oblique strain component.

A better way (e.g., Cloud, Cesarz, and Leeak 1979) is to print on the specimen surface a "three-way" grating, which is quite arbitrary in nature and which is easy to create. It allows the determination of all three required displacements from one specimen, so that the entire strain field can be established without resorting to a second specimen. Obtaining high-contrast fringes is, of course, a bit more difficult than it is for one- or two-way gratings.

The three-way grating can usually be made by extending whatever method has been selected as most suitable for the problem at hand. For example, a three-way grating can be created in photoresist by exposing it three times through a single master grill, the grill being rotated 45° between exposures. The resulting photoresist grating can then be etched, plated, or filled with ceramic for high-temperature studies. Figure 8.12 is a photomicrograph of a grating that was manufactured in this way.

It is important to recognize that no care need be exerted to register the third grating with the other two. One should also note that the third grating is, in principle, not even necessary. If it is not used, then a submaster analyzer grill of different pitch has to be used for the 45° direction, and the sensitivity will be different for that axis. This approach is used in moire interferometry (Chapter 13) to obtain the three components.

Forming the fringe patterns with this grating is no different from normal procedures, except that the fringes will be of lower contrast. To separate the three families and give best contrast, a one-way grill should be used as the analyzer. Direct superimposition works well with fairly coarse gratings. A better way is to use high-resolution photography to record the specimen grating in the unloaded and loaded states and then to create the fringe patterns by superimposing the grating photographs with a submaster grill in an optical spatial filtering system (Chapter 11). Extremely sharp fringe patterns can be obtained in this way.

Yet another approach is to photograph each component grill of the three-way grating using a slotted "tuned aperture" in the camera, as described in sections 8.3 and 11.5. The directional and frequency selectivity of the tuned imaging aperture is exploited to create three separate grating photographs from the single three-way grating. The slots in the lens are aligned with one of the specimen grills for the first photograph. Then the assembly is rotated to align with the second grill for the second photograph, and so on for the third. This method can be used in real time with a master grating in the camera, in double exposure, or by direct grating photography and subsequent superimposition with or without optical spatial filtering.

8.8 Sensitivity of geometrical moire

A major difficulty with the geometrical moire technique is obtaining and attaching gratings and grids that are fine enough in pitch to be useful in measuring strains of the magnitudes that occur in elastic deformation of high-modulus materials such as metals. Attention is focused on the basic sensitivity requirements as well as on some ways to reduce the severity of the requirements or to avoid meeting them.

The grating pitch required for a given strain sensitivity is easily calculated. Take ϵ_s to be the smallest strain to be measured. Let h be the maximum fringe spacing that can be tolerated with the apparatus or procedure employed. The grating pitch required is p, so $1/p$ will be the grating density or spatial frequency; it is customary to specify spatial frequency rather than pitch as the basic grating parameter. Use of the equations developed in section 7.3 leads to

$$\frac{\partial N}{\partial x} = \frac{1}{h}$$

$$\epsilon_s = \frac{p}{h}$$

(8.8)

As a numerical example, take $\epsilon_s = 5 \times 10^{-5}$, which is marginally acceptable sensitivity for elastic strains in metals. Suppose also that one fringe per inch is needed in the fringe photographs, a spacing that is greater than optimum with ordinary data processing methods. The required pitch for this example works out to be 5×10^{-5} inches, or a grating density of 20,000 lines/in.

Gratings and grids of density great enough for small-strain analysis are manufactured, obtained, and handled with difficulty. An obvious solution is to modify the technique in some way to allow the use of less-dense gratings. The following is a list of several such techniques, along with short descriptions.

Low-modulus materials. The moire specimens are made of low-modulus material such as plastic or rubber. Then the deformations and strains will be greater by the ratio of the elastic moduli of the model and prototype. The effect is to increase the acceptable minimum strain to be measured (ϵ_s in eq. 8.8). The scaling of model to prototype strains follows essentially the same laws of similarity as were developed for photoelasticity, as long as the deformations are not so large as to change the geometry of the model.

Fractional fringe orders. Fractional orders of moire fringes can be measured. Such a procedure increases the number of data points, which increases the acceptable maximum fringe spacing (h in eq. 8.8). Careful examination of the optics laws for moire analysis suggests that fractional orders can be established by light intensity measurement. Sciammarella (1965) has pursued the idea extensively. Others have used similar ideas in recent years with computerized optical processing, with apparent good results (Voloshin et al. 1986). Anyone contemplating this approach should examine carefully the paper by McKelvie (1986) in which, among other significant contributions, the relationships between light intensity and fringe order are worked out from fundamental concepts, with results that are not exactly what are suggested from simple considerations.

Fringe multiplication. The so-called fringe multiplication techniques are another way of increasing the acceptable fringe spacing to reduce the required grill pitch. These methods have become important with the development of the laser and the improved understanding of optical data processing. One such technique, for example, allows the use of coarse specimen gratings (below 20 lines/mm) to obtain sensitivities equivalent to that achieved with gratings of hundreds of lines/mm. Such procedures are discussed in Part IV.

Moire interferometry. For truly remarkable sensitivities in moire work, such as are required for measuring small strain in ceramics and metals, moire interferometry is the answer. This subject is taken up in Part V.

A conclusion is that elastic strain analysis by geometrical moire is appropriate for low-modulus materials, such as soft tissue, wood, elastomers, and some plastics. It is also useful for investigations on metals that are deformed plastically. As we shall see in the next chapter, geometric moire is also valuable for creating contour maps of objects, and in certain investigations of fluid flow.

8.9 Effects of pitch mismatch

A difference of pitch between the model and master grills will cause moire fringes to form even though the model is unstrained. These fringes usually appear as an initial pattern that would contribute serious error if it were not eliminated in the data analysis. On the other hand, the effects of pitch mismatch can be turned to advantage as a device for increasing the capability of the moire technique.

A clear understanding of pitch mismatch results from study of equation 7.6. Rotation effects have already been discussed, so take the alignment mismatch to be zero. If we let

$\theta = 0$

p be the original pitch for the model grating

q be pitch of the master grating

δp be change in p caused by strain (this is the quantity that must be measured accurately)

then equation 7.6 becomes

$$\left(\frac{p + \delta p - q}{p}\right)x = Nq \tag{8.9}$$

$$\left(\frac{p - q}{p}\right)x + \left(\frac{\delta p}{p}\right)x = nq \tag{8.10}$$

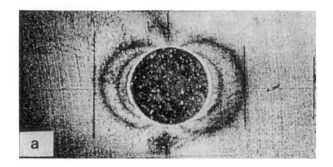

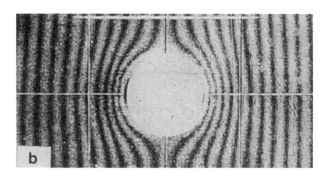

Figure 8.13. Sample geometrical moire pattern for a deformed body taken (a) without mismatch and (b) with a small pitch mismatch (patterns optically filtered) (Cloud 1980).

Let

$$\frac{p - q}{p} = \rho = \text{mismatch ratio} \qquad (\text{may be } + \text{ or } -) \qquad (8.11)$$

and notice that

$$\frac{\delta p}{p} = \epsilon_x$$

so equation 8.9 becomes

$$\rho x + \epsilon_x x = Nq$$

$$\epsilon_x = \frac{Nq}{x} - \rho \tag{8.12}$$

This result can be written in terms of the gradient of moire fringe order as was shown in Chapter 7 and used in subsequent discussion:

$$\epsilon_x = q \frac{\partial N}{\partial x} - \rho \tag{8.13}$$

Equation 8.13 shows how to eliminate the effects of initial pitch mismatch. In context, the pitch mismatch ratio ρ is essentially the spacing of moire fringes in the initial pattern observed before straining the model. In other words,

$$\rho x = N_i = \text{initial fringe order} \tag{8.14}$$

The result just derived also demonstrates that an initial mismatch can be used to increase the number of fringes in a moire measurement. Such an increase of fringe order sometimes makes the fringes easier to locate and count, and subsequently it facilitates the determination of the space derivatives of fringe order. This idea can be especially valuable where strain gradients are small and where fringes are widely spaced.

Note that it is not absolutely necessary to know the grating pitches in order to calculate and eliminate the pitch mismatch effect. "Before strain" and "after strain" fringe records (photographs) can be used directly with equations 8.13 and 8.14. As an added possibility, these equations can be used in reverse to determine the unknown pitch of any grating if a grating having known pitch is available. The concepts are also important when a lens is used to image the model grill in the plane of the master grill, since it is tedious to establish exactly the magnification ratio beforehand.

Emphasis must be given the fact that pitch mismatch does not increase the sensitivity of the moire process. The sensitivity is governed by the pitches of the gratings. True sensitivity multiplication can be obtained by optical data processing (see Chapter 11) of the grating images. Mismatch helps in providing more data points; it is primarily an aid to interpolation of the fringe patterns.

Sample fringe patterns with and without pitch mismatch are shown in Figure 8.13.

References

Austin, S., and Stone, F. T. (1976). Fabrication of thin periodic structures in photoresist. *Applied Optics*, 15, 4: 1070–4.

Burch, J. M., and Forno, C. (1975). A high sensitivity moire grid technique for studying deformation in large objects. *Optical Engineering*, 14, 2: 178.

Chiang, F. P. (1989). Moire methods of strain analyses. In *Manual on Experimental Stress Analysis*, 5th ed., Eds. J. F. Doyle and J. W. Phillips, pp. 107–35. Bethel, CT.: Society for Experimental Mechanics.

Cloud, G. (1976). Slotted apertures for multiplying grating frequencies and sharpening fringe patterns in moire photography. *Optical Engineering*, 15, 6: 578–82.

Cloud, G. L. (1980). Simple optical processing of moire grating photographs. *Experimental Mechanics*, 20, 8: 265–72.

Cloud, G., and Bayer, M. (1988). Moire to 1370°C. *Experimental Techniques*, 12, 4: 24–7.

Cloud, G., Cesarz, W., and Leeak, J. (1979). A true whole-field strain rosette. *Proc. 8th Congress of International Measurement Confederation*, Moscow, USSR: S7–27.

Cloud, G., Radke, R., and Peiffer, J. (1979). Moire gratings for high temperatures and long times. *Experimental Mechanics*, 19, 10: 19N–21N.

Dally, J. W., and Riley, W. F. (1991). *Experimental Stress Analysis*, 3rd Ed. New York: McGraw-Hill.

Dantu, P. (1964). Extension of the moire method to thermal problems. *Experimental Mechanics*, 4, 3: 64–9.

Holister, G. S., and Luxmoore, A. R. (1968). The production of high-density moire grids. *Experimental Mechanics*, 8, 5: 210–16.

Kiesling, E. W. (1966). *Nonlinear, Quasi-Static Behavior of Some Photoelastic and Mechanical Model Materials*. Ph.D. dissertation. East Lansing, MI: Michigan State University.

Kiesling, E. W., and Pindera, J. T. (1969). Linear limit stresses of some photoelastic and mechanical model materials. *Experimental Mechanics*, 9, 8: 337–47.

Luxmoore, A. R., and Hermann, R. (1970). An investigation of photoresists for use in optical strain analysis. *Journal Strain Analysis*, 5, 3: 162.

Luxmoore, A. R., and Hermann, R. (1971). The rapid deposition of moire grids. *Experimental Mechanics*, 11, 5: 375.

McKelvie, J. (1986). On the limits to the information obtainable from a moire fringe pattern. *Proc. Spring 1986 Conference on Experimental Mechanics*, pp. 971–90. Bethel, CT: Society for Experimental Mechanics.

Parks, V. J. (1987). Geometric Moire. *Handbook on Experimental Mechanics*, Ed. A. S. Kobayashi, Ch. 6. Englewood Cliffs: Prentice-Hall.

Post, D. (1965). The moire grid-analyzer method for strain analysis. *Experimental Mechanics*, 5, 11: 368–77.

Post, D., Han, B., and Ifju, P. G. (1994). *High Sensitivity Moire: Experimental Analysis for Mechanics and Materials*. New York: Springer-Verlag.

Sciammarella, C. A. (1982). The moire method – a review. *Experimental Mechanics*, 22, 11: 418–32.

Sciammarella, C. A. (1965). Basic optical law in the interpretation of moire patterns applied to the analysis of strains – part I. *Experimental Mechanics*, 5, 5: 154–60.

Voloshin, A. S., Burger, C. P., Rowlands, R. E., and Richard, T. S. (1986). Fractional moire strain analysis using digital imaging techniques. *Experimental Mechanics*, 26, 3: 254–8.

Zandman, F. (1967). The transfer-grid method: a practical moire stress-analysis tool. *Experimental Mechanics*, 7, 7: 19A–22A.

9

Moire mapping of slope, contour, and displacement

This chapter describes techniques for using geometric moire to measure out-of-plane displacement and slope, and also for mapping the contours of three-dimensional objects (Chiang 1989; Parks 1987; Theocaris 1964). The ideas are illustrated by an example from biomechanics, which has been a major area of application.

9.1 Shadow moire

The shadow method of geometric moire utilizes the superimposition of a master grating and its own shadow (Takasaki 1970; Takasaki 1973). The fringes are loci of points of constant out-of-plane elevation, so they are essentially a contour map of the object being studied. In studies of deformable bodies, the method can be used to measure out-of-plane displacements or changes in displacement. To understand the creation of fringes and to be able to interpret them, consider the optical system shown in the conceptual sketch of Figure 9.1.

A master grating of pitch p is placed in front of an object that has a light-colored nonreflective surface. The combination is illuminated with a collimated beam at incidence angle α. Observation is at normal incidence by means of a field lens that serves to focus the light to a point, where a camera or an eye that is focused on the object is located.

The incident illumination creates a shadow of the grating on the surface of the specimen. The grating shadows are elongated on the specimen by a factor that depends on the inclination of the surface, and they are shifted laterally by an amount that depends on the incidence angle and the distance w from the master grating to the specimen. Examination of the sketch shows, for example, that the apparent lateral shift of a grating shadow is given by $\delta = w \tan \alpha$, where w is the z-component of displacement.

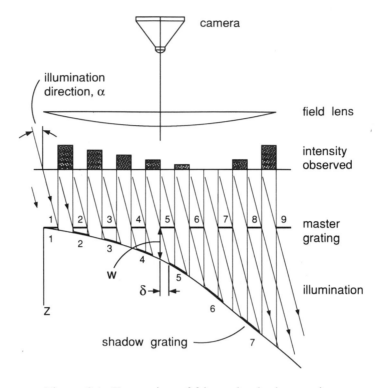

Figure 9.1. Formation of fringes in shadow moire.

The shadows are observed through the spaces in the master grating. At some locations in the composite, the shadows line up with the dark lines of the master, and these areas will appear light. In other areas, the shadows will be aligned with the spaces in the master, and these areas will appear dark. The sketch shows one complete cycle of this effect, from light to dark and back to light. Note that the elongation of lines in the shadow do not affect the composite picture for viewing along the normal.

The relationship between fringe order and z-displacement is derived in a manner similar to that used for in-plane displacement. Examination of Figure 9.1 shows that a complete fringe cycle takes place when, in this case, the shadows of seven grating lines are spread over the expanse of eight lines in the original grating. In general terms, the shadows of m grating lines are spread over the expanse of $m + 1$ lines of the grating for one fringe order. Let w be the change in the z-component of distance between master grating and specimen over the same expanse. Simple trigonometry gives

$$[(m + 1) - m] \frac{p}{w} = \tan \alpha$$

In more useful form, the change in z corresponding to one fringe order is

$$w = \frac{p}{\tan \alpha} \tag{9.1}$$

If there are N fringes between two specific locations on the image taken with normal incidence, then the change in displacement between those two points is

$$w = \frac{Np}{\tan \alpha} \tag{9.2}$$

In practice, it is uncommon to use a field lens to obtain true normal incidence. In that case, equation 9.2 will not be exact except on the viewing axis. The error will be small if the viewing distance is large compared with the specimen breadth. Similar statements apply if noncollimated light is used for illumination.

If viewing is along some direction other than the normal, the sensitivity of the fringes to displacement will be altered. For example, assume that viewing is from a direction opposite the normal to the angle of incidence (clockwise from the normal in the sketch); then the effect of the displacement on the fringes will be exaggerated, so sensitivity is increased. The relationship between fringe order and displacement is easily developed for this more general case, with the following result:

$$w = \frac{Np}{\tan \alpha - \tan \beta} \tag{9.3}$$

where α is the incidence angle
              β is the viewing angle
              p is the grating pitch
              N is the moire fringe order
              w is the axial distance from grating plane to object

Note that the relative signs of the angles are important. If illumination and viewing are from opposite sides of the normal, then β is taken as negative.

Equation 9.3 can also be used to estimate the errors caused by noncollimated illumination or when a field lens is not used to establish normal viewing. One merely examines the equation for the range of angles that are imposed by the particular setup.

One of the major applications of shadow moire is in contour mapping of the human body, with the object of detecting asymmetries that indicate certain infirmities (e.g., Adair, Van Wijk, and Armstrong 1977). In some of these biomechanics studies, the necessarily large reference grating is created by merely stretching strings across a framework.

9.2 Reflection moire

A moire method for measuring the slope or rotation, or (more precisely) the change of slope of structural components, was developed by Ligtenberg (1954) and later utilized and improved by several other investigators (e.g., Bradley 1959; Kao and Chiang 1982). This approach is used in the study of plates in bending and similar problems. Moire methods that give contour, such as the shadow method, can be used for obtaining the curvatures that are important in studies of plates. The direct measurement of slope offers a distinct advantage, however, in that it eliminates one differentiation of experimental data.

The basic experimental arrangement for reflection moire measurement of a plate or membrane is shown in Figure 9.2. The plate to be studied is polished on one side so as to act as a mirror; it is then mounted in a holder that provides correct boundary conditions and facility for the required load. At distance d from the plate a moire master grating is erected. Errors are reduced if the grating is curved; but a flat grating is often used, and the resulting small errors are eliminated in other ways. The grating has a hole at its center, and a camera is set up behind the aperture. The camera is aimed at the plate, but it is focused on the virtual image of the grating as it is reflected in the polished plate. If the camera aperture is small enough that the depth of focus is large, then the plate itself will also be imaged by the camera lens.

A first exposure is made with the specimen in its initial state, which is taken to be the flat, unloaded position in the sketch. In this case, the grating element at point Q is reflected from point P on the plate, and this grating element appears superimposed on the image of point P. The plate is then deformed so that it acts as a curved mirror. Point P on the plate has moved to point P′, which is, to a reasonable approximation, imaged to the same location on the film as point P. Because of the curvature, grating element Q′ is now superimposed on this image point in the second exposure. The effect in the image is that the reflected grating has shifted, thereby producing a moire pattern in the double-exposure image.

Rather than analyze the grating shift and the resulting moire pattern at the image plane, it is convenient to refer it to the real plate dimensions. If θ_x is the change of slope in the x-direction ($\partial w/\partial x$) at point P as it moves to P′, then the law of reflection indicates that the line PQ in the sketch rotates through angle $2\theta_x$ as it moves to position P′Q′. The apparent relative shift δ of the grating between exposures is found to be

$$\delta = d[\tan(\alpha + 2\theta_x) - \tan \alpha] \tag{9.4}$$

If p is the pitch of the grating, then at the point in question there will be produced a moire fringe of order N, where $N = \delta/p$. Notice that any effect of the deflection on the apparent relative shift of the grating in the

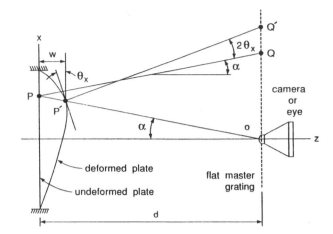

Figure 9.2. Reflection moire arrangement for measuring changes of slope.

double-exposure photograph is ignored. Although such an assumption is appropriate for typical plates subjected to bending, it may not be correct for membranes or shells.

The distance d from camera to specimen is ordinarily large compared with plate dimensions. In that case, the paraxial approximation pertains, and the equation relating slope and fringe order is

$$\theta_x = \frac{Np}{2d} \qquad (9.5)$$

The error caused by use of this simplified equation can be ascertained for a given arrangement by the use of the more exact equation 9.4. It can usually be brought within acceptable limits by making d large and using a lens of long focal length in the camera. Otherwise, a good bit of the error in the paraxial approximation can be eliminated by using a grating that is curved to a cylindrical shape. The gratings are often just ruled on posterboard, and they are easily mounted on curved rails sawn from plywood. Ligtenberg (1954) shows that the optimum radius of curvature of the grating is about $3.5d$.

For complete analysis of the plate or other object, the slope in the y-direction θ_y must also be measured. The exception is when symmetry is absolutely established, and θ_y in one quadrant corresponds to θ_x in another quadrant. Independent observation of θ_y requires that either the specimen or the grating be rotated $90°$ about the optical z-axis of the system and that a second double-exposure photograph be made.

The grating pitch required for a given arrangement and a specified sensitivity is easily determined from equation 9.5. For example, if d is 2 m and if one fringe is to represent 10^{-3} rad, then the pitch is 4 mm. Such a reference grating can be fabricated by ruling the lines with a pen onto a sheet of posterboard.

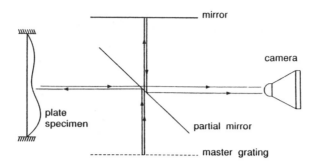

Figure 9.3. Optical arrangement for real-time reflection moire analysis of slope.

The fringe contrast in all types of double-exposure moire negatives tends to be marginal. Nonlinear high-contrast processing is helpful. Fringe visibility can be greatly enhanced by using optical spatial filtering as described in detail in Chapters 10 and 11.

An improved but more complex optical system, which will give a moire pattern in real time so that it can be examined and then recorded with a single exposure, is shown in Figure 9.3 (Parks 1987). Several other improvements of the technique have been developed.

9.3 Projection moire

A third method for using moire for out-of-plane displacement measurement or contour mapping involves projecting the reference grating on the specimen by means of, for example, a slide projector (Der Hovanesian and Hung 1971). The analysis is quite similar to that of shadow moire. One distinct advantage of projection moire is that only a small transparency of the reference grating is required. Alternatively, a projected grating can be created by interference of coherent light, as will be discussed at the end of this section. Shadow moire, on the other hand, requires a master grating that is as large as the object being studied. Another advantage of projection moire is that, if the surface is not initially flat, its out-of-plane displacement can be determined from only one double-exposure photograph. With shadow moire, one would need to calculate the difference in contours given by two separate fringe photographs. In fact, the difference between contours of two entirely distinct objects is easily determined by projection moire.

Figure 9.4 shows a simplified optical arrangement for analysis of the z-component of displacement by projection moire. The grating pitch and surface displacement are greatly exaggerated for analysis, and the viewing distance is foreshortened. This setup emphasizes both the similarities and the differences between projection moire and reflection moire. Surface S_1 is the

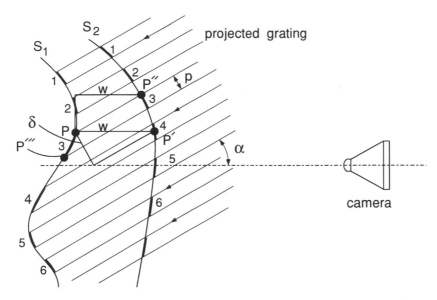

Figure 9.4. Optical arrangement for projection moire analysis of contour or changes of z-displacement.

specimen in its initial state. A grating of pitch p is projected obliquely onto the surface from a device that keeps the pitch constant in space. This condition is approximated by a slide projector placed far away from the specimen; it can be achieved exactly by using interference of coherent plane waves to create the grating, as will be described. Observation is parallel to the optic axis everywhere. This condition implies that a field lens is used, with the camera placed at its focus. An approximation is that viewing or photography is from a distance that is much larger than the breadth of the specimen. Both of these paraxial approximations can often be achieved in practice for an acceptable level of error.

A first exposure is recorded, after which the specimen is deformed to its second state, represented on the sketch by surface S_2, and a second exposure is taken. For the moment, consider only the case in which a typical point P on the surface is caused to move axially a distance w to point P′. The effect of lateral shift of the specimen is considered later. Examination of the sketch shows that as P undergoes axial movement w, it moves across the projected grating by the amount $\delta = w \sin \alpha$. Utilize now what has been learned about fringe formation in geometric moire; the fringe order at point P in the double-exposure photograph will be $N = \delta/p$. Making appropriate substitutions and inverting the result gives the axial displacement in terms of the fringe order as

$$w = \frac{Np}{\sin \alpha} \qquad (9.6)$$

where w is the axial displacement
             N is the moire fringe order
             p is the grating pitch at the specimen
             α is the incidence angle of the projected grating

The preceding result is fundamental in the interpretation of projection moire fringe patterns. It is subject to the limitations imposed by the assumptions of paraxiality.

Note that, unlike shadow moire but in common with reflection moire, it is essential that the camera be able to resolve the grating lines and to create an image with high contrast. Also, the projection system must be able to project the grating with sufficient contrast. The optics need to be quite good, but not superb, because the gratings for projection moire often have a pitch on the order of only one line/millimeter or less at the specimen, depending on the sensitivity sought. If the grating lines are not imaged with good contrast, then fringe visibility will suffer.

Consider now the effect of lateral motion of the observed point P. Suppose the point moves to location P″ in the sketch. The important fact to remember is that, in the double-exposure image, the laterally shifted point P″ is no longer superimposed onto the image of the original point P. Rather, its place is taken by a different point, such as P‴ on the specimen, which moves to the location P′ after deformation. The moire fringe order created at image point P shows just the change in elevation between point P and the new point P′. In general, the moire fringes in projection moire are not necessarily associated with the "before" and "after" states of the same point. The same is true of reflection moire, although it is not often mentioned because that method is usually employed for the study of plates that are not allowed to shift sideways. This characteristic of projection moire can be a disadvantage or an advantage, depending on application.

One positive way in which this peculiarity can be exploited is in measuring the difference in contours between two distinct objects. The first exposure is recorded using the first object, and then the second object is put into its place for the second exposure. The moire fringes indicate the contour difference. If the first object is a flat surface, then the fringes are a normal contour map of the object. Otherwise, the mismatch between the objects is established. One obvious application of this idea is in quality control of, say, stamped metal or molded components. A master prototype can be used for one exposure, and deviations in a production sample measured by this comparison process.

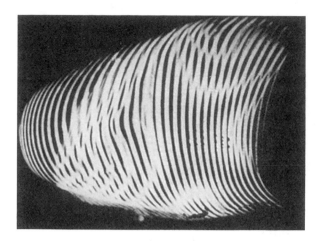

Figure 9.5. Moire fringes depicting contour differences between the human foot and a shoelast (Der Hovanesian and Hung 1975).

Although the approach is proven to be practical, it seems not to be used in industry. Der Hovanesian and Hung have pursued this contour difference technique in elegantly simple experiments for several years (e.g., Der Hovanesian and Hung 1971; Der Hovanesian and Hung 1975). Figure 9.5 is an example of the results obtained for a shoe-fitting application. In general, the projection moire approach is appropriate for a wide variety of experimental applications. In view of its simplicity, it seems surprising that it is not more widely used.

In practice, it is not usually practicable to adhere completely to the paraxial conditions. If a slide projector is used to generate the reference grating, then the grating will be diverging in space even if the projector is at considerable distance from the object. For large objects, a field lens is prohibitively expensive, so the camera is actually at some finite distance from the object. These practicalities mean that equation 9.6 is accurate only on the optic axis of the moire system. To properly interpret the fringes for off-axis points, or to assess the error in the results obtained by the simple analysis, the fringe–displacement relationship must be obtained for finite projector and camera distances. The approach is similar to that used earlier, except the trigonometry is more complicated. The problem has been solved in detail (Khetan 1975); only the results are reproduced here, with minor notation changes. Figure 9.6 identifies the surfaces and the variables involved.

The developed relationships are as follows:

$$w = w_d - w_0 \tag{9.7}$$

where w is the displacement of P during deformation
          w_0 is the initial z coordinate of point P
          w_d is the final z coordinate of point P displaced to P$'$

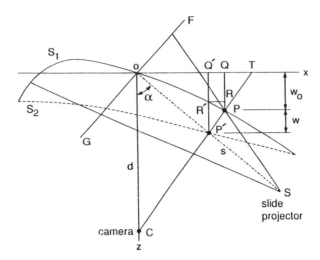

Figure 9.6. Geometry for projection moire with finite projector and camera distances (based on Khetan 1975).

Individual relationships of w_d and w_0 are as follows:

$$w_d = w_0 + \frac{Np}{\sin \alpha} \left[\left(1 - \frac{2w}{s} \cos \alpha - \frac{Np}{s \tan \alpha} \right) \right.$$

$$\left. - \frac{x}{s \sin \alpha} \left(\frac{s}{d} \cos \alpha - \cos 2\alpha \right) \right] \tag{9.8}$$

$$w_0 = d - x_{a0}$$

where N is the fringe order (number of grating lines in cone P'SP)
         p is the pitch of projected grating lines on plane FG (assumed parallel
            to the y-axis)
         s is the distance from projector to subject reference plane
         d is the distance from camera to subject reference plane
         x is the x coordinate of point P observed by the camera lens
         α is the angle between camera and projector optical axis
         x_a is the x coordinate of point P'
         x_{a0} is the x coordinate of point P
         x is the x_a
         n is the number of lines in cone OSP

Some approximations are incorporated into the derivation of equation 9.8. Khetan states that it provides results within 1% of the exact analysis. He also develops simplified equations for specific applications.

The use of double exposures to create the fringe patterns in projection moire is subject to the usual problem of marginal fringe visibility. Fringe contrast can often be improved by high-contrast nonlinear processing of the negative. Another approach for improving fringe quality is to utilize coherent optical spatial filtering of the moire grating photographs, as presented in

Chapters 10 and 11. In some cases, both spatial filtering and nonlinear chemical processing are employed to gain optimal results.

Projection moire can be accomplished in real-time mode. The idea is to record the first exposure of the specimen onto a high-contrast photographic plate having a glass substrate for stability. The single exposure is developed and then placed back into the camera plateholder in exactly its original position. As the specimen is deformed, the fringes will be created at the photoplate, which is now serving as a master grating. The fringes can be viewed as they form, and they can be photographed by placing a second camera behind the first.

Time-average analysis of periodic vibrations by projection moire has also been accomplished (Der Hovanesian and Hung, 1971; Harding and Harris 1983). The concept is very similar to that used in time-average holographic interferometry or speckle interferometry as discussed in later chapters, so it will not be developed here. The approach would be indicated when relatively large amplitude vibrations of flexible objects are to be studied. The holographic and speckle techniques tend to fail when deformations are larger than several wavelengths of light.

The only special apparatus needed for projection moire is a good quality grating of a size that can be placed into a slide projection. This requirement is best met by purchasing a Ronchi ruling of the correct size and spatial frequency from a supplier of laboratory apparatus or optical components. An acceptable alternative is to rule a grating onto a large piece of posterboard and photograph it with a 35-mm camera.

A quite different and highly satisfactory approach for projecting the grating in space is to utilize the interference of two coherent beams having plane wavefronts (Harding and Harris 1983). This approach to creating a grating was discussed in section 2.10. Either mirrors or a Fresnel biprism can be used with an expanded laser beam to form the three-dimensional grating structure. This method of creating the projected grating has two distinct advantages. The first is that the grating is everywhere in focus. No matter how convoluted the object, or how large its deformation, the grating lines remain sharp. The second advantage is that, with proper setup, the grating pitch is constant throughout the field. This is equivalent to having a slide projector at infinite distance from the object. The only real disadvantage of the method is that the grating and the photographic images will be contaminated by laser speckle. If small apertures are used, the speckle size might approach the width of the grating lines, which would seriously degrade moire fringe quality.

9.4 Projection moire application in biomechanics

An interesting example of the use of projection moire to analyze the *changes* in contour of an object is drawn from the field of biomechanics of the human

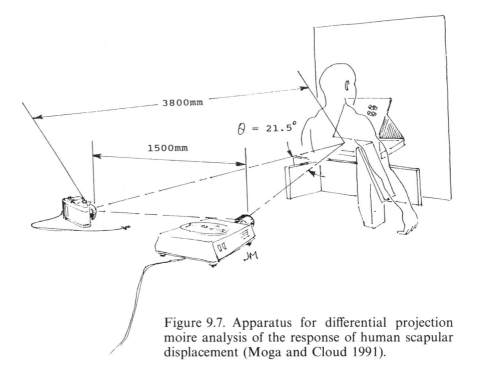

Figure 9.7. Apparatus for differential projection moire analysis of the response of human scapular displacement (Moga and Cloud 1991).

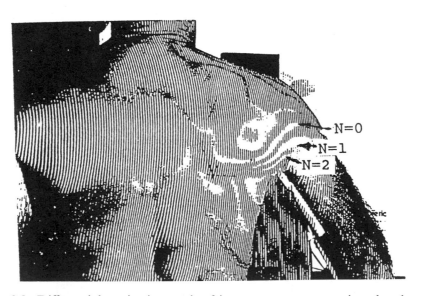

Figure 9.8. Differential projection moire fringe pattern representing the change in topography of the human scapular mechanism with muscular effort; result for 20% adduction of upper arm (Moga and Cloud 1991).

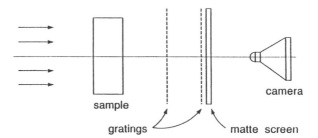

Figure 9.9. Apparatus for
moire analysis of gas flow.

body (Moga and Cloud 1991). The objective of the study was to quantify the
movement of the scapular mechanism as a function of adduction (tension
toward the body) of the upper arm. This application is, then, one of differential
moire in a process that occurs in several discrete steps. It is different from
other biomechanical applications of moire topography, which are directed
only toward the determination of absolute contours in a single state of the
body. In either case, the advantages of moire methods for biomechanics
studies are clear. It is noncontacting, remote, nontraumatic, and nonstressful;
and the sensitivity can easily be adjusted to appropriate levels. Figure 9.7 is
an isometric schematic of the experimental arrangement.

Figure 9.8 is a sample of one of the moire fringe patterns obtained by
double exposure between the at-rest state and a specified degree of muscle
contraction effort expressed as a fraction of the subject's maximum voluntary
contraction. Such fringe patterns were recorded for several levels of muscular
effort in various subjects. They were analyzed to obtain plots of out-of-plane
displacement as a function of percent of maximum voluntary contraction.

9.5 Moire in fluid mechanics

Not much attention is being given in this book to interferometric measurements on fluids. In many cases, the extension to fluid mechanics studies is
obvious. One common modification for work on gas flow is that the optical
system must be changed so that the light passes through the sample volume
instead of being reflected or scattered. The geometric moire technique is no
exception, even though not truly interferometric. Figure 9.9 illustrates an
arrangement of the sort that has been used (Keren et al. 1981) for studying
aspects of gas flow.

The operational principle is that the refractive index of the gas varies
throughout the flow field as a result of pressure/density gradients. The
refractive index variations cause distortions of the image of the grating that
is placed on the far side of the sample volume. This distorted image is
superimposed with a reference grating by one or another of the methods
common to geometric moire. The product is a system of moire fringes
indicative of pressure or temperature distributions.

References

Adair, I. V., Van Wijk, M. C., and Armstrong, G. W. D. (1977). Moire topography in scoliosis screening. *Clinical Orthopaedics and Related Research*, 129: 165–71.

Bradley, W. A. (1959). Laterally loaded thin flat plates. *Journal of Engineering Mechanics Division, Am. Soc. Civil Engineers* (Oct. 1959); also with discussion in *Trans. American Society of Civil Engineers*, paper no. 3194 (1960).

Chiang, F. P. (1989). Moire methods of strain analyses. In *Manual on Experimental Stress Analysis*, 5th ed., Eds. J. F. Doyle and J. W. Phillips, pp. 107–35. Bethel, CT.: Society for Experimental Mechanics.

Der Hovanesian, J., and Hung, Y. Y. (1971). Moire contour-sum, contour-difference, and vibration analysis of arbitrary objects. *Applied Optics*, 10, 12: 2734–8.

Der Hovanesian, J., and Hung, Y. Y. (1975). Moire contouring using new scanned ruling method. *The Engineering Uses of Coherent Optics*, Proceedings, Ed. E. R. Robertson. Cambridge University Press.

Harding, K. G., and Harris, J. S. (1983). Projection moire for vibration analysis. *Applied Optics*, 22, 6: 856–61.

Kao, T. Y., and Chiang, F. P. (1982). Family of grating techniques of slope and curvature measurements for static and dynamic flexure of plates. *Optical Engineering*, 21, 4: 721–42.

Keren, E., Bar-Ziv, E., Glatt, I., and Kafri, O. (1981). Measurements of temperature distribution of flames by moire deflectometry. *Applied Optics*, 20, 24: 4263–6.

Khetan, R. P. (1975). *Theory and Applications of Projection Moire Methods*. Doctoral dissertation. Stony Brook, NY: State University of New York.

Ligtenberg, F. K. (1954). The moire method, a new experimental method for determining moments in small slab models. *Proc. Soc. Experimental Stress Analysis*, 12, 2: 83–98.

Moga, P. J., and Cloud, G. L. (1991). A study of scapular displacement by differential moire methods. *Proc. 1991 Conference on Experimental Mechanics*. Bethel, CT: Society for Experimental Mechanics, 736–50.

Parks, V. J. (1987). Geometric Moire. *Handbook on Experimental Mechanics*, Ed. A. S. Kobayashi, Chap. 6. Englewood Cliffs: Prentice-Hall.

Takasaki, H. (1970). Moire topology. *Applied Optics*, 9, 6: 1457–72.

Takasaki, H. (1973). Moire topology. *Applied Optics*, 12, 4: 845–50.

Theocaris, P. S. (1964). Moire fringes of isopachics. *Journal of Scientific Instruments*, 41, 3: 133–8.

PART IV

Diffraction theory, optical processing, and moire

10

Diffraction and Fourier optics

The interference phenomenon and its place in optical measurement have, to this point, been the dominant themes. This chapter introduces and develops in detail the second anchor point in optics, diffraction by an aperture. After discussion of diffraction theory, several simple but important examples are presented; then the idea of optical spatial filtering is explained. These concepts are important in the remainder of the book.

10.1 Overview and problem identification

One of the oldest and most fundamentally important problems in optics is to predict the nature of the light field at any distance and direction from an illuminated aperture having arbitrary shape and perhaps containing certain optic elements in the form of a lens, a grating, or some kind of a filter. This problem is important because it provides an understanding of the formation of images by optical components, and it leads to ways of predicting, specifying, and measuring the performance of optical systems. Diffraction theory also leads to a conception of certain optical components as Fourier transforming devices, and it gives us the theoretical basis of whole-field optical data processing. The recording and analysis of moire gratings, the process of holography and holographic interferometry, and the methods of speckle interferometry and speckle photography all can be understood as diffraction processes.

The diffraction problem is complex and has not yet been solved in generality. The classic solutions rest on some severe assumptions that are not altogether realistic and logical. Even with the simplified solutions, the calculation of the optical field for arbitrary or complex apertures is forbidding. It testifies to the brilliance of the devisers of these solutions that their results describe and predict with considerable accuracy what is observed.

The oldest solution, known as the Huygens–Fresnel construction, rests on the assumption by Huygens in 1678 that the problem is equivalent to a simpler one comprising an array of infinitesimal point sources spread across the aperture. This assumption might not seem reasonable, but the resulting solution, which was obtained by Fresnel in 1818, is indeed the correct one, although a component known as the obliquity factor could not be derived. This solution was, and remains, a standard approach to explaining diffraction phenomena. Perhaps its outstanding attributes are that it is mathematically straightforward and the physical process is easy to visualize.

The solution of the diffraction problem was accomplished with more mathematical rigor in 1882 by Kirchhoff, who treated it as a boundary value problem. That this solution gave the correct result, including the part missing in the Huygens–Fresnel approach, might seem surprising since it was later shown that Kirchhoff's assumptions about the boundary values were not consistent. The theory was extended and refined by Kottler and by Sommerfield using mathematical theory developed by Rayleigh so as to get around the boundary value discrepancy.

When these solutions are said to give the correct result, it must be remembered that they rest on certain approximations that limit validity. The practical significance of these limitations is that the optical field must not be observed too close to the aperture and that the aperture must be larger than the wavelength of the light.

The Rayleigh–Sommerfield approach is physically the most realistic of the approximate diffraction theories. Completely rigorous solutions are few, the notable exception being the exact solution by Sommerfield describing diffraction by a perfectly conducting thin plate. The Kirchhoff solution is, on the other hand, simpler and sufficient for most purposes. Neither solution is easy to perceive as a physical process, but either can be shown equivalent to the Huygens construction for simple cases, so the benefits of visualization are not lost entirely.

Much of this development appears in a standard reference (Born and Wolf 1975), although it is not easily extracted. Another text (Goodman 1968) does not press through the entire derivation but is useful to those whose main concern is utility. The treatment here follows the pattern elucidated in the extraordinary lectures of the late Walter Welford of Imperial College, London (unpublished lecture notes 1976–77). The lectures and notes of Robert Haskell (1971) have also been valuable. Because of the length of the integral expressions that are encountered, the derivation is in terms of complex amplitude (recall section 2.6). After the manner of Professor Welford, the integral equations are left in the form containing complex fractional superscripts, because they are most easily understood in that form, and the relationships between geometric space and frequency space remain relatively clear.

The next section is a summary of the Kirchhoff solution. Some of the details pertaining to the implications of the various assumptions are left out for the sake of brevity and so as not to obscure the physical essentials.

10.2 General diffraction theory

10.2.1 The Kirchhoff integral

The problem is to describe the light field in terms of its complex amplitude in the vicinity of a point P, which is illuminated by radiation that passes through an aperture on its way from source Q. Figure 10.1 illustrates the general problem. The distribution of light on the side of the aperture opposite the source of radiation will not be the same as would be seen if the aperture were not in place. This is because the optical system truncates the wavefront from the source, so the wavefront is modified through diffraction effects.

Kirchhoff converted this problem to one of boundary values by assuming that the point P is contained in a large dark vessel having the aperture as its opening. Assumptions about the complex amplitude and its normal derivative at the aperture and on the dark side of the vessel were then combined with the Helmholtz equation to yield a solution. Attention is given here to the formulation of the Helmholtz equation and then to the assumptions and approximations that led Kirchhoff to the diffraction integral solution. Subsequently, additional approximations, which lead to especially simple and useful solutions, are considered.

Maxwell's equations (section 2.2) for a homogeneous medium specify that both the electric vector $\mathbf{E}$ and the magnetic vector $\mathbf{H}$ satisfy the wave equation. Scalar diffraction theory assumes that the $\mathbf{E}$ and $\mathbf{H}$ vectors can be treated independently. If attention is confined to monochromatic waves and the usual definition of complex amplitude is used,

$$U = A(x, y, z)e^{i(\mathbf{k} \cdot \mathbf{r} - \phi_0)} \tag{10.1}$$

then U must satisfy the wave equation, with given boundary conditions,

$$\nabla^2 U + k^2 U = 0 \qquad k = \frac{2\pi}{\lambda}$$

To obtain the Helmholtz–Kirchhoff equation, recall Stokes's theorem for two well-behaved functions U_1 and U_2 defined in a volume V bounded by surface S, as shown in Figure 10.2. Stokes's theorem relates the surface and volume integrals of the functions,

$$\iiint_V (U_1 \nabla^2 U_2 - U_2 \nabla^2 U_1)\, dv = \iint_S (U_1 \nabla U_2 - U_2 \nabla U_1) \cdot \mathbf{n}\, ds \tag{10.2}$$

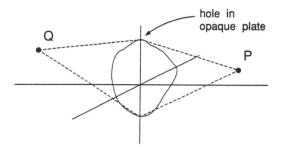

Figure 10.1. The diffraction problem.

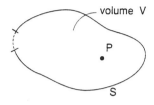

Figure 10.2. Region of integration for Stokes's theorem.

where **n** is the unit vector normal to the surface. Because

$$\nabla U_1 \cdot \mathbf{n} = (\mathbf{grad}\ U_1) \cdot \mathbf{n} = \frac{\partial U_1}{\partial n},$$

the right-hand side of Stokes's equation becomes

$$\iint\limits_S \left(U_1 \frac{\partial U_2}{\partial n} - U_2 \frac{\partial U_1}{\partial n} \right) ds \qquad (10.3)$$

If both functions are solutions of the wave equation as specified, the left side of Stokes's equation is zero, leaving just

$$\iint\limits_S \left(U_1 \frac{\partial U_2}{\partial n} - U_2 \frac{\partial U_1}{\partial n} \right) ds = 0 \qquad (10.4)$$

At this point, the unknown complex amplitude inside the surface is chosen to be of the form

$$U_2 = \frac{e^{ikr_2}}{r_2} \qquad (10.5)$$

where r_2 is measured from point P. The reason for this step is that the function at the particular (but general) point P is sought, so it is convenient to work in terms of spherical waves centered at P. Now, because P might be a singular point, such as a focus of a lens installed in the aperture, it must be excluded from the domain of integration by surrounding it with a sphere of radius ϵ whose surface we designate S_2 and whose outward normal is $\mathbf{n}_2$, as shown in Figure 10.3.

For the moment, designate the original surface by S_1. The integral can be

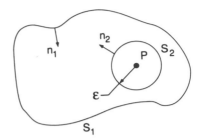

Figure 10.3. Integration
surfaces for Helmholtz–
Kirchhoff formula.

extended over the whole of the region between S_1 and S_2, which means that
the integral of equation 10.4 must range over S_1 and S_2 with the adopted
expression for U_2,

$$\iint_{S_1}\left[U_1\frac{\partial}{\partial n_1}\left(\frac{e^{ikr_2}}{r_2}\right)-\left(\frac{e^{ikr_2}}{r_2}\right)\frac{\partial U_1}{\partial n_1}\right]dS_1$$

$$+\iint_{S_2}\left[U_1\frac{\partial}{\partial n_2}\left(\frac{e^{ikr_2}}{r_2}\right)-\left(\frac{e^{ikr_2}}{r_2}\right)\frac{\partial U_1}{\partial n_2}\right]dS_2=0 \qquad (10.6)$$

Notice that $\partial/\partial n$ on surface S_2 is just $\partial/\partial r_2$, so equation 10.6 can be rewritten
as

$$\iint_{S_1}\left[U_1\frac{\partial}{\partial n_1}\left(\frac{e^{ikr_2}}{r_2}\right)-\left(\frac{e^{ikr_2}}{r_2}\right)\frac{\partial U_1}{\partial n_1}\right]dS_1$$

$$=-\iint_{S_2}\left[U_1\left(\frac{ik}{r_2}e^{ikr_2}-\frac{1}{r_2^2}e^{ikr_2}\right)-\frac{e^{ikr_2}}{r_2}\frac{\partial U_1}{\partial n_2}\right]dS_2 \qquad (10.7)$$

The second integral is a surface integral evaluated on S_2, so

$$r_2=\epsilon \qquad \text{and} \qquad dS_2=\epsilon^2\,d\Omega$$

where $d\Omega$ is the element of solid angle subtended by dS_2. These substitutions
leave for the second integral,

$$\int\left[U_1 e^{ik\epsilon}\left(\frac{ik}{\epsilon}-\frac{1}{\epsilon^2}\right)-\frac{e^{ik\epsilon}}{\epsilon}\frac{\partial U_1}{\partial n_2}\right]\epsilon^2\,d\Omega$$

$$=-4\pi\epsilon^2 U_1\frac{e^{ik\epsilon}}{\epsilon}\left(ik-\frac{1}{\epsilon}\right)+\epsilon\int e^{ik\epsilon}\frac{\partial U_1}{\partial n_2}\,d\Omega \qquad (10.8)$$

If ϵ is small, U_1 cannot vary much over the surface S_2 because it is
"well-behaved." Which is to say that though U_1 might be large, its normal
derivative $\partial U_1/\partial n$ is small. Thus, the integral remaining in equation 10.8 is
finite, and the expression containing the integral tends to zero as ϵ
approaches zero. As ϵ is diminished, the first expression in equation 10.8

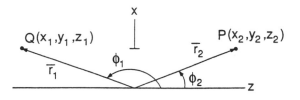

Figure 10.4. Geometry of source and aperture for deriving the Kirchhoff integral.

tends toward,

$$4\pi U_1|_{\text{at } r_2 = 0} = 4\pi U_P \tag{10.9}$$

These results replace the right side of equation 10.7, so the complex amplitude at the point P can be written,

$$U_P = \frac{1}{4\pi} \iint_S \left[U_1 \frac{\partial}{\partial n} \left(\frac{e^{ikr_2}}{r_2} \right) - \frac{e^{ikr_2}}{r_2} \frac{\partial U_1}{\partial n} \right] ds \tag{10.10}$$

This equation is the Helmholtz–Kirchhoff formula giving the complex amplitude at a point in terms of the values of complex amplitude on a surface surrounding the point.

To extend the preceding equation to create a solution of the problem of diffraction by a screen, Kirchhoff introduced the following assumptions about the boundary values of U_1:

1. $U_1 = 0$ and $\partial U_1/\partial n = 0$ on the dark side of the surface except at the aperture.
2. U_1 takes the same values in the aperture as it would if the surface were not in position.

Therefore, if the nature and position of the radiation source is known in relation to the aperture, and if the character of the aperture is known, then U_1 and $\partial U_1/\partial n$ are known everywhere on the surface, and so the Helmholtz–Kirchhoff integral can be evaluated.

The first assumption implies that the integral over the whole surface is zero except for the part functioning as an aperture. Accordingly, subsequent integrations need extend only over the aperture. This assumption may be questioned and is one aspect of the theory that was refined by Sommerfield.

Suppose then, that U_1 comes from a point source Q at distance r_1 and inclination ϕ_1 from the origin of coordinates, which is now taken to lie within the aperture, as shown in Figure 10.4. This adoption of a point source is not restrictive because an extended object can be viewed as a collection of such sources.

For the light incident on the aperture screen,

$$U_1 = A \frac{e^{ikr_1}}{r_1} \tag{10.11}$$

$$\frac{\partial U_1}{\partial n} = A \frac{e^{ikr_1}}{r_1} \left(ik - \frac{1}{r_1} \right) \cos \phi_1 \tag{10.12}$$

Taking a result that was used in developing equation 10.7,

$$\frac{\partial}{\partial n} \left(\frac{e^{ikr_2}}{r_2} \right) = \frac{e^{ikr_2}}{r_2} \left(ik - \frac{1}{r_2} \right) \cos \phi_2 \tag{10.13}$$

Points P and Q can be safely confined to regions far enough removed from the diffraction screen so that,

$$\begin{aligned} 1/r_1 &\ll k \\ 1/r_2 &\ll k \end{aligned} \tag{10.14}$$

Substituting these values into the Helmholtz–Kirchhoff formula (eq. 10.10) gives,

$$U_P = -\frac{Aik}{4\pi} \iint\limits_{ap} \frac{1}{r_1 r_2} e^{ik(r_1 + r_2)} (\cos \phi_1 - \cos \phi_2) \, ds \tag{10.15}$$

$$= -\frac{Ai}{2\lambda} \iint\limits_{ap} \frac{e^{ik(r_1 + r_2)}}{r_1 r_2} (\cos \phi_1 - \cos \phi_2) \, ds \tag{10.16}$$

This result is the Kirchhoff integral (also called the Fresnel–Kirchhoff formula) giving the complex amplitude anywhere in the region on the side of a diffracting screen opposite a point source of light. Although derived for pure monochromatic (and coherent) light, it holds for nonmonochromatic, noncoherent illumination if the final result is put in terms of irradiance.

10.2.2 Inclusion of a transparency and the plane wavefront case

A small extension of the thinking outlined above extends the usefulness of the Kirchhoff integral. A common situation in optical data processing situations is that the diffracting screen contains some sort of "transparency" that modifies the amplitude, phase, or intensity of the illumination beam as it passes through the aperture. An easy way to handle such a problem is to define an aperture transmittance function $T(\xi, \eta)$, which might be complex, so that the complex amplitude U_1 at the aperture is the complex amplitude of the source times the transmittance function. ξ and η are coordinates in the aperture plane. The derivation remains unchanged except the $T(\xi, \eta)$ enters the integral as a multiplier,

$$U_P = -\frac{Ai}{2\lambda} \iint_{ap} T(\xi, \eta) \frac{e^{ik(r_1 + r_2)}}{r_1 r_2} (\cos \phi_1 - \cos \phi_2) \, d\xi \, d\eta \qquad (10.17)$$

There is considerable to be said about the development of the transmittance function for specific classes of problems. An important distinction must be established between diffraction systems operating with incoherent illumination and those using coherent light. In the first case, T operates on the irradiance rather than on complex amplitude. In many practical situations the transparency can be treated as a device that modifies either the amplitude or phase of the incoming light. A harmonic amplitude grating, for example, could be represented as

$$T(\xi, \eta) = T_0 + T_1 \cos \frac{2\pi\xi}{d} \qquad (10.18)$$

If the aperture contains refraction elements (lenses), a similar approach can be employed except that the transmittance function contains only phase terms that change with distance from the optical axes. An alternative approach is to view the lens as a device that changes the distance r_1 to the point source. As was stated earlier, more complicated input complex amplitudes can be viewed as coming from an array of point sources, and one need only sum the output complex amplitudes.

A minor digression is in order at this point. One special case is easily treated and is of considerable importance. Suppose that an aperture containing a transparency is illuminated normally by a plane monochromatic wavefront. In that case, the Kirchhoff integral can be rederived by taking

$$U_1 = Ae^{ikz}e^{i\phi_0}T(\xi, \eta)$$

$$\frac{\partial U_1}{\partial n} = Aike^{ikz}e^{i\phi_0}T(\xi, \eta) \qquad (10.19)$$

The diffraction integral for this special case is developed from equation 10.10.

$$U_P = \frac{1}{4\pi} \iint \left[Ae^{ikz}e^{i\phi_0}T(\xi, \eta) \frac{e^{ikr_2}}{r_2} \left(ik - \frac{1}{r_2} \right) \cos \phi_2 \right.$$

$$\left. - \frac{e^{ikr_2}}{r_2} Aike^{ikz}e^{i\phi_0}T(\xi, \eta) \right] d\xi \, d\eta \qquad (10.20)$$

Again, take $1/r_2 \ll k$.

$$U_P = \frac{ik}{4\pi} \iint_{ap} \left[A e^{ikz} e^{i\phi_\circ} T(\xi, \eta) \frac{e^{ikr_2}}{r_2} (\cos \phi_2 - 1) \right] d\xi \, d\eta \qquad (10.21)$$

$$= \frac{ik}{4\pi} \iint_{ap} U_1 \frac{e^{ikr_2}}{r_2} (\cos \phi_2 - 1) \, d\xi \, d\eta \qquad (10.22)$$

or

$$= \frac{ik}{4\pi} A e^{ikz} e^{i\phi_\circ} \iint_{ap} T(\xi, \eta) \frac{e^{ikr_2}}{r_2} (\cos \phi_2 - 1) \, d\xi \, d\eta \qquad (10.23)$$

Equation 10.22 suggests again that one can operate with only the complex amplitude of the light from the aperture without much concern for the nature of the source. Both of the last two equations indicate more clearly than those before that any diffraction process, including optical imaging, can be described as a linear integral transformation. It is shown in the next section that certain reasonable approximations reduce the diffraction integral to a Fourier transform. This simplification is carried out for the general Kirchhoff integral, rather than on the special case resulting in equation 10.23. With the general result in hand, some special cases are then discussed briefly. An interesting and useful exercise is to perform similar operations on the specific case involving a transparency illuminated by a plane wavefront.

10.3 The Fresnel and Fraunhofer approximations

At this point, some simplifications are adopted in order to develop the important approximate equivalents to the general solution. The Kirchhoff diffraction integral (eq. 10.16) proves unwieldy when evaluated for even simple apertures. The reason is that r_1, r_2, and the obliquity factors $\cos \phi_1$ and $\cos \phi_2$ vary as the location of area element ds varies. That is, general relationships between these variables must be found before the integration can be performed. The resulting integral often proves intractable. It is convenient to accept some approximations that simplify the integral and also supply reasonably accurate solutions for important classes of diffraction problems. Fortunately, these special cases include almost all practical situations.

The basic idea behind these simplications is to write approximations for the position variables r_1 and r_2 and observe that certain terms can be dropped and others moved outside the integral sign when r_1 and r_2 are large in comparison with the aperture dimension. Understanding the process is easier if attention is confined to the two-dimensional situation. Extensions to three

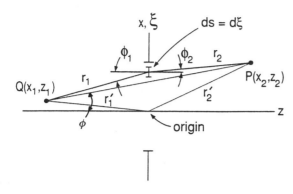

Figure 10.5. Diffraction geometry in two dimensions.

dimensions are readily derived and intuitively obvious. Figure 10.5 shows the geometry of the problem.

First, observe that if points P and Q are far removed from a small aperture, then the factor $1/r_1 r_2$ that appears inside the integral can be approximately replaced by the constant $1/(r'_1 r'_2)$, where the primed distances are to the origin of coordinates that lies in the plane of the aperture rather than to the location of $d\xi$. Likewise, the factor $(\cos \phi_1 - \cos \phi_2)$ in the integral is replaced by the constant $2 \cos \phi$, where ϕ is the angle between the normal to the aperture and the line QP. The Kirchhoff integral for this case becomes

$$U_P \approx -\frac{iA}{\lambda} \frac{\cos \phi}{r'_1 r'_2} \int_{ap} e^{ik(r_1 + r_2)} \, d\xi \tag{10.24}$$

Note that the r_1 and r_2 appearing in the exponent cannot be replaced by r'_1 and r'_2 because the exponential oscillates rapidly as area element $d\xi$ explores the aperture. However, further simplification results from adoption of reasonable approximations to r_1 and r_2 in the exponential within the integrand. Observe that

$$r_1 = [(x_1 - \xi)^2 + z_1^2]^{1/2}$$
$$= (x_1^2 - 2x_1\xi + \xi^2 + z_1^2)^{1/2}$$
$$= z_1\left(1 + \frac{x_1^2}{z_1^2} + \frac{\xi^2}{z_1^2} - \frac{2\xi x_1}{z_1^2}\right)^{1/2} \tag{10.25}$$

likewise,

$$r_2 = z_2\left(1 + \frac{x_2^2}{z_2^2} + \frac{\xi^2}{z_2^2} - \frac{2\xi x_2}{z_2^2}\right)^{1/2}$$

Now, if $z_1 \gg |x_1 - \xi|$, then the binominal expansion,

$$(1 + \epsilon)^{1/2} = 1 + \frac{\epsilon}{2} + \cdots$$

can be used to write

$$r_1 \approx z_1\left(1 + \frac{x_1^2}{2z_1^2} + \frac{\xi^2}{2z_1^2} - \frac{x_1\xi}{z_1^2}\right)$$

$$= z_1 + \frac{x_1^2}{2z_1} + \frac{\xi^2}{2z_1} - \frac{x_1\xi}{z_1} \tag{10.26}$$

and

$$r_2 \approx z_2 + \frac{x_2^2}{2z_2} + \frac{\xi^2}{2z_2} - \frac{x_2\xi}{z_2}$$

Substitution of these approximations into equation 10.24 yields

$$U_P = \frac{-iAk}{4\pi}\frac{\cos\phi}{r_1'r_2'}\int_{ap} T(\xi)e^{ik\left[z_1 + \frac{x_1^2}{2z_1} + \frac{\xi^2}{2z_1} - \frac{x_1\xi}{z_1} + z_2 + \frac{x_2^2}{2z_2} + \frac{\xi^2}{2z_2} - \frac{x_2\xi}{z_2}\right]}\,d\xi \tag{10.27}$$

or

$$U_P = \frac{iAk}{4\pi}\frac{\cos\phi}{r_1'r_2'}e^{ik(z_1+z_2)}e^{\frac{ik}{2}\left(\frac{x_1^2}{z_1}+\frac{x_2^2}{z_2}\right)}\int_{ap} T(\xi)e^{\frac{ik\xi^2}{2}\left(\frac{1}{z_1}+\frac{1}{z_2}\right)}e^{-ik\left(\frac{x_1\xi}{z_1}+\frac{x_2\xi}{z_2}\right)}\,d\xi$$

$$\tag{10.28}$$

It is worth noting at this point that the quantities x_1/z_1 and x_2/z_2 just specify the angular deviations of the incident and diffracted waves. We shall allow these quantities to stay in this slightly unwieldy but physically meaningful form.

Equation 10.28 is the Fresnel approximation to the Kirchhoff integral. Most or all of the constants outside the integral can be absorbed in a single complex constant. A common form for the result is

$$U_P = Ce^{\frac{ik}{2}\left(\frac{x_1^2}{z_1}+\frac{x_2^2}{z_2}\right)}\int_{ap} T(\xi)e^{\frac{ik\xi^2}{2}\left(\frac{1}{z_1}+\frac{1}{z_2}\right)}e^{-ik\left(\frac{x_1}{z_1}+\frac{x_2}{z_2}\right)\xi}\,d\xi \tag{10.29}$$

Notice that the integral is beginning to look like an integral transform. This simplified form is very important, but it is still more unwieldy than is necessary for most applications.

A simpler and more useful approximation requires the more stringent condition that the quantity,

$$e^{\frac{ik\xi^2}{2}\left(\frac{1}{z_1}+\frac{1}{z_2}\right)}$$

be near unity so that it may be dropped. This condition is met if

$$\frac{\pi\xi^2}{\lambda}\left(\frac{1}{z_1}+\frac{1}{z_2}\right) \ll \frac{\pi}{4}$$

Because ξ_{max} is one-half the aperture width w, the limitation translates to

$$\frac{1}{z_1} + \frac{1}{z_2} \ll \frac{\lambda}{w^2} \tag{10.30}$$

The implication is that the source point and/or the observing point is far removed from the aperture. This condition is very severe in physical terms, but it is satisfied closely enough in many optical systems. Section 10.4 shows how this restriction can lead to difficulties in correlating the predictions of diffraction theory with results of simple experiments. The restriction is almost necessary, however, because the more general equation is very difficult to apply even for simple cases.

After imposition of the severe small aperture condition just given, equation 10.29 reduces to a simplified form known as the Fraunhofer approximation,

$$U_P = Ce^{\frac{ik}{2}\left(\frac{x_1^2}{z_1} + \frac{x_2^2}{z_2}\right)} \int_{ap} T(\xi)e^{-ik\left(\frac{x_1}{z_1} + \frac{x_2}{z_2}\right)\xi} \, d\xi \tag{10.31}$$

where C is the complex constant defined in connection with equation 10.29. We see now that the integral is a Fourier transform of the aperture transmittance function $T(\xi)$, and the Fraunhofer equation can be written as

$$U_P = Ce^{\frac{ik}{2}\left(\frac{x_1^2}{z_1} + \frac{x_2^2}{z_2}\right)} F\{T(\xi)\}_{f=\frac{1}{\lambda}\left(\frac{x_1}{z_1} + \frac{x_2}{z_2}\right)} \tag{10.32}$$

In the above notation, $f = (1/\lambda)(x_1/z_1 + x_2/z_2)$ gives the "spatial frequency" or dimension parameter in transform space. A useful observation is that the x/z terms are angular deviations from the optic axis, the approximation for small angles having already been imposed.

As pointed out in section 10.2.2, a case of special interest in holography and optical data processing is when the transparency represented by $T(\xi)$ is illuminated by a plane monochromatic (coherent) wavefront. Rigorous treatment of this case would start with equation 10.21 and follow through with approximations similar to those of this section. An approach that is less rigorous, in that it doesn't deal with the terms that were absorbed into the complex constant but gives the correct result in an agreeable way, is to merely let $z_1 \to \infty$ in equation 10.30. The result is

$$U_P = Ce^{\left(\frac{ik}{2}\frac{x_2^2}{z_2}\right)} \int_{ap} T(\xi)e^{-ik\frac{x_2\xi}{z_2}} \, d\xi$$

$$= Ce^{\left(\frac{ik}{2}\frac{x_2^2}{z_2}\right)} F\{T(\xi)\}_{f=\frac{x_2}{\lambda z_2}} \tag{10.33}$$

This equation is the usual form of the Fraunhofer diffraction integral. Both this version and the previous one are important! The distribution of light at a point in the far field is seen to be the Fourier transform of the aperture function, evaluated at the field point, times an exponential containing the coordinates of the point. The result is identical to that obtained by the Huygens–Fresnel construction except for the exponential function. A similar process of letting $z_1 \to \infty$ will produce the equivalent Fresnel diffraction equation for the function U_P in the near field from equation 10.29, but the result is much more complex, and it is difficult to apply for useful practical problems.

Equation 10.33 and its Fresnel equivalent are more powerful than might be apparent at first. The transmittance function in these equations actually represents the complex amplitude of the light emitted by the aperture when it is illuminated by a plane wave. The amplitude and phase of the illumination beam were factored out as constants. If the illuminating beam is not a plane wave, then equation 10.33 may be written as

$$U_P = C e^{\left(\frac{i\pi}{\lambda} \frac{x_2^2}{z_2} \right)} \int U(\xi) e^{\frac{i2\pi}{\lambda} \frac{x_2}{z_2} \xi} \, d\xi \qquad (10.34)$$

In this form, $U(\xi)$ is the aperture transmittance function times the complex amplitude of the beam that is incident on the aperture. This understanding can be developed more soundly starting with equation 10.20. The resulting form of the diffraction equation facilitates handling complicated optical systems consisting of a succession of diffraction screens and lenses by successive application of a fairly simple integral or successive Fourier transforms.

Rigorous derivations of these general forms of the simplified diffraction integral require going back to the equations in section 10.2.2 and carrying through the Fresnel and Fraunhofer approximations. Another approach is to utilize the fact that an arbitrary incident wave can be viewed as the superposition of several of the point sources used in the Kirchhoff formulation.

10.4 Diffraction by a clear aperture

To illustrate both the power and the shortcomings of the Fraunhofer approximation of general diffraction theory, it will be applied with care to a case that is, at first glance, trivially simple. It is actually fairly complex; it is also useful and important in many ways. Several sections of this text will refer to it implicitly or explicitly, and careful study of it is recommended.

This problem involves a plane wavefront falling at normal incidence upon a

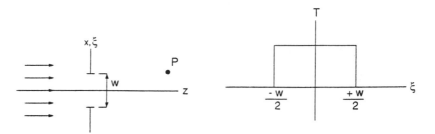

Figure 10.6. Geometry for diffraction by a clear aperture in two dimensions.

Figure 10.7. Transmittance function for a clear aperture.

clear aperture of diameter w. A two-dimensional analysis will suffice; this does not imply that the aperture is merely circular, although such an assumption is often put to good use. Figure 10.6 illustrates the geometry. The transmittance function for this problem is shown in Figure 10.7.

The mathematical expression for this rectangular pulse (in space) is

$$T(\xi) = \text{rect}\left(\frac{\lambda}{w}\right) \equiv \begin{cases} 1 & \text{for } |\xi| \leq \dfrac{w}{2} \\ 0 & \text{otherwise} \end{cases} \tag{10.35}$$

Because the illumination is a plane wavefront, this transmittance can be substituted directly into equation 10.33,

$$U_P = C e^{\frac{ik}{2}\frac{x_2^2}{z_2}} \int_{-\frac{w}{2}}^{+\frac{w}{2}} e^{-ik\left(\frac{x_2}{z_2}\right)\xi} \, d\xi \tag{10.36}$$

The integral is the Fourier transform $F\{T(\xi)\}$ of the aperture function, which is evaluated as follows,

$$F\{T(\xi)\} = \left[\frac{e^{-ik\left(\frac{x_2}{z_2}\right)\xi}}{-ik\left(\dfrac{x_2}{z_2}\right)} \right]_{-\frac{w}{2}}^{+\frac{w}{2}} \tag{10.37}$$

$$= \frac{\sin\left(kw\dfrac{x_2}{2z_2}\right)}{k\left(\dfrac{x_2}{2z_2}\right)} \tag{10.38}$$

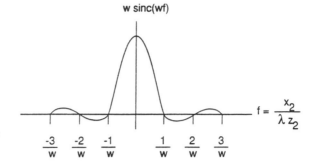

Figure 10.8. Shape of sinc function for an aperture of width w.

Note that $kwx_2/2z_2$ is just πf if we switch variables according to the usual practice with $k = 2\pi/\lambda$ and

$$f = \frac{x_2}{\lambda z_2} \tag{10.39}$$

Recall from the previous section that this parameter is spatial frequency, which is the distance dimension in transform space. With this substitution, the transform becomes

$$F\{T(\xi)\} = \frac{\sin \pi wf}{\pi f} = w \, \mathrm{sinc}(wf) \tag{10.40}$$

A diagram of the form of the sinc function appears in Figure 10.8.

The intensity or irradiance in the far field is found by computing $U_P U_P^*$, and thus it involves the square of the sinc function. This quantity must be multiplied by the constant and exponential that were outside the transform integral to get the intensity. The square of the sinc function is shown in Figure 10.9; it helps in visualizing the light distribution in the far field. We see that the light distribution that will appear on a screen placed at some distance from the aperture consists of a bright central spot flanked in this two-dimensional case by alternating light and dark spots. The three-dimensional analog is a circular aperture, which will produce a central patch surrounded by concentric light and dark rings. This pattern, called the Airy disc, is easy to create by shining laser light on a small pinhole. Figure 10.10 shows a typical pattern of this type.

The diameter of the central spot for a typical case is calculated as follows, assuming that the diameter is twice the distance from the origin to the zero point at $1/w$ in the frequency domain of the figure. Remembering the dimension parameter for the abscissa of the plot in Figure 10.9, we can write

$$d = \text{diameter} = 2x_2 \quad \text{at a given } z_2$$

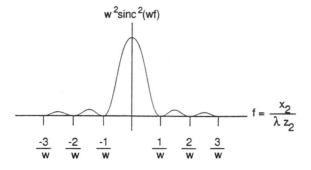

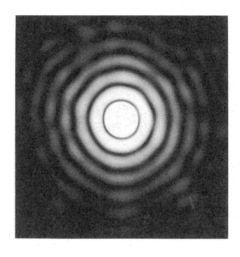

Figure 10.9. Shape of the square of the sinc function for an aperture of width w.

Figure 10.10. Diffraction pattern showing the Airy disc for a circular aperture.

where x_2 and z_2 are developed from the fact that the minima are reached at $f = 1/w$. So

$$2\left(\frac{x_2}{\lambda z_2}\right) = 2\left(\frac{1}{w}\right)$$

and

$$d = 2x_2 = \frac{2(\lambda z_2)}{w} \tag{10.41}$$

Take, for example, $\lambda = 0.6\ \mu m$ and $w = 0.1$ mm, and place the screen at $z_2 = 1{,}000$ mm. The diameter of the Airy disc is then 12 mm.

If the same setup is used but the aperture is decreased to 0.01 mm, then the diameter of the central spot in the diffraction pattern increases to 120 mm.

These two examples illustrate an interesting inverse reciprocity between aperture size and the spread in the diffraction pattern. The smaller the aperture, the larger the area illuminated in the diffraction observing plane.

It is educational to look at the two extreme cases. Suppose that the aperture is made extremely small, to the point that it almost vanishes. One would expect that the illumination in the plane of viewing the diffraction pattern would be spread over an area approaching infinity. We cannot pass to the limit in the laboratory, but experimental observations with decreasing apertures imply the truth of our expectation. Mathematically, too, the experiments are given support. In the limit, the vanishingly small pinhole is represented by the impulse function, otherwise known as the Dirac delta. This function is represented as

$$\delta(x - a) = 0 \quad \text{at } x \neq a \qquad \text{and} \qquad \int \delta(x - a) \, dx = 1 \qquad (10.42)$$

That is, the delta function is zero everywhere but at the point where the impulse is located and is such that the area under the function is unity.

Now, it is easy to establish that the Fourier transform of the delta function is just unity for all values of f in the transform plane. One must not forget the parameters that multiply the transform to get the complex amplitude, but the implication is that the illumination in the diffraction viewing plane is spread over a large area.

The other extreme case raises more of a problem. What if the aperture were "large"? The symmetry of the problem suggests that only a very small spot in the transform viewing plane would be illuminated. This expectation seems, at first glance, to be supported by diffraction theory. The Fourier transform of a constant extending to infinity in aperture space is just the Dirac delta function at the origin of the transform plane. As the aperture expands, the sinc function gets narrower and approaches the impulse function.

This theory seems contrary to our experience. If we go into the laboratory and illuminate a hole having a diameter of, say, 10 mm, and we set up a screen at 1,000 mm behind the hole, then we will see a bright area on the screen that corresponds roughly to the usual shadow of the hole, with some fuzziness around the edges if we look carefully. Yet, the diffraction theory predicts for this case an illuminated patch on the screen of only 0.12 mm. If the aperture diameter is increased to 100 mm, then the theory predicts an illuminated spot of only 0.012 mm.

The apparent contradiction between experiment and theory here is caused by the fact that we have misused the Fraunhofer equation. To be specific, the condition laid down by equation 10.30 has been violated. In this equation, $1/z_1 = 0$ because we are using a plane wave. We must have, for the Fraunhofer equation to give valid results, $z_2 \gg w_2/\lambda$. For the 10-mm aperture, $z_2 \gg 1.6 \times 10^5$ mm. In other words, to validly test the theory for this aperture, the viewing screen must be a kilometer or so away from the aperture. For a 100-mm aperture, the screen would have to be at about 100 km for the Fraunhofer restriction to be met! This aspect of the Fraunhofer approximation is not widely appreciated.

To analyze the large aperture situation, one must go back to the "near-field" solution given by the Fresnel equation (eq. 10.29). This equation is difficult to work with even for the simple aperture function under discussion. Remember also that the Fresnel equation is itself an approximation, which was developed by simplification of obliquity factors appearing in the Kirchhoff integral. These factors would be important if the screen were "close" to a "large" aperture.

One reason for discussing this large-aperture difficulty in some detail, perhaps tediously so, is that it addresses a problem that is inherent in most treatments of moire interferometry and the reconstruction of holographic images. This problem seems always to be ignored. In fact, typical developments of the holography equations merely develop the equation of the transmitted wave right at the grating plane, and then "infer" what it must be downstream. Both holographic and moire processes involve the illumination of large pieces of diffraction gratings (50–200 mm broad) and subsequent prediction of where the diffracted beams go in the near distance opposite the grating. Application of the Fraunhofer integral relationship to this problem, which is obviously one of diffraction, implies that the beams converge to a small spot. This is never seen in the laboratory, and now we know why. The laboratory is not big enough! Given the scales involved, the collimated incident beam, when diffracted, is essentially still collimated.

Another way to utilize the predictions of the Fraunhofer approximation in this context of a large aperture is to view the predicted delta function as a direction specifier that tells where the broad wavefront will be headed. Alternatively, one may think of the large aperture as an assembly of small apertures that produces a corresponding assembly of sinc functions in the near field. The structure of the diffracted beam is not adequately predicted in this way, but the directional relationships come out right. This approach leads one full circle back to the Huygens construction.

The discussion of this large-aperture case leads to the question of how one might use diffraction to create optical Fourier transforms of spatial signals that extend over a broad aperture within a laboratory of finite size. The answer now seems obvious. The required physical distance to the viewing screen can be shortened by using a lens, which causes the near-collimated diffracted beam to converge more quickly. In other words, the lens allows one to get around the severe limitation of the small-aperture restriction.

Some other specific diffraction problems that are useful in understanding various optical measurement techniques shall now be discussed. Most of these use the preceding results in one way or another.

10.5 Diffraction by a harmonic grating

Consider a special diffraction problem that is important in several types of moire measurement and also in holographic interferometry. We seek to understand the ways in which light is affected by a simple grating. The grating

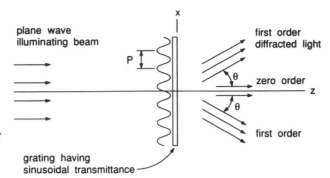

Figure 10.11. Diffraction of a plane wave by a simple harmonic grating.

consists of closely spaced parallel lines. For simplicity, begin with the case of normal incidence of a narrow plane wavefront upon a grating whose amplitude transmittance varies sinusoidally. Figure 10.11 shows schematically what happens. The light is not all transmitted directly through the grating. Some of it is channeled into two beams that deviate from the main beam by angle θ. The deviation angle is a function of the wavelength of light and the spatial frequency (line density) of the grating.

The grating amplitude transmittance function is, with reference to Figure 10.11,

$$T(\xi) = Q - R \cos\left(\frac{2\pi\xi}{p}\right) \tag{10.43}$$

where p is the pitch of grating
          R is the transmittance variation
          Q is the average transmittance

Before substituting this into a diffraction integral, rewrite the cosine as

$$\cos\left(\frac{2\pi\xi}{p}\right) = \frac{e^{i(2\pi\xi/p)} + e^{-i(2\pi\xi/p)}}{2} \tag{10.44}$$

Substitute equation 10.44 into 10.43, and then use that expression in the Fraunhofer diffraction equation 10.33. Then, to save rewriting the expressions that appear outside the integral in equation 10.33, define

$$C' = C e^{\frac{ikx_2^2}{2z_2}} \tag{10.45}$$

Also use $k = 2\pi/\lambda$. The integral for the complex amplitude exiting the grating can then be divided into three parts of the following form,

$$\frac{U_P}{C'} = Q \int_{-\frac{w}{2}}^{+\frac{w}{2}} e^{-ik\frac{x_2}{z_2}\xi} \, d\xi - \frac{R}{2} \int_{-\frac{w}{2}}^{+\frac{w}{2}} e^{-ik\left(\frac{x_2}{z_2} - \frac{\lambda}{p}\right)\xi} \, d\xi - \frac{R}{2} \int_{-\frac{w}{2}}^{+\frac{w}{2}} e^{-ik\left(\frac{x_2}{z_2} + \frac{\lambda}{p}\right)\xi} \, d\xi$$

$$\tag{10.46}$$

All three of these integrals are identical in form to the one discussed in section 10.4, so only a little more need be said.

The first integral is just the diffraction by the clear aperture with an extra factor Q, which is a measure of the overall transmittance of the grating, attenuating the beam. Of course, the whole beam complex amplitude is attenuated over its extent by the factor C'. This beam is the one that travels along the z-axis, and it is called the 0-order beam. Section 10.4 discussed the nature of this beam in detail. If the extent of the grating and the beam cross section are substantial, then this beam is nearly collimated over any reasonable distance.

Now take up the second integral in equation 10.46. This, somewhat surprisingly perhaps, is the same as the clear aperture integral with a spatial frequency shift of $1/p$ appearing in the exponential. The spatial frequency modification just imposes a coordinate shift in the transform plane. For the small aperture, the transform is the sinc function centered at the coordinate $f = f_0$ in the transform viewing plane. To figure out where that is, take

$$f = f_0 = \frac{x_2}{\lambda z_2} = \frac{1}{p}$$

which gives the location of the sinc function along the line from the center of the aperture with

$$\frac{x_2}{z_2} = \frac{\lambda}{p} \qquad (10.47)$$

If the total aperture of the grating is large, then the transform is the delta function,

$$\delta(f - f_0) \qquad (10.48)$$

which corresponds to a bright patch at location $f = f_0$ in the transform viewing plane.

Again, making use of the understanding developed in section 10.4, we know that the delta function is not realized within a reasonable distance, so we actually see a broad wavefront traveling along the line specified by equation 10.47. This beam is deviated on the plus side of the z-axis and is termed the $+1$ diffraction order.

Identical arguments are brought to bear on the third integral of equation 10.46. The results are the same, except that the sinc function or the delta function is at $f = -f_0$, meaning that the beam is on the negative side of the z-axis. This one is called the -1 diffraction order.

Another potential problem appears here. If the diffraction angle is θ, then the preceding results tell us that $\tan \theta = x_2/z_2 = \lambda/p$. The commonly accepted diffraction grating formula is $\sin \theta = \lambda/p$. Why the discrepancy? There is no contradiction or error if we recall again the limitations of the Fresnel

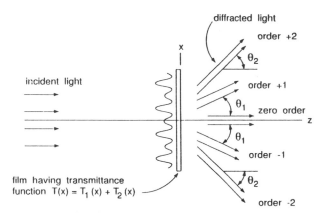

Figure 10.12. Formation of higher diffraction orders by a grating having two Fourier components.

approximation, which are carried on to the Fraunhofer diffraction integral. To obtain equation 10.27, the distance r to the field point was, to some degree, replaced by its z-coordinate. It should not be surprising that the distinction between sine and tangent is lost. When the diffraction is needed in subsequent sections, the form with the sine function will be used.

If the grating is not a simple sinusoidal one, there will be more than two diffracted beams. In fact, there will be two diffracted waves or orders for each Fourier component of the grating. These are easy to develop using the process just outlined. Figure 10.12 illustrates the idea when the grating transmittance is the superposition of two sine waves. This idea is used to advantage in Fourier processing of moire gratings to improve sensitivity, as is discussed in Chapter 11. The diffraction equations show that the deviation depends on the wavelength of light. If more than one wavelength is present, the diffraction grating will separate them as a prism does. The grating is a spectrum analyzer, and it is extensively used as such in spectrophotometry and spectroscopy.

The case in which the illuminating wave is oblique to the aperture containing the grating will not be discussed. It is important in, for example, holographic reconstruction and moire interferometry. The results are intuitively sensible, and they will just be presented where they are needed, as in Chapter 13.

Finally, the use of wavefronts that are not planar will also not be described mathematically. This situation often happens in holography, but an exact solution is not required in order to understand the process.

10.6 The lens as Fourier analyzer

Because diffraction at an aperture gives an optical rendition of the Fourier transform of spatial signals, the possibility arises of using optical transforms in various ways, particularly in modification of images by spatial filtering in

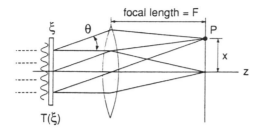

Figure 10.13. Creation of an optical Fourier transform with spatial signal aperture in front of lens.

the transform plane. One would like to have a good way of looking at these transforms in a way that is conducive to such manipulation.

The major problem, as was made clear in section 10.4, is that the true transform appears only in the far field. In the near field, the Fresnel equation does not give a true transform, in that the integral is contaminated with extra exponential terms. The far-field picture for a broad aperture is too far away to be of much use.

The answer is to bring the far-field picture of the optical Fourier transform closer to the aperture. As usual, this task can be performed by a lens. Several lens arrangements can be employed. Two of these will be considered from different viewpoints, one more physically meaningful and the other more mathematically rigorous.

First, consider the system shown in Figure 10.13 in an intuitive way, using the results of the diffraction analysis for a simple grating. The development of sections 10.4 and 10.5 showed that the beam coming from the grating is essentially collimated, and it is deviated from the z-axis by the diffraction angle θ, which is dependent on grating spatial frequency $(1/p)$ and the wavelength. The light strikes the lens, where it is refracted. The property of a lens that is important is that it will cause a collimated beam (plane wavefront) to converge to a point at the so-called back focal plane. The distance from the plane to the principal plane of the lens is the focal length of the lens, here symbolized by F. The x-coordinate of the focal spot for the inclined rays clearly depends on the diffraction angle θ.

To see how this works out mathematically, at least for small deviations and in two dimensions only, start with equation 10.33. Let $T(\xi)$ be the aperture spatial signal, take for convenience of notation $x_2/z_2 = \tan \theta = \theta$, and also ignore the quadratic multiplier preceding the integral to get the complex amplitude going into the lens,

$$U = \int_w T(\xi)e^{-ik\theta\xi}d\xi \qquad (10.49)$$

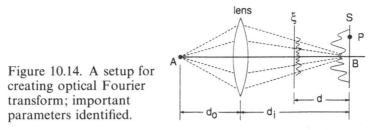

Figure 10.14. A setup for creating optical Fourier transform; important parameters identified.

The definition of focal length implies that $x = \theta F$. Substitute x/F for the θ and also use $k = 2\pi/\lambda$ in equation 10.49 to find

$$U_P = U(x) = \int T(\xi)e^{-i2\pi\left(\frac{x}{\lambda F}\right)\xi}\, d\xi \qquad (10.50)$$

This is a Fourier transform as before; but now, importantly, the scaling factor in the transform plane is modified by the focal length of the lens. The spatial frequency metric is $f = x/\lambda F$ rather than the $x_2/\lambda z_2$ that appears in the Fraunhofer equation.

To give this idea some numerical illustration, go back to the example involving a 100-mm aperture containing a grating. Rather than needing many kilometers to find the delta function representing the transform, we now need put the transform viewing screen at a distance equal to the focal length of the lens, probably only a meter or less.

An important distinction must be established. It is often said that a lens is a Fourier transforming device. It is indeed, but so is any other aperture. Diffraction at any aperture is a Fourier transforming process, whether a lens is present or not. The lens makes it possible to create the transform in a useful way for large-aperture systems.

Examination of a second optical transform setup from a second viewpoint will serve several purposes. It will confirm the observations made above but with a more mathematically rigorous development. It will illustrate another way to use the lens to make the transform accessible. Finally, this system happens to be one for which the more general Fresnel diffraction integral gives good results without heavy calculations. The questions surrounding the use of the Fraunhofer approximation are thereby dispelled. This development follows closely that of Haskell (1971).

The optical scheme is shown in Figure 10.14. The development will be confined to two dimensions in order to keep the equations shorter. The intuitive extension to three dimensions is correct if radial distance is used in place of x, and so on. Notice that the signal transparency aperture is now opposite the lens from the light source. The light comes from a point source at A, so a spherical wavefront approaches the lens at distance d_0 from the source. The distance may be taken as very large to create the plane wavefront case. The effect of the lens is to multiply the entering complex amplitude by a complex transmittance function so that the light is caused to converge to the back focal plane at distance d_i, where the image of the source point is located. The signal $T(\xi)$ is placed in its aperture at distance d from the back focal plane. The task is to predict what is seen in the back focal plane.

Although the results are not really needed for the diffraction development, this is a good place to explain the effect of a lens in terms of complex amplitude. The amplitude transmittance function for a lens can be developed from its known effect on a light from a point source at the focal length F of the lens. The complex amplitude falling on the lens will be

$$U_-(x) = Ae^{\frac{ik}{2F}x^2} \tag{10.51}$$

This incident light is multiplied by the lens transfer function $\psi(x, y)$ to get the outgoing complex amplitude $U_+(x, y)$, which must be a plane wavefront,

$$U_+(x) = U_-(x)\psi(x) = Ae^{\frac{ik}{2F}x^2}\psi(x, y) = \text{const} \tag{10.52}$$

From which the complex transmittance function or transfer function for a lens is determined to be

$$\psi(x) = e^{-i\frac{k}{2F}x^2} \tag{10.53}$$

Getting back to the diffraction system, we can take the spherical wave exiting the lens and converging on B in the focal plane as

$$U_+(x) = Ae^{-i\frac{k}{2d_i}x^2} \tag{10.54}$$

The minus sign in the exponential implies that the wavefront is converging. Just left of the signal input plane, the illuminating spherical wave, still centered at B, is similar, but the distance is now just d and the lateral coordinate is ξ.

$$U_-(\xi) = Ae^{-i\frac{k}{2d}\xi^2} \tag{10.55}$$

Just right of the signal plane, the complex amplitude has been multiplied by the signal amplitude transmittance function, to yield,

$$U_+(\xi) = Ae^{-i\frac{k}{2d}\xi^2}T(\xi) \tag{10.56}$$

From this point, there are two ways to proceed. The first is to use the form of the Fresnel equation for the case in which the illuminating wavefront is plane. Take $1/z_1$ in equation 10.29 to be zero, as was done for the Fraunhofer equation, to obtain

$$U_P(x_2) = Ce^{\left(\frac{ik}{2}\frac{x_2^2}{z_2}\right)} \int_{ap} T(\xi)e^{\frac{ik}{2z_2}\xi^2} e^{-ik\frac{x_2}{z_2}\xi} d\xi \tag{10.57}$$

Then use the idea, demonstrated in equation 10.34, that the non-plane wavefront integral can be found by just substituting the complex amplitude of the actual illuminating wave times the transmittance function in place of the transmittance function alone as it appears in the plane wave integral. Substitution into the general integral involving a point source (eq. 10.29) is not correct; it must first be reduced to the plane wave case. Carrying out the substitution of equation 10.56 into 10.57 yields, after realizing that z_2 in the equation is just the distance d in the optical setup and x_2 is now coordinate s in the transform plane,

$$U_P(s) = Ce^{\frac{ik}{2}\frac{s^2}{d}} \int T(\xi)e^{-ik\frac{s}{d}\xi} d\xi \tag{10.58}$$

The striking feature of this development is that the extra term appearing in the Fresnel integral has been canceled by the term representing the effect of the lens on the illuminating beam. The result is, except for the obliquity term appearing outside the integral, an exact Fourier transform of the input spatial signal. It is not necessary to resort to the more severe Fraunhofer approximation in order to get a true Fourier transform. The scaling relationship between input spatial frequency f and distance s in transform space is $f = s/\lambda d$.

A simpler approach to obtaining the same result is to look at the general Fresnel integral (eq. 10.29) and realize that the effect of a lens, as it changes a diverging beam to a converging one, is to mathematically move the illuminating source to the focal plane. In the equation, therefore, $1/z_1 = -1/z_2$, and so the extra term in the integral vanishes. Also, $x_1 = 0$ for the system chosen. The general Fresnel equation is reduced to exactly the form of equation 10.58.

The reverse of the system shown in Figure 10.14 – that is, with a point source and the spatial signal aperture preceding the lens – can also be analyzed using the Fresnal integral. The development is considerably longer and gives similar results, except that the input transmittance function is multiplied by a scaling factor that depends on distances in the optical arrangement.

Because this process of creating an optical Fourier transform is important, it needs to be summarized and illustrated in practical terms. Consider the situation shown in Figure 10.15 where collimated (usually) coherent light passes through some optical signal $f(x, y)$, which often takes the form of a

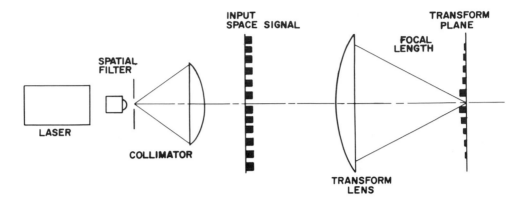

Figure 10.15. Creation of an optical Fourier transform of input signal in form of transparency (filter and collimator are optional).

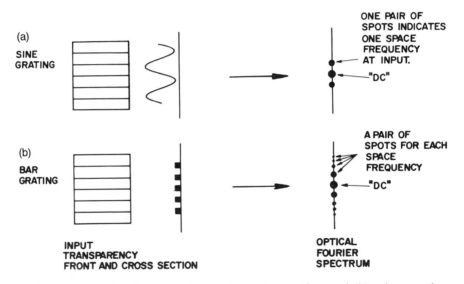

Figure 10.16. Two simple space signals, (a) a sine grating and (b) a bar grating, and their optical Fourier transforms (Cloud 1976).

transparency having a transmittance function that depends on the space coordinates. The modulated light beam then passes through a simple lens. At the back focal plane of the lens (the focus for the undisturbed light beam), there will be a diffraction pattern whose intensity distribution is related to the square of the amplitude of the Fourier transform of the input signal.

If the input is a sinusoidal grating, for example, as pictured in the first case shown in Figure 10.16a, the transform plane will exhibit three bright patches. The central dot corresponds to the uniform field or "DC" component of the input. The other two dots indicate the spatial frequency content of the input signal, with radial distance in the transform plane representing spatial frequency in the input plane. If the input signal is a "square wave" bar and space grating, as in Figure 10.16b, there will be in the transform plane a row of dots whose positions and brightnesses indicate the presence and importance of various harmonics of the fundamental space frequency at the input. A two-dimensional grid input will generate a Fourier spectrum at the transform plane that is a two-dimensional array of dots corresponding to the two-dimensional Fourier transform.

10.7 Optical spatial filtering

The preceding sections have developed the idea that any aperture causes an integral transformation of the light field impinging on the aperture. For the case in which the aperture contains a lens, the transform will be visible at the focal plane of the lens. If the lens receives some sort of spatial signal in the form of a transparency that is illuminated with coherent light, then the optical Fourier transform of the spatial signal will be created. Observation and interpretation of this transform field is most easily done for certain arrangements of the spatial signal plane, the transforming lens, and the observation plane.

The practical significance of these findings is expanded greatly by the addition of two more ideas. The first is that the spatial frequency content of the original input optical signal can easily be modified in the Fourier transform plane. The second idea is that a second lens can be used to perform a second transformation, which is an inverse transform, to regenerate the original signal, perhaps now modified by having its spatial frequency content changed.

To execute these ideas, place another lens behind the transform plane of the systems just shown. The image of the original input may be cast on the screen. Such a system is illustrated in Figure 10.17. The second lens forms the inverse transform to recover the original signal. It is possible and often useful, however, to modify the frequency content of the optical image at the Fourier transform plane before completing the inverse transform. This task can be accomplished by blocking or otherwise changing some portion of the light distribution at the transform plane. Such a procedure is called spatial filtering, coherent optical data processing, or optical Fourier processing.

A fundamental example of optical Fourier processing is shown in Figure 10.18. Here, the input signal (a) is a two-dimensional grid of crossed lines that produces a two-dimensional array of dots at the transform plane (b). All the dots except the central horizontal row are blocked by a suitable screen with a slit (c), which is placed in the transform. The inverse transform created by the second lens is found to be a simple grating of vertical parallel lines (d).

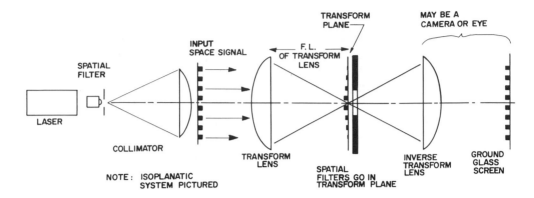

Figure 10.17. Optical system for spatial filtering in the Fourier transform plane and the creation of an inverse transform of a filtered image.

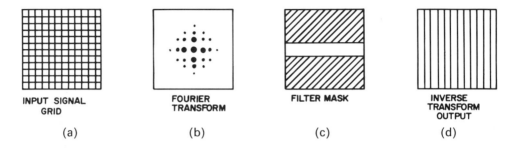

INPUT SIGNAL GRID	FOURIER TRANSFORM	FILTER MASK	INVERSE TRANSFORM OUTPUT
(a)	(b)	(c)	(d)

Figure 10.18. Example of optical spatial filtering to create a bar grating from a grid of dots or crossed lines (Cloud 1980).

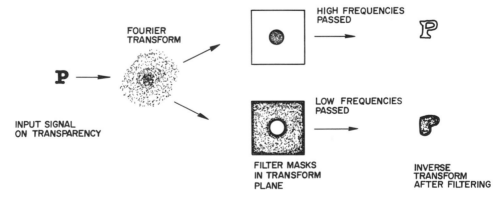

Figure 10.19. Example of optical spatial filtering for image modification (Cloud 1980).

The horizontal family of lines is suppressed by the optical filter, which has removed all the light rays required to image the horizontal lines. The potential usefulness of such a process is very great.

Figure 10.19 illustrates what can be done in modifying the appearance of a single printed character by blocking either the high-frequency or low-frequency components. One can imagine many applications in image enhancement, noise reduction, character recognition, and optical coding. A fundamental advantage is that the procedure is a simultaneous analog treatment of a whole optical field. The entire scene or message can be hidden in code, modified by removing portions of its frequency content, or recovered in one operation. A specific application of these ideas for enhancing the utility of the moire technique will be discussed in the next chapter.

10.8 The pinhole spatial filter

A particularly useful example of optical spatial filtering (a process) is the pinhole spatial filter (a device).

The purpose of and need for this device are easy to demonstrate. Often, a laser beam must be expanded to illuminate a scene that is much larger in breadth than the diameter of the beam. A lens having a short focal length, such as a microscope objective, can be used to expand the laser beam. The problem with this solution is that the diverging illuminating beam will contain a multitude of spurious diffraction and interference fringes. These lines and blotches will be transmitted to the hologram, moire pattern, or whatever else the system is being used for. At best the effect will be only unsightly; in many cases it will interfere with the measurement. An example that illustrates the problem appears in Figure 10.20, which is a photograph of a laser beam as expanded by a microscope objective.

One way to clear the expanded beam of spurious fringes is to place a pinhole aperture of approximately 10 to 25 μm diameter at the focus of the microscope objective. Figure 10.21 shows the system. Figure 10.22 is an example of a laser beam when it has been expanded by a pinhole spatial filter.

The function of this device can be explained in two ways, both based on the diffraction theory and examples appearing in earlier portions of this chapter. The first and simplest is to view the pinhole which is illuminated by the concentrated light at the focus of the microscope objective as providing a very narrow impulse function (delta function) at the aperture plane of a diffraction system. The Fourier transform appearing at a distance, but not very distant because the aperture is so small, is just the wide, uniform rectangular function.

A better view is that the pinhole spatial filter uses optical spatial filtering to eliminate high-frequency noise components at the transform plane of the

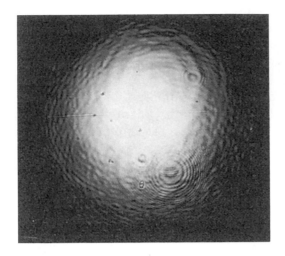

Figure 10.20. Laser beam expanded by a microscope objective, showing spurious fringes.

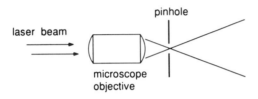

Figure 10.21. Pinhole spatial filter for creating a clean expanded laser beam.

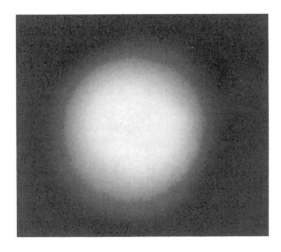

Figure 10.22. Laser beam expanded by a pinhole spatial filter, using the same microscope objective as was used in Figure 10.20.

microscope lens. Refer to Figure 10.21. The parallel laser beam enters the transform lens to be converged at the transform plane. The energy contained in the optical noise will appear off-axis at this plane. The pinhole stops the noise components but allows the low-noise DC components to pass through. Given the short focal length and the small aperture, there is no need for a second lens unless the convergence of the emerging beam must be modified, as into a plane wavefront.

Pinhole spatial filters, often called just "spatial filters," are manufactured and sold by several commercial organizations. The adjustment of one of these can be very frustrating without some instruction. The proper procedure is to first set up the optical system without the pinhole in place. Put a white card about 10 in. away from the pinhole holder. Then insert the pinhole into its holder and back the lens away from the pinhole until some light gets through to the card. This puts the pinhole within the cone of light. Center it by adjusting the pinhole holder laterally until the exiting pattern is nicely symmetrical. It should be an Airy disc (see section 10.4). Then, advance the objective toward the pinhole until the light disc starts to slide off sideways. Recenter the disc by slowly adjusting the pinhole holder laterally some more. Repeat this process until the outer fringes of the disc disappear and the pattern on the screen looks like a large glowing ball. At that position, the pinhole is centered at the exact focus of the objective. What has happened is that the pinhole was first located within the light cone where it is broad. Then the pinhole is tracked back down the cone until it arrives at the apex. Do not disturb the apparatus even minutely once the spatial filter is adjusted, or the process will have to be repeated from the beginning.

References

Born, M., and Wolf, E. (1975). *Principles of Optics*, 5th ed. New York: Pergamon Press.

Cloud, G. L. (1976). Lasers in engineering education. *Engineering Education*, 66, 8: 837–40.

Cloud, G. L. (1980). Simple optical processing of moire grating photographs. *Experimental Mechanics*, 20, 8: 265–72.

Goodman, J. W. (1968). *Introduction to Fourier Optics*. New York: McGraw-Hill.

Haskell, R. E. (1971). *Introduction to Coherent Optics*. Rochester, MI: Oakland University.

11

Moire with diffraction and Fourier optical processing

In this chapter, diffraction and spatial filtering theory are put to good use in moire measurement of displacements, with significant gains in sensitivity and flexibility. Before we describe the method, some additional development of diffraction by superimposed gratings is necessary. Improvement of moire results by performing spatial filtering during the grating photography is also explored.

11.1 The basic idea

Although useful moire fringe patterns can be obtained by direct superposition of the grating photographs with one another or with a submaster grating, as is discussed in Chapter 8, such a simple procedure does not yield the best results, nor does it exploit the full potential of the information that is stored in a photograph of a deformed specimen grating. Increased sensitivity, improved fringe visibility, and control of the measurement process can be had by utilizing some of the basic procedures of optical data processing.

Three related physical phenomena are important in developing an understanding of moire fringe formation and multiplication by superimposing grating photographs in a coherent optical analyzer. The first of these phenomena is the diffraction of light by a grating, or more accurately, by superimposed pairs of gratings having slightly different spatial frequencies. The second is the interference fringe patterns that are produced in the diffraction orders by interference of two beams that come together at small relative inclination. The third important phenomenon is that a simple lens acts as a Fourier transformer or spectrum analyzer and offers the possibility of performing filtering operations on space-dependent optical signals in a manner analogous to the treatment of time-dependent vibration and electrical signals. Actually, these concepts are not independent from each other; they are manifestations of fundamental interference and diffraction processes.

11.2 Diffraction by superimposed gratings

The theory of moire fringe formation by superimposing two diffraction gratings of nearly equal pitch and orientation has been presented in considerable detail in a book by Guild (1956). His ideas have been extended, refined, and demonstrated within the context of moire strain analysis in a series of papers by Chiang (1969), Cloud (1978, 1980), Post (1967, 1968, 1971), Post and McLaughlin (1971), and Sciammarella (1969).

Recall first from section 10.5 the effect of a single diffracting screen upon a beam of light. If a single, narrow monochromatic beam is made to pass normally (normal incidence is chosen for simplicity; oblique incidence may be used) through (or reflect from) a sinusoidal amplitude or phase grating, the beam will be divided into three parts at the grating. Figure 11.1 shows this behavior schematically. The first part, called the zero order, is an undisturbed portion of the beam that passes directly through the grating. The other two parts, called first orders, deviate symmetrically on either side of the zero order at an inclination that depends upon the spatial frequency of the grating and the wavelength of light as well as the incidence angle, which has been taken to be zero for this discussion.

Now, consider what happens when a narrow collimated beam passes through two sinusoidal gratings of slightly different spatial frequencies, as illustrated in Figure 11.2. The three diffracted beams or rays from the first grating are each divided into three more rays by diffraction at the second grating. The diffracted rays shown in Figure 11.2 are numbered according to the order of diffraction at each grating. For example, the ray marked C + 1, F − 1 was diffracted into the + 1 order at the coarse grating and then into the − 1 order at the fine grating. If the grating spatial frequencies are nearly equal, or if one spatial frequency is nearly an integral multiple of the other, then the diffracted rays will naturally fall into clusters or groups. Five distinct ray groups will appear in the case shown. The center group is a single attenuated version of the incident beam. Each of the extreme orders contains a single ray that has been diffracted by each of the two gratings in succession. The intermediate ray groups numbered + 1 and − 1 are those of interest in moire work. Each contains two rays, one of which has been diffracted at the first grating only, and the other having been diffracted at the second grating. These two rays in the intermediate group are nearly parallel because the spatial frequencies of the two gratings are nearly equal. Now, if the two rays can be made to overlap, as by a lens or imaging system, and if they both came from a single light source that has a coherence length great enough for interference to be possible, then the two rays forming the group will interfere with one another. Figure 11.3 shows how the ray groups can be refracted by a lens and made to converge to a spot or patch in what actually is the Fourier transform plane.

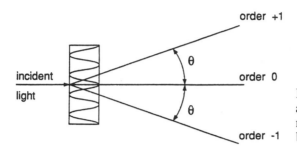

Figure 11.1. Diffraction of a narrow collimated monochromatic light beam by a sine grating.

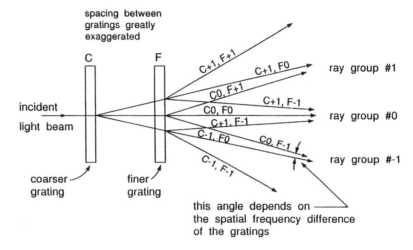

Figure 11.2. Diffraction of light by two superimposed sine gratings having slightly different spatial frequencies.

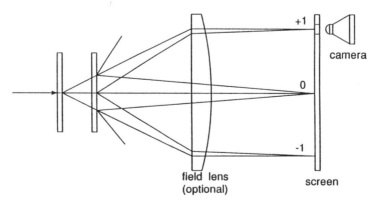

Figure 11.3. One way to form a two-beam interference fringe pattern by light diffracted through two sine gratings having slightly different spatial frequencies (Cloud 1980).

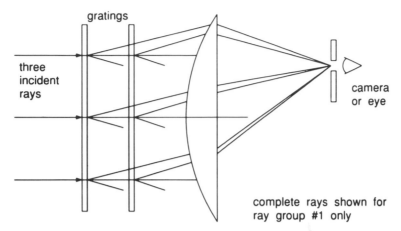

Figure 11.4. Diffraction of a wide collimated beam by two sine gratings to form whole-field interference pattern (Cloud 1980).

The interference is that of two coherent beams impinging on a surface with a small difference of incidence angle. It is a classic example of two-beam interferometry, as was studied in Chapter 2. For a given wavelength of light, the small angular difference between the two beams is a measure of the spatial frequency difference between the two gratings. The interference fringe pattern is a function of the angular difference. The result is an interference pattern indicative of the pitch and orientation differences of the two diffraction gratings. In short, it is the moire pattern of the two gratings for the area subtended by the incident beam. This discovery, which might seem a bit startling at first, is fundamental to both intermediate-sensitivity moire and high-sensitivity moire interferometry.

For moire strain measurement, it is necessary to illuminate the whole field of the two gratings by coherent collimated (usually) light. In this case, a whole field of rays will be diffracted by the first grating, and a second field by the second grating, as pictured in Figure 11.4. A field lens is placed in the diffracted beams to decollimate the rays and converge them at a focus. In general, the rays diffracted at the first grating will focus at a point slightly displaced from the focus of the rays diffracted at the second grating. If they are close enough to overlap, then an interference pattern is produced. A more useful procedure is to use another lens and screen (a camera or, carefully, an eye) to construct images of the two grating fields with the light contained in the ray group. Essentially, two images are constructed so as to lie atop one another. Because coherent light was used, the two images interfere with one another, the degree of interference depending mainly on the relative displacement of the two focal spots. That, in turn, depends on the relative inclinations of the two sets of rays coming from the diffraction gratings. The image in the camera displays, then, a pattern of interference fringes that are indicative of

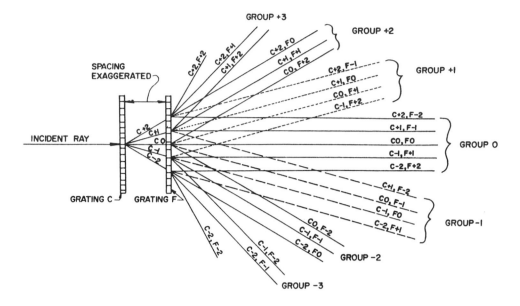

Figure 11.5. Diffraction of a narrow beam by two bar and space gratings to form ray groups containing higher diffraction orders (Cloud 1980). Notes: (1) For simplicity, spatial frequencies of gratings C and F are taken nearly equal. (2) Diffraction orders beyond ± 2 are not included. (3) Numbers on diffracted rays indicate orders at first and second gratings (e.g., C+2, F−1 means ray portion deviated into +2 order at grating C and then into −1 order at grating F).

the local spatial frequency and orientational differences between the two gratings.

Only minor extensions of these basic ideas are needed to understand the use of moire gratings in practical measurement situations.

The first complication is that the gratings tend to vary in pitch and orientation from point-to-point in any strain field of practical interest. One need only apply the reasoning previously outlined to each elemental area of the whole field. The result, clearly, is a set of fringes that vary in direction and spacing from point to point in the field.

The second complication is more difficult to analyze but is important because it leads to methods for enhancing sensitivity in moire experiments – a critical factor. In general it is neither wise nor possible to work with sinusoidal gratings. There will exist, therefore, higher-order diffractions at each of the two gratings. The number of orders produced from a single ray by each grating depends mainly on the sharpness of the grating or the degree to which it approaches a rectangular wave periodic structure. One finds in such a situation that each group of near-parallel rays consists of several individual rays corresponding to different orders of diffraction at each grating. Figure 11.5 illustrates this behavior.

DATE 3/10

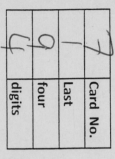

Surname	First four letters
A	W	A	N

Initials
M	RS

Card No.	Last four digits
7	1	9	4

Take your book to the Issue
machine in the Self
Reservations area

Follow the on-screen
instructions to issue your
book

Check your receipt for the
date that book should be
returned/renewed

Check your library account in
MUSE each day

If you have any problems
with your account please
speak to a member of staff

DATE

	Surname
	First
	four
	letters

	Initials

	Card No.
	Last
	four
	digits

Take your book to the Issue machine in the Self Reservations area

Follow the on-screen instructions to issue your book

Check your receipt for the date that book should be returned/renewed

Check your library account in MUSE each day

If you have any problems with your account please speak to amember of staff

For this work, the complex cases do not need to be considered in great detail. It is sufficient to observe that the basic diffraction and interference model still applies. In general the interference at the image will involve more than two component images or beams. In practice, the higher-order diffractions are attenuated to the point where only the basic two beams in each ray group are of any consequence. That is not to say that the higher-order groups are not useful.

There is one important related fact that holds true if the two gratings are of nearly the same spatial frequency. Each ray group in this case corresponds to a higher diffraction order, which corresponds in turn to a grating frequency that is a multiple of the basic grating frequency. The image formed by any ray group contains a moire pattern corresponding to grating frequencies equal to the diffraction order (or group number) times the fundamental specimen grating frequency. This important concept is the basis of multiplying the moire sensitivity when the two gratings must be of the same base pitch. All one need do is use the light in one of the higher-order ray groups to form the image and its fringe pattern. The implication is that rather coarse specimen gratings can be employed in a moire experiment, and then, through use of high-order diffraction groups, a sensitivity corresponding to finer gratings can be obtained. Such a multiplication technique can yield several-fold increases of moire sensitivity with accompanying simplification of the experiments owing to the necessity of handling only relatively coarse specimen gratings.

A third and even more useful extension of the basic concepts arises when the gratings are grossly different in spatial frequency – that is, when one grating frequency is a multiple of the other plus a small additional mismatch, which might be imposed deliberately and/or be the quantity that is to be determined. In such a situation, the diffractions are somewhat more complicated, as is the makeup of each of the diffracted ray groups. Figure 11.6 illustrates what happens where the second grating frequency is three times the frequency of the first. Several extraneous rays are omitted from each group in this figure.

The basic idea of forming an interference pattern with the rays forming a given group still applies. The question arises as to what such an interference pattern means in terms of the frequency and orientation differences between the two gratings. A general interpretation can be very complicated. An important simplification is that, by design and because of the natural attenuation of high diffraction orders, only two of the component rays in any useful ray group will interact to form a visible fringe pattern. Examination of the two main components in (for example) ray group 3 in the case pictured produces the answer to the interpretation question. These two rays correspond to the first diffraction order at the fine grating and the third order at the coarse grating. The image formed with these two groups will be the same as

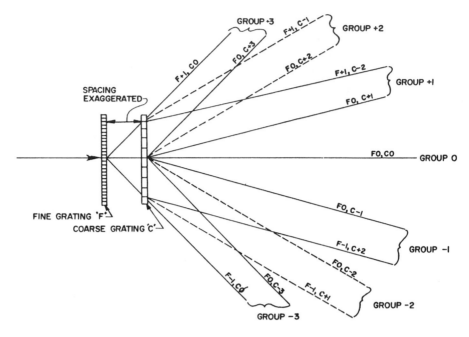

Figure 11.6. Diffraction by two superimposed gratings, one having a spatial frequency three times that of the other; only the important component rays are shown (Cloud 1980). Notes: (1) Diffraction orders beyond ± 2 are not included. (2) Numbers on diffracted rays indicate orders at first and second gratings (e.g., $C+2$, $F-1$ means ray portion deviated into $+2$ order at grating C and then into -1 order at grating F). (3) Diffraction angles are shown larger than encountered in practice.

the image produced by two gratings having nearly equal frequencies at three times the fundamental frequency of the coarse grating. The moire interference fringes in the image will correspond to the fringes that would be produced by two fine gratings. This conclusion is supported by careful theoretical analysis. The effects of the remaining beams will be to increase the background noise in the fringe pattern, perhaps to the point of obscuring the moire fringes.

A striking feature of the situation just discussed is that the higher-order moire fringe patterns in the camera image are identical, except for background noise and overall brightness, no matter which ray group is used to form the image. Stated another way, the sensitivity is not increased by going to a higher diffraction order, as is the case when two similar gratings are superimposed. A given group, however, can have superimposed moire patterns that correspond to different orders and therefore different sensitivities. It is possible, and good practice, to utilize whichever group gives the best fringe visibility, without further worry about the meaning of the fringes. The best group will usually be the one for which the diffractions are simplest, such as happens for group 3 in the illustration.

This case, where one grating frequency is an integral multiple of the other, has striking implications for moire measurement of the type currently under discussion. It allows the use of a coarse specimen grating that is easily applied and photographed. When a specimen grating is superimposed with a finer grating, the moire fringe pattern that appears is the same as would be created by two fine gratings. That is, a coarse specimen grating gives a measurement sensitivity equivalent to that of a much finer grating. The potential sensitivity multiplication is greater than can be obtained using high-order groups from two coarse gratings.

11.3 Optical Fourier processing of superimposed moire gratings

Another productive approach to understanding the creation of moire fringes by superimposing specimen and master gratings, or their replicas, in a coherent optical system is based on the fact that a simple lens acts as a Fourier transforming device. This idea was thoroughly developed in Chapter 10, and we need only apply the findings to the specific problem at hand.

In the elementary situation under study here, two superimposed gratings are placed in an optical data processing system. A Fourier spectrum of the gratings is created at the transform plane. All but one of the bright dots (actually two or more bright dots close together) are eliminated by a pinhole in a dark mask. The light in this one dot is used to form an image on a screen by the second lens – this lens and screen combination being an ordinary camera. The image is constructed, then, of light that carries with it information about the periodic structure of the two gratings for whatever fundamental space frequency has been chosen by the placement of the pinhole. The only rays that get through the pinhole are those that have been modulated by the gratings at a single space frequency, which may be the fundamental grating frequency or one of its harmonics.

A distinctive feature of this approach is that it considers the output image of the gratings to consist of a desirable signal plus a great deal of other information. The important signal, which is the moire pattern, is made visible by sifting it from all the extra information. There is a certain latitude in selecting the information that best suits a given purpose. Selection of higher-frequency components will give moire sensitivity multiplication, for example, when the input gratings are of similar space frequencies. All the flexibility discussed in the context of the diffraction model has its parallel in the Fourier filtering model.

Having two explanatory models of the same physical process raises the question of which one is correct, or are both faulty? Actually, the two explanations are not different in basic concepts. The difference is one of emphasis. In the diffraction model, the diversion of portions of the incident beam of light is the important feature. With the Fourier processing approach, attention centers on the transfer characteristics of an aperture that happens to have a lens in it, given an optical signal that is already generated by passing light through a transparency. Of course, the lens would not work correctly if the transparency did not divert portions of the incident beam by diffraction. So the combining and rationalizing of the two approaches can be pursued to a final consistent model. The price to be paid for this nicety is a small increase of complexity. Further study of the problem will not contribute to the goals of this work, so we shall abandon it with one final observation. As so often is the case with optical processes, the uniting physical phenomenon is that of interference. This property of light is what makes visible for study those minute differences of propagation directions or wavefront shapes that are the physical manifestations of important processes such as diffraction and double refraction.

11.4 Creation of moire fringe patterns

In engineering applications of moire analysis, one of the gratings is typically a specimen grating or its replica, and the other grating is some sort of master or submaster. Alternatively, the gratings may both be replicas of the specimen grating for two different states of the specimen.

As was pointed out in the discussion of geometric moire, it is not easy to obtain good moire fringe patterns by superimposing, directly or optically, a specimen grating with a master grating. The problem gets worse if, because of the application environment, the specimen grating must be recorded, imaged, or transferred by a lens. Such processes always reduce contrast and add optical noise to the grating replicas, and fringe visibility is reduced when these replicas are directly or optically superimposed with a master grating. The deterioration of fringe contrast and visibility can usually be tolerated if coarse gratings are used, but sensitivity of the measurement is then severely limited. When fine gratings are employed to gain sensitivity, the fringes obtainable may not have tolerable levels of visibility.

The preceding sections have outlined in general terms and in different ways an optical filtering approach that can be used to construct moire fringe photographs from two gratings. The advantages of such an approach in terms of fringe visibility and reduction of contamination by optical noise are readily apparent. Multiplication of measurement sensitivity beyond the limitations normally imposed by coarse specimen gratings can be obtained. The primary data are permanently stored for all specimen states, and the data can be processed at leisure in the laboratory to give optimum results with control of pitch mismatch. The approach has been shown to be especially effective in hostile environments. Apparent disadvantages include the requirement for

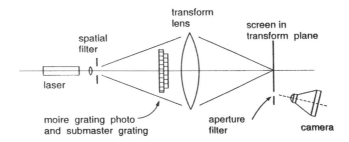

Figure 11.7. Creation of moire patterns using optical processing on photograph of specimen grating.

grating photography and the added complications in primary and secondary data reduction.

To see how the method works, without too much obscurative detail, refer to Figure 11.7, which is a schematic of a Fourier optical processing system for retrieving moire fringe patterns from a photograph of a specimen grating and a submaster grating that is a photograph of a master. At the focal plane of the lens, there will appear an array of bright dots, which are the ray groups. These ray groups will be immersed in a sort of background halo, which is the optical transform of all the optical noise and unwanted information in the two gratings. A screen with an aperture that is small enough to pass only one ray group is placed in the transform plane. An imaging system is placed behind the screen so as to construct an image with only the energy that gets through the aperture. This imaging system, which can be the eye, a camera, or a video capture device, should be focused at the apparent input grating plane as seen through the first lens. The image seen is formed only with the radiation corresponding to the spatial frequency characteristic of that ray group. All the rest is eliminated, including the optical noise radiation. The result is a moire fringe pattern of improved contrast and visibility, at least. Additional improvements such as incorporation of pitch mismatch, sensitivity enhancement, and optimization of fringe contrast are attained by changing submasters and/or ray groups.

The next chapter treats this entire moire procedure in more specific detail, and some sample results are included.

11.5 Grating photography with slotted apertures

Many moire techniques, including those just described, require imaging and/or photography of fine grating structures. The opposing requirements of large field and high resolution stretch the capabilities of even the finest lenses. The resulting grating photographs are often of low contrast, and they may carry considerable noise. Optical spatial filtering of the type described in the preceding section will improve the signal-to-noise ratio, and it offers the possibility of doing some sensitivity multiplication. Another possibility is to perform some spatial filtering during grating photography by modifying the lens aperture so its spatial frequency response is biased toward the grating frequency. By biasing the lens response to multiples of the grating frequency, the effective frequency is multiplied, giving multiplied moire sensitivity (Cloud 1976).

Forno (1975) and Burch and Forno (1975) demonstrated that slotted apertures can be used to tune a camera lens to give sharpened photographs of grating structures, improve depth of field, and enhance the response of a photographic system to certain spatial frequencies that might be contained in a random pattern. The potential applications of such procedures, which are based on well-known principles of optics, are many. The work cited seems directed mainly toward measurement of deformation and strain through elegant but simple improvements of the moire and speckle techniques.

Slotted camera apertures can also be used for multiplication of grating spatial frequency in moire photography. Sensitivity of the measurement, which is often a serious problem when measuring strains with the moire method, can be increased several fold. Depth of field is increased, and a camera lens of ordinary quality can be used to photograph the high-frequency gratings. The method gives improved photographic rendering of grating lines when sensitivity multiplication is not needed. The photography of two-dimensional arrays (grid and dot patterns) is simplified. These improvements are apart from any gains derived from coherent optical processing of the moire photographs (Cloud 1976).

In one set of experiments, gratings of frequency 39 lines/mm (1,000 lines/in.) were applied to the specimen using ordinary photoresist techniques. As a result of processing deficiencies, the specimen gratings were far from perfect. Plastic deformation during the testing process caused significant damage to the gratings.

Even with the obvious flaws, it proved possible to photograph the specimen gratings in reflected light without special apertures. The resulting moire patterns were useable, except that fringe definition was seriously degraded in the regions of high plastic deformation, where the resist coating was damaged. With slotted apertures, the specimen gratings proved sharp enough so that emphasis of the second harmonic with a properly tuned lens produced photographs of specimen gratings equivalent to 80 lines/mm, which is the highest frequency harmonic theoretically obtainable with the lens used. Some details of these experiments are presented to illustrate the capabilities of the slotted aperture idea.

A micrograph of a typical photograph of a poor specimen grating, taken at $f/11$ in white light without any sort of optical filtering, is reproduced in Figure 11.8. Grating frequency in the film is 30 lines/mm. Negative photographs of this sort were used extensively in moire studies. Single exposures can be superimposed on one another to produce observable fringes. They

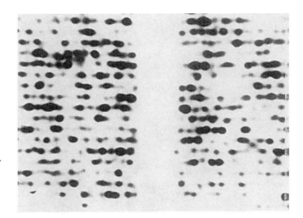

Figure 11.8.
Photomicrograph of
negative photograph of
poor specimen grating
obtained with no filter
mask (Cloud 1976).

have been used in various combinations in the optical filtering setup. Although double-exposure plates produce barely observable fringes, the patterns are clear when they are observed in the data processor; and, in fact, higher-order patterns with the appropriate fringe multiplication can be observed.

Forno (1975) summarizes the information required to design slotted aperture filter masks that will tune a given camera for photographic emphasis of particular space frequencies. These same principles can be used to obtain photographs containing gratings having frequencies that are multiples of the fundamental frequency in the specimen grating if (1) the higher frequencies are present in the structure being photographed and (2) the camera and film are capable of responding to those frequencies. The practical significance of requirement (1) is that the specimen grating be sharp and of high contrast – that is, it should resemble a ruling more than a simple sine grating. The limits imposed by condition (2) are easily calculated for any camera–film combination.

To improve the results of the studies mentioned, slotted apertures were designed to fit behind the iris diaphragm inside the lens. Slot sizes and locations were calculated to tune the lens to spatial frequencies of 30 and 60 lines/in. in the image plane for green light. Because the masks were to be placed near the iris, it was necessary to account for magnification of the mask by the rear lens element. This magnification was found to be 1.09.

In establishing a slot width, which governs the band pass of the tuned lens, one must account for the maximum and minimum grating frequencies that will be encountered in the photography of the strained grating. Typically, masks were designed for a fairly broad band pass, giving a strain response of more than $\pm 10\%$. Dimensions of the two masks used are shown in Figure 11.9. The results reported here were obtained with apertures that were cut from cardboard with a pocket knife and then painted black.

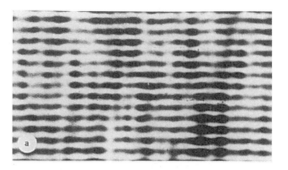

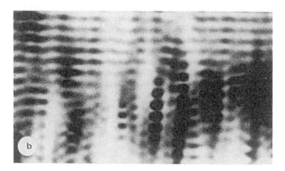

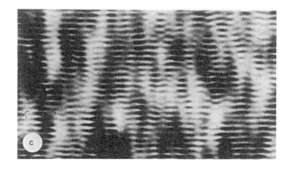

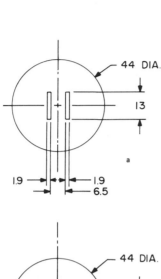

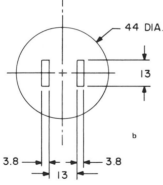

Figure 11.9. Slotted aperture masks used in photographing moire gratings: (a) mask for 39 lines/mm; (b) mask for 79 lines/mm (Cloud 1976).

Figure 11.10. Photomicrographs of negative photographs of a specimen grating obtained with slotted aperture masks and white light: (a) in a small strain region with an aperture having 39 lines/mm center frequency at specimen, (b) same as (a) but in a region damaged by large strain, (c) in a small strain region with an aperture having 79 lines/mm center frequency at specimen (Cloud 1976).

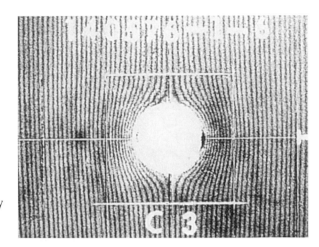

Figure 11.11. Moire pattern obtained with a sensitivity increase of 2 obtained solely by using a slotted aperture to multiply grating spatial frequency (Cloud 1976).

Figure 11.10 shows micrographs of typical photoplates of the same specimen used for Figure 11.8 but taken with the apertures installed in the camera lens. Comparison of Figures 11.10a and 11.10b with Figure 11.8 suggests the degree of improvement that can be expected from using filter masks in photographing moire specimen and master gratings. Especially important is the delineation of the grating in the areas where it cracked and flaked because of the plastic deformation of the specimen. The comparison is more suggestive if the specimen is not flat and normal to the optical axis. The depth of focus for the unmasked lens is so small that the grating will not be resolved over the whole specimen.

Figure 11.10c illustrates the grating frequency multiplication that can be obtained with slotted apertures, even with a lens having a rather low upper-frequency limit. The grating in Figure 11.10c is 60 lines/per mm on the film, which is equivalent to 78 lines/mm on the specimen. Athough this grating shows local nonuniformities, it produces very good moire fringes when superimposed with an appropriate submaster. The striations in the background of these grating photographs are traceable to irregularities in the moire master grating, the specimen surfaces, and the sprayed resist coating.

The gains in fringe pattern quality from using aperture masks are evident when fringe multiplication is carried out in the processor. Figure 11.11 was derived by superimposing a specimen grating made with a slotted aperture tuned to 60 lines/mm and a submaster made by ordinary photoreduction. The measurement sensitivity is thereby increased by a factor of 2. Higher multiplications are possible with a lens having wider aperture so that the slots can be further apart. Sensitivity multiplications of 4 have been obtained by combining slotted aperture photography with optical Fourier processing (Cloud 1976).

The slotted apertures are also useful in isolating just one grating from a two- or three-way grid for moire rosette investigations (see Chapter 8).

The price for the advantages gained are mainly in loss of light. Exposure will be longer and stability requirements more severe. The use of bright strobe lighting is highly recommended for photography with these modified apertures.

References

Burch, J. M., and Forno, C. (1975). A high sensitivity moire grid technique for studying deformations in large objects. *Optical Engineering*, 14, 2: 178.

Chiang, F. P. (1969). Techniques of optical spatial filtering applied to the processing of moire-fringe patterns. *Experimental Mechanics*, 9, 11: 523–6.

Cloud, G. L. (1976). Slotted apertures for multiplying grating frequencies and sharpening fringe patterns in moire photography. *Optical Engineering*, 15, 6: 578–82.

Cloud, G. L. (1978). *Residual Surface Strain Distributions Near Fastener Holes Which Are Coldworked to Various Degrees*. Air Force Material Lab Report AFML-TR-78-1-53. Ohio: Wright Aeronautical Labs.

Cloud, G. L. (1980). Simple optical processing of moire grating photographs. *Experimental Mechanics*, 20, 8: 265–72.

Forno, C. (1975). White-light speckle photography for measuring deformation, strain, and shape. *Optics and Laser Technology*, 217.

Guild, J. (1956). *The Interference Systems of Crossed Diffraction Gratings*. Oxford: Clarendon Press.

Post, D. (1967). Analysis of moire fringe multiplication phenomena. *Applied Optics*, 6, 11: 1938.

Post, D. (1968). New optical methods of moire fringe multiplication. *Experimental Mechanics*, 8, 2: 63.

Post, D. (1971). Moire fringe multiplication with a nonsymmetrical doubly blazed reference grating. *Applied Optics*, 10, 4: 901–7.

Post, D., and MacLaughlin, T. F. (1971). Strain analysis by moire-fringe multiplication. *Experimental Mechanics*, 11, 9: 408.

Sciammarella, C. A. (1969). Moire-fringe multiplication by means of filtering and a wave-front reconstruction process. *Experimental Mechanics*, 9, 4: 179–85.

12

Procedures of moire analysis with optical processing

This chapter brings together many of the concepts discussed in the preceding five chapters to develop a moire technique that is superior to simple geometric moire but simpler than moire interferometry. The details of the procedures are presented in considerable detail. Many of the techniques described here, including specifics of using pitch mismatch, reproducing gratings, differential processing, and digital fringe reduction apply equally well to geometric moire and moire interferometry. In fact, many of the general ideas are useful in other areas of interferometry.

12.1 Introduction

A moire technique that incorporates spatial filtering has several attractive aspects. The fundamental idea is to take advantage of the sensitivity multiplication and noise reduction offered by optical Fourier processing of moire grating photographs, which are recorded for various states of a specimen. Sensitivity of the method can be controlled *after* the experimental data are recorded, within limits that are between those of geometric and interferometric moire. The method also is very flexible in that any two specimen states can be compared easily. The original data are permanently recorded for leisurely study later. Certain common errrors are automatically eliminated. Fringe visibility is usually much improved over that obtained by any method of direct or optical superimposition. Finally, the method is useful in difficult environments.

For orientation purposes, a short summary of a typical but quite specific procedure is presented first. Recognize that many variations in the procedure are possible, and the method is easily adaptable to whatever resources are at hand.

In the sections following the procedure summary, some specific details of technique are given. These details should be viewed as suggestive examples.

Most of them are drawn from investigations conducted by the author and colleagues (e.g., Cloud 1978, 1979, 1980a, 1980b, Cloud and Paleebut 1984, 1992; Cloud and Sulaimana 1981) and they are meant to be neither all-inclusive nor exclusive.

12.2 Outline of the technique

The steps in intermediate sensitivity moire using Fourier optical processing are listed here. Because these steps extend from specimen preparation through data reduction, the list appears rather lengthy. Some of the steps are, of course, common to other moire procedures, especially geometric moire.

1. Master gratings, typically of 1,000 lines/in. (lpi) (39.4 lines/mm), are obtained and reduced photographically to create a set of working submasters of various grating frequencies including the fundamental frequency.
2. Gratings of the fundamental frequency are applied to the specimen, typically using photoresist and contact printing.
3. Fiducial marks and identity labels are applied to the specimen surface.
4. The specimen is placed in the loading system and the grating photographed at low magnification, say between 1:1 and 2:1.
5. The specimen is loaded to the first load.
6. The grating, now deformed with the specimen, is photographed again.
7. The photographic plate of the undeformed grid is superimposed with a submaster grating having a spatial frequency of one to three (sometimes higher or lower, depending on the sensitivity desired) times the frequency of the photographed specimen grating, plus or minus a small frequency mismatch.
8. The assembly is placed in a coherent optical processor and adjusted to produce the correct baseline (zero strain) fringe pattern at the processor output. This fringe pattern is photographed.
9. Steps 7 and 8 are repeated with the deformed grating photoplate.
10. Steps 7–9 are repeated as desired with other submaster gratings to produce fringe patterns having different pitch mismatch and, in some cases, different sensitivity multiplication factors.
11. Steps 5–10 are repeated for each specimen load state that is of interest in the experiment.
12. The fringe patterns are enlarged and printed.
13. The prints of the fringe patterns are sorted and coded for identification during the data analysis procedure.
14. The locations of the fringes along the desired axes are determined and recorded as arrays of numbers showing fringe order and location from a specified fiducial mark; this is best done on a digitizing tablet if available.
15. The arrays of fringe data are converted to plots, and differentiations yield strain; computerized reduction is appropriate.

An important characteristic of this procedure is that it operates in a differential mode, which, as a rule, is best in any experiment. To understand this remark, notice that fringe data are collected for the specimen in each load state, including a baseline data set at zero or beginning load. A requirement is that nothing but the load is changed between creation of the successive data sets. The data sets are then always processed in pairs, often, but not always, including the baseline data. There are several advantages to this procedure, these being typical of differential measurements. For one, the effects of any residual initial fringe patterns, whether accidental or deliberately induced as by pitch mismatch, are automatically eliminated. There is no false initial strain. Another distinct advantage is that any two of the specimen states can be easily compared because the grating at each state is permanently stored on glass photoplates. Such a capability is very useful when the specimen is loaded in several irreversible stages, as in studies of plastic deformation or fracture.

12.3 Master and submaster gratings

As mentioned in the chapter on geometric moire, master gratings are best purchased from suppliers. For the purposes described here, gratings of 1,000 lines/in. were obtained from Graticules Ltd. (Tonbridge, TN9 1RN, Kent, England), and grids (mesh) in stainless steel were obtained from Buckbee-Mears (St. Paul, MN).

The making of 1:1 copies of moire gratings for submasters and for use in printing the grating onto the specimen can be accomplished in several ways. Good contact copies of the master have been made by a method similar to that used by Chiang (1969) on Kodak high-resolution plate (HRP) using a Durst enlarger head with a 150-mm Schneider lens at $f/5.6$. A sketch of this setup is shown in Figure 12.1.

This arrangement has several noteworthy features. First, the light rays falling on the master and photo plate are not collimated as is often thought necessary. It is not a source of error here because of the small equivalent source size and the intimacy of the contact between the master and the photoemulsion. No spacer is needed between the two to eliminate the diffraction lines, which appear if monochromatic light is used. The emulsion of the HRP should be placed in contact with the emulsion side of the master and held by scale weights on the enlarger easel. Only a thin film of index-matching fluid (xylene) is needed between the two to reduce the effects of possible lenticulation in the master. Diffraction lines do not appear because white light from the enlarger head is used. This procedure originated as a means of reducing exposure times, which still will amount to about 8 min with a lengthy 12-min development of the HRP in D-19. The resultant ratio

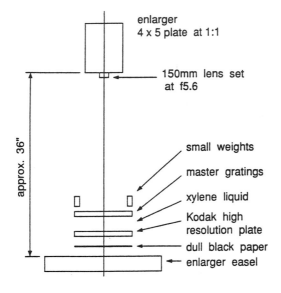

Figure 12.1. Schematic of a setup for contact printing of submaster grating.

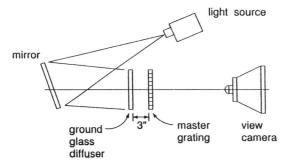

Figure 12.2. Sketch of a setup for photographically producing a submaster grating of high spatial frequency.

of line width to space width in the grating copies is 0.7, which is a bit on the low side for the best moire work. The submasters so obtained will be quite good within the limitations imposed by the master grating.

Direct photographic reproduction is a good choice for manufacture of the several submaster gratings having various spatial frequencies required for optical data processing of the specimen photographs. In a typical experiment, several each of gratings having spatial frequencies of 743, 783, 797, 1,488, 1,535, 2,200, 2,225, 2,288, 2,999, and 3,049 lines/in. were produced. These values are 1, 2, and 3 times the fundamental spatial frequency of the specimen grating photographs (1,000 lines/in. divided by magnification used), plus or minus various frequency mismatches. Figure 12.2 shows a sketch of an apparatus used in the absence of an optical bench. In this setup, the master grating is held in a laboratory clamp base and backlit with light from a Kodak carousel slide projector. A ground glass plate is placed about 3 in. behind the grating to scatter the incident light. The lens and camera are those used in photographing the specimen grating; they are described in section 12.5. The whole setup rests on a holography table that uses airbag suspension for vibration isolation.

Camera-to-subject grating distances can be estimated by calculation and finalized by trial and error to give the sought-for submaster grating frequency on the photographic plate. Focus of the grating image is very critical in such a situation because of the extreme resolution and contrast required of the system. An ordinary camera ground glass is much too coarse and may not be exactly in the film plane. A good technique is to replace the ground glass by a developed and fixed unexposed film plate of the same types as are to be used in the photography. This focus plate is held in an ordinary 4 × 5 plate holder with the separator removed, and the assembly is placed in the camera back and carefully seated. All the plateholders should be checked to see that their critical dimensions are uniform. Critical focusing can be accomplished with a hand-held magnifier (50 × or so), which is adjusted so that the emulsion of the focus plate is in focus when the magnifier is held against the back side of the plate. Image sharpness and parallax observations can both be used as focus criteria, dye (magic marker) marks having been put into the emulsion of the focus plate for the purpose. In some cases, a previously produced contact copy of the grating is placed in the plateholder, and the sharpness of the moire fringes produced on this plate is used as the focus indicator as well as an indicator of magnification. A better technique is to mount a microscope directly to the camera back, as is done in high-resolution photography of integrated circuits.

In theory, it is best to use the maximum lens opening for greatest sharpness and resolution in such a demanding situation, but trials are required to establish the best aperture for a given lens. Vignetting and light falloff in the extremes of the field might prove serious enough, however, to make it difficult to obtain a completely satisfactory exposure at maximum aperture. For this reason, aperture settings of $f/8$ and up to $f/16$ have been found most satisfactory for the optics used by the writer. The slight softening of the grating is compensated by the more uniform illumination of the image. Some dodging is often used in the grating photography to reduce gross gradations in image density.

Monochromatic light is usually preferred for high-resolution photography such as this. Experiments along these lines have shown that image degradation caused by the very long exposures required for typical monochromatic sources can be greater than the gain in using such sources, especially if the lens is well corrected for color.

Both Kodak high-resolution plate and Kodak 649-F spectroscopic plate (often used for holography) have been used for photoreduced submasters. The higher-speed holographic films such as Agfa 8E75 and l0E75 have given good results. Grain effects are noticeable with the 649-F with a heavy 7–8 min development in D-19.

Experiments have been conducted to show that the slotted aperture technique described in section 11.5 can be extended to achieve improved photography of moire gratings. This improvement will not be pursued here.

Performance of the submaster gratings should be checked by observing their diffraction efficiencies as they are produced. Both the gross transmittance and the diffraction performance of the submaster must be "complementary" to these characteristics of the specimen grating photographs in the optical data processing stage. For this reason, several different photoplates of each submaster spatial frequency should be produced. Exposure and development times are varied in order to produce submasters having different properties. In general, the denser submasters prove more useful with typical specimen photos.

To some extent, the variation of density over the extent of the submaster plate, which results from cosine^4 light falloff, proves useful in optical data processing. It tends to counteract the normal Gaussian distribution of light in the expanded laser beam to give a near-uniform field in the fringe photographs.

12.4 Specimen gratings

Methods for printing moire grills and gratings onto the specimen have been explored in considerable detail by several investigators (e.g., Cloud and Bayer 1988; Cloud, Radke, and Peiffer 1979; Holister and Luxmoore 1968; Luxmoore and Hermann 1970, 1971; Straka and Pindera 1974). There are many solutions to the problem. The choice of technique for creating specimen gratings depends to a degree on the material and shape of the specimen. The photoresist methods discussed later are appropriate for specimens that are flat, or nearly so, and that are made of metal or (most) plastics, including composites. The photoresist approach to creating gratings on specimens is fairly simple, requires minimal special equipment, and offers the possibility of baking and/or etching the grating to make it resistant to damage.

A typical photoresist is Shipley AZ1350J marketed by Shipley Co. (Newton, MA). This particular resist is formulated for applying acid resist coatings to aluminum substrates, and its solids content is comparatively high at 30%. The companion dye, thinner, cleaner, and developer should be purchased with the resist.

12.4.1 Application of photoresist

It is desirable for moire work, as with most other photoresist usage, that the resist coating be thin and uniform. Common application methods include spinning, dipping, spraying, wiping, and roller coating. The dipping and wiping techniques have been found deficient in that they tend to leave some buildup and sagging near the boundaries, that is, in the region of greatest interest. The spraying and spinning methods often prove best. Only the spraying technique is described here.

In order that the photoresist wet the surface and spread to a uniform coating, it is essential that the specimen surface be chemically clean. A procedure almost exactly the same as that commonly used for cleaning aluminum surfaces in preparation for the bonding of electrical resistance strain gages serves quite well, except that the surface is best polished before beginning. Initial brisk solvent cleaning is followed by treatment with the two common strain–gage applications materials, which, evidently, are weak solutions of phosphoric acid and sodium hydroxide, respectively. The specimen is rinsed in a spray of distilled water after this cleaning in order to assure the removal of any surface deposits of NaOH, which would affect the resist. (Resist developers are essentially weak solutions of NaOH.) Cleaning only with solvents works well if care is taken. This approach is most useful for recleaning a specimen after removal of a faulty sprayed resist coating. As a final cleaning step, the wet specimen can be rinsed in a copious spray of fresh reagent grade anhydrous methyl alcohol and allowed to air dry. This whole procedure is quickly and easily completed, and the specimen will dry for coating very soon. Furthermore, the specimen is left slightly chilled, which helps retard drying of the sprayed photoresist until it has a chance to smooth and flatten.

Extensive testing has established a balance of resist–thinner–dye proportions, air pressure, airbrush nozzle opening, spraying distance, and brush motion. The values arrived at represent a workable combination, but probably not the best one.

The proportions arrived at through trial are, by volume, 30 parts AZ1350J, 20 parts AZ thinner, and 1 part dye (if used). The best air pressure is about 40 psi, and it is very important that the air be dried and the pressure regulated.

A workable spraying procedure calls for laying the clean and dry specimen horizontally or inclined at about $10°$. The airbrush containing the resist is held about 18 in. (45 cm) from the specimen and pointing to one side of it. Flow of the atomized resist is begun and allowed to stabilize for about $\frac{1}{2}$ to 1 sec, after which the spray is quickly shifted onto the specimen. At the range and the air flow used, coverage is wide enough so that it is not necessary to sweep the brush, although small oscillations seemed to aid in giving good coverage while helping to settle the operator's nerves. This procedure eliminates problems connected with trying to overlap airbrush strokes. Another acceptable technique involves holding the brush about 6 in. from the specimen with about 25 psi pressure and covering the specimen with a few sweeping strokes, starting closest to the operator. An overlap of $\frac{1}{3}$ to $\frac{1}{2}$

the fan width is used. This method gives coatings that tend to have some large-scale thickness variations. With either approach, it is absolutely necessary to start the spray well before bringing it to bear on the specimen, as some coarse droplets are expelled at the beginning of flow. Coating thickness is controlled by spraying time once the nozzle opening is set. Minimum coatings should be about 2–5 μm thick; 10–15 μm (approximately 0.0005 in.) seems better.

Because of the volatile and mildly toxic nature of the solvents and cleaners employed, all the cleaning and coating procedures should be conducted in a laboratory fume hood.

After the coatings dry to touch, a process requiring only a minute or so, they can be placed in a small laboratory oven for drying at a temperature of about 25°C for 20–30 min. The coated specimens should then be placed inside padded, light-tight boxes to await exposure and development of the grating image.

12.4.2 Printing grating onto specimen

The moire grating can be printed into the photoresist coating on the specimen by a simple contact printing procedure in which a grating submaster is clamped to the specimen and the assembly exposed to ultraviolet light from a mercury lamp. The procedure closely follows that described by Holister, Hermann, and Luxmoore in the references cited earlier. Additional useful information on producing fine periodic structures is contained in a paper by Austin and Stone (1976).

Clamping of the submaster to the specimen can be accomplished with ordinary spring-type clothespins or small glue clamps. The assembly then is placed in front of an unfiltered mercury lamp, with caution taken to protect the eyes and skin. Exposure times depend on coating thickness, lamp power, and distance, and they are best established by trial. Typically, they are 1–5 min. Lenticular effects in the submaster can be reduced by using a 50% aqueous glycerin solution between the submaster emulsion and the photoresist. Usually, the fluid can be eliminated, probably because lenticulation of the thin submaster emulsion is not great and/or the data processing procedure is quite forgiving of noisy gratings. The newly exposed photoresist is developed according to manufacturer's instructions in the standard Shipley AZ developer diluted with water.

A variation of the procedure can be used to create a two-dimensional grid pattern of dots (as contrasted with the usual array of continuous parallel lines) through a two-step exposure. The submaster is arranged as usual for one-half the total required exposure; it is then rotated 90°, reclamped, and the resist exposed for the remaining time.

One aspect of the behavior of photoresist deserves further comment. It is possible to balance exposure time and coating thickness to produce specimen

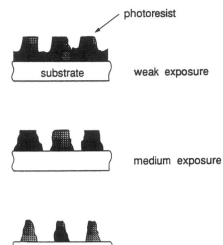

Figure 12.3. Sketches of typical cross sections of specimen gratings in photoresist for various degrees of exposure.

gratings that will photograph more sharply than those ordinarily thought of as "good" gratings. The phenomenon utilized is that incomplete exposure and development leaves "debris" between the unexposed grating lines. Figure 12.3 illustrates varying degrees of this behavior. It turns out that the debris tends to scatter and absorb the incident light, giving high-contrast dark lines against the smooth unexposed lines. It seems wiser, therefore, to use a thick coating (0.0005 in. or more) and not try to cut through to base metal in the exposure and development. This conclusion may not apply for some types of surfaces, especially if the finish has a matte structure.

The green dye often used with photoresists is useful only in indicating coating uniformity and in checking the exposure. It bleaches rapidly when exposed to the light required for grating photography. It may be eliminated without ill effect.

12.5 Photography of specimen gratings

It is critical to the intermediate sensitivity moire technique that the specimen grating be photographed with enough resolution and contrast that the potentials of optical Fourier processing can be fully exploited. The processing does largely eliminate the effects of optical noise, but sensitivity multiplication and sharp fringe patterns are most easily obtained when the grating replicas show good contrast, regardless of the noise.

Several different setups for accomplishing the high-resolution photography of the specimen grating have been utilized. Good results came from an arrangement such as is sketched in Figure 12.4. One camera used successfully is a 4 × 5 Burke and James "Orbit" monorail that is stiffened with angle

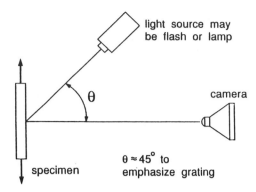

light source may
be flash or lamp

θ

camera

θ ≈ 45° to
emphasize grating

specimen

Figure 12.4. Sketch of apparatus for photographing a specimen grating.

iron and weighted with lead blocks and some steel plate. The lens is a Goerz Red Dot Artar apochromat having a focal length of $9\frac{1}{2}$ in. (2.4 cm) and maximum aperture $f/9$. Another sucessful setup incorporates a Horseman 4×5 camera, which is also stiffened. The lens is a Schneider 50-mm Retinar-Xenar or a Zeiss 120-mm S-Planar that is optimized for finite conjugates. Typically, the camera is set at full extension to give an image of the specimen that is magnified by a factor of about 1 to 1.3. These setups sometimes are mounted on heavy pedestals to photograph specimens in a testing machine.

Focus of the specimen image is very critical in this high-resolution situation. The ground glass of the camera is not satisfactory for this critical work because it is too coarse and because such focus plates are often not exactly in the photoemulsion plane. For focusing, a blank plate of the thickness and type used in the photography is developed and fixed and then mounted in a 4×5 plateholder that has the separator removed. The image of the specimen in the emulsion is examined with a magnifier adjusted to focus in the emulsion plane while the magnifier base rests on the opposite side of the film plate. The best method is to mount an ordinary microscope to the camera back. The image of the specimen grating should be checked over the whole area of interest for maximum sharpness and contrast. As a check on focus, parallax between the grating image and the marks (scratches and magic marker lines) can be used. Zero parallax means correct focus. This focus procedure is rather tedious, but it need not be repeated as long as the photographic system is not disturbed.

Although a monochromatic filter is usually specified for this type of high-resolution photography, it is often not practicable. When available light is so low that contrast and definition are lost with the filtered light and the films used, white light gives better results, provided the lens is well corrected for color.

For many applications, the light source shown in Figure 12.4 is best replaced with a flashlamp. Focus is done with a continuous source, then the flash is used in recording for improved vibration rejection.

The Kodak high-resolution plate, the backed Kodak spectroscopic 649-F plate, or holography plates may be utilized. The HRP material often proves somewhat slow for the illumination available, so the holography media are best. The glass photoplates are placed in regular 4 × 5 plateholders that have been checked for matching critical dimensions.

The lenses are used mostly at apertures of $f/9$ and $f/11$. Higher f-numbers will give reduced light falloff and exposure reduction in the extremes of the field. Also, the frequency response of typical lenses is such that there is softening of the contrast of the 1,000 lpi grating at the highest apertures. The lower f-numbers must be used for proper resolution and the exposure adjusted for the best compromise over the area of interest. Dodging can be utilized to even up the unequal exposure of the plate caused by the oblique illumination and cosine^4 light falloff.

Exposure determination is largely by trial and error. Minor correction is possible during plate development. Crude measurements of light intensities at the camera back can be obtained with an ordinary light meter. Much time and film are saved if a sensitive photometer probe is obtained for measuring light level at the film plane during focus. It also helps in adjusting illumination and in calculating exposure.

It is worth noting here that the angle of incident illumination is chosen by trial to give the best contrast in the grating image. Shadows in the three-dimensional grating structure can be exploited to enhance visibility of the grating.

The development of the plates should be monitored by very short examination under red safelight; many holographic films are red sensitive. Development is stopped when plate density reaches about 50%, the value that seems to give the best fringe visibility in the optical data processor.

12.6 Formation of moire fringe patterns

Although useful moire fringe patterns can be obtained by direct superposition of the grating photographs with one another or with a submaster grating, such a simple procedure does not yield the best results, nor does it exploit the full potential of the information which is stored in a photoplate. Increased sensitivity and control of the measurement process can be had by utilizing some of the basic procedures of optical data processing that were discussed in Chapters 10 and 11.

We have been discussing a multitude of technical details, so let us indulge in a very brief summary of the two related physical phenomena that are important in developing an understanding of moire fringe formation and

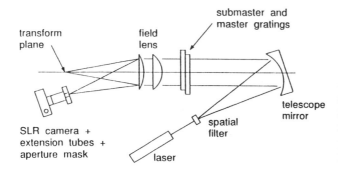

Figure 12.5. Schematic of an optical processing system used for obtaining moire fringe photographs from specimen grating photoplates; compact folded configuration.

multiplication by superimposing grating photographs in a coherent optical analyzer. The first of these phenomena includes the diffraction of light by a grating (or, more accurately, by superimposed pairs of gratings having slightly different spatial frequencies) and the related multiple-beam interference patterns produced in the diffraction orders. The second important phenomenon is that a simple lens acts as a Fourier transformer or spectrum analyzer and offers the possibility of performing filtering operations on space frequency components in a manner analogous to the treatment of vibration and electrical signals. Actually, either phenomenon is sufficient by itself for understanding moire processes, from slightly different points of view.

One optical system typical of those devised for this purpose is pictured schematically in Figure 12.5. This one is unique in that the optical system is folded to fit into very limited space. The light source used for such devices is typically a helium–neon laser of 10 to 50 mW output. The laser beam passes through a simple spatial filter that converts it to a moderately clean diverging beam. This beam is directed to a spherical astronomical telescope mirror of 4 in. diameter. The mirror folds the beam back along the optical axis of the processor to compensate for lack of a long enough space, and it also collimates the expanded beam. The moire submaster grating and the photoplate of the specimen grating are placed emulsion sides together in the optical path normal to the light beam, and they are clamped and held to an optical mount by spring-type clothespins or clamps. After passing through the photoplate, the beam, now containing diffraction components, is decollimated by the simple lenses (most often only one) acting in series.

The focal plane of the system is found by trial with the data plates removed. In this plane, which is the transform plane of the field lens combination, is placed a black paper screen containing a hole of approximately 3/32 in. diameter. The screen can actually be contained in the filter mount of the 35-mm camera lens, which is mounted to the camera body with extension rings so as to provide close focus. The camera is mounted on a swinging bar so that the hole in the filter mask can be made to coincide with the chosen ray group, a series of which appear as bright spots in the diffraction pattern.

Selection of the proper bright patch must be accomplished with the camera pointed so as to focus an image of the specimen grating plate on the camera film. The whole assembly is best mounted on a simple optical bench.

The camera is focused in the apparent plane of the data plates as seen through the field lens. After proper adjustment of the grating photoplates, a moire pattern will be visible in the image of the specimen seen in the camera viewfinder. After final adjustment of the plates to eliminate rigid-body rotation effects in the fringe pattern, the pattern is photographed. The fringe photograph negatives can be made with any good film. Exposures are usually not critical, and development is normal. If desired, a view camera can be substituted for the 35-mm camera, and instant films can be employed for quick results. For quantitative use, the fringe patterns are best printed in 8×10 size.

It is at this point that the flexibility of the optical data processing procedure becomes useful. The baseline (zero strain) and deformed grating data are permanently stored on glass photographic plates. It is possible to superimpose these plates with different submaster gratings in order to gain maximum useful sensitivity multiplication and to improve subsequent fringe reading and data analysis by optimizing the spatial frequency mismatch of the superimposed gratings. Fringe contrast will be much higher than that obtained with direct superimposition. Submasters having density and diffraction characteristics that balance those of the specimen gratings can be chosen by trial at this stage to optimize fringe visibility. Also, the best ray group can be chosen.

As a practical example, it worked out in one experiment that the specimen grating photographs had a spatial frequency of 762 lpi (30 lines/mm), which is the frequency obtained when a specimen grating of 1,000 lines/in. is magnified 1.3 times. These plates could be superimposed with submasters of around 2,200 lines/in. to get a sensitivity multiplication of 3, or of 1,542 lines/in. for a multiplication of 2. The various mismatches were chosen to yield the closest fringe spacing obtainable with good fringe visibility, and the signs of the mismatches were chosen to fit the situation in the areas of interest. Most grating photoplates were processed with at least three mismatch levels, and sometimes with more than one sensitivity multiplication factor for checking purposes and because it was often not possible to assess the quality of a dense fringe pattern through the camera viewfinder, which was used without a magnifier.

Three samples of the moire fringe patterns that were obtained by this procedure are reproduced in Figures 12.6–12.8.

12.7 Reduction of fringe data

A moire interference fringe is a locus of points for which the in-plane displacement of the specimen surface normal to the grating lines, plus pitch

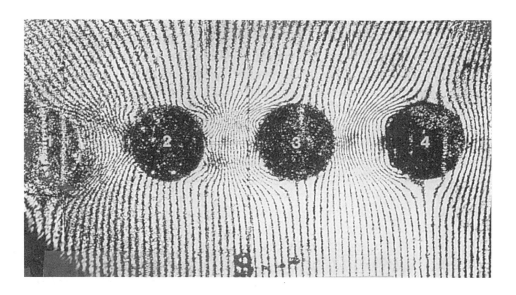

Figure 12.6. Moire pattern showing the strain field in the vicinity of a row of holes that were coldworked in the order 1, 2, 4, 3 (Cloud and Sulaimana 1981).

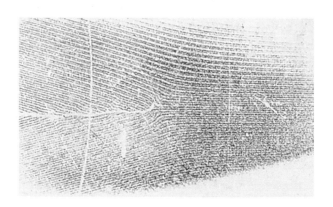

Figure 12.7. Moire fringes showing vertical strain near a crack and the crack-opening displacement taken during a fatigue test of a titanium alloy at 600°F; the grating is vacuum deposited aluminum.

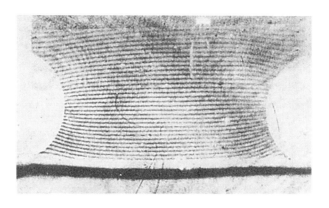

Figure 12.8. Moire pattern showing axial strain under load at 1204°C using a 40-lines/mm nickel mesh grating embedded in ceramic; this type of grating has been used up to 1370°C (Cloud and Bayer 1988).

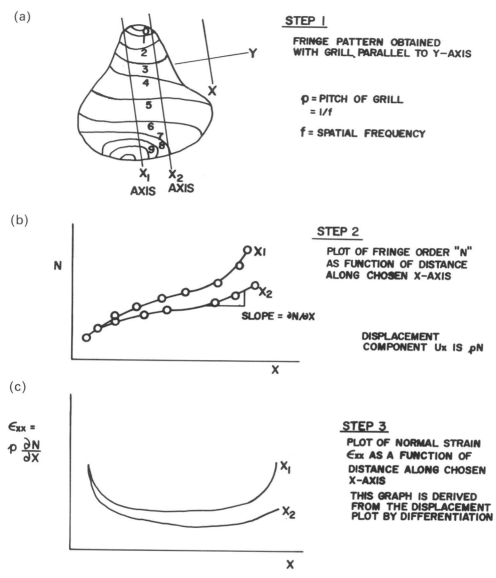

(a)

STEP I

FRINGE PATTERN OBTAINED
WITH GRILL PARALLEL TO Y-AXIS

p = PITCH OF GRILL
    = $1/f$

f = SPATIAL FREQUENCY

(b)

STEP 2

PLOT OF FRINGE ORDER "N"
AS FUNCTION OF DISTANCE
ALONG CHOSEN X-AXIS

SLOPE = $\partial N/\partial X$

DISPLACEMENT
COMPONENT U_x IS pN

(c)

$\epsilon_{xx} = p\dfrac{\partial N}{\partial X}$

STEP 3

PLOT OF NORMAL STRAIN
ϵ_{xx} AS A FUNCTION OF
DISTANCE ALONG CHOSEN
X-AXIS

THIS GRAPH IS DERIVED
FROM THE DISPLACEMENT
PLOT BY DIFFERENTIATION

Figure 12.9. General steps involved in reducing moire fringe photographs to obtain displacement and strain.

mismatch if it exists, is a constant multiple of the pitch (period) of the grating. Figure 12.9 shows in graphic form the general process of reduction of the fringe patterns to obtain strain. The steps will now be outlined.

Begin as shown in Figure 12.9a by drawing lines on the moire pattern to delineate the axes along which strain is desired. A picture of an in-plane displacement component can be constructed by plotting moire fringe order along a given axis in the specimen as shown in Figure 12.9b, subtracting the pitch mismatch (baseline, typically zero-load) fringe orders, and multiplying the remainder by the inverse of the grating spatial frequency and the fringe multiplication factor (sensitivity multiplication) used in the optical data processing.

Once a map of a displacement component is obtained, the corresponding strain component can be generated by calculating the derivative of displacement with respect to the appropriate space variable, as suggested by Figure 12.9c. For example, if the grating lines are aligned with the y-axis in the specimen, then the x component of displacement (u_x) is given by the moire fringe pattern. A plot of u_x along any x-axis is easily constructed. The derivative of u_x with respect to x ($\partial u_x/\partial x$) is the x component of the normal strain (ϵ_x). Other displacement and strain components are obtained by logical extension of this idea. As pointed out in Chapter 8, it may not be wise, however, to develop the shear strain information by using the cross partial derivatives $\partial u_x/\partial y$ and $\partial u_y/\partial x$ because these quantities are sensitive to errors caused by relative rigid-body rotations of the two gratings. If shear strain is to be measured, then the best procedure is to obtain a measurement of normal strain along a third axis inclined to the x and y systems. The strain rosette equations can then be used with the three normal strain components to find shear strain ϵ_{xy}. That the normal strain is not affected by moderate relative rotation of the gratings was demonstrated in Chapter 8.

The reduction of moire fringe data to obtain strains can be performed by graphical methods, by numerical techniques, or by a combined approach. When planning a data reduction process, it is well to remember that differentiation of experimental data is necessary. Such differentiation is an inefficient process in that it is laborious, often gives poor results if not done with extreme care, and an estimate of errors is difficult to obtain. Careful graphical processing, although tedious, is probably as dependable as any approach because it allows constant critical study of all intermediate results.

For most investigations, the volume of moire data makes numerical processing highly attractive. A numerical reduction computer plotting scheme has been developed (Cloud 1978) that incorporates most of the advantages of high-speed computing while retaining desirable features of the graphical approach, such as allowing examination of intermediate results and the introduction of a user-chosen small degree of data smoothing. The algorithm uses spline functions and curve fitting. Space does not permit inclusion of the routines in this book, but they can be copied from the references, or they will be supplied by the author upon request.

The remainder of this chapter deals with the digitizing of moire fringe data and the computerized reduction of such data for some different situations. The availability of some sort of manual digitizing tablet and at least a personal digital computer is presumed.

The sort of manual digitization that is discussed here is, of course, quite

tedious. Several ideas for automating the process, including fully computerized image acquisition and fringe reduction, have been tested, and some
have appeared in commercially available software. The numbering of fringe
orders in a convoluted pattern, where some of the fringes may be broken or
folded, or immersed in optical noise, is something that machines do not do
very dependably. The human eye is very good at tracking a fringe through
noise and at finding the proper center of a fuzzy line. The automated
technology is rapidly improving in performance and price, so we may not
need to do manual digitization for many more years.

12.7.1 *Digitizing fringes for normal strain parallel to the chosen axis*

The steps involved in manually digitizing and reducing the moire fringe
photos to obtain the normal strain parallel to the axis of interest for a
situation incorporating sensitivity multiplication and pitch mismatch interpolation are summarized pictorially in Figure 12.10. This figure represents
the reasoning just outlined applied to a specific problem and including both
at-load and baseline fringe patterns.

Recall from section 12.5 that the results of the moire grating photography
and optical data processing are enlargements showing the moire fringes in
the area, along with various identifying labels and fiduciary marks (Fig.
12.10a). To obtain normal strain along an axis (call it the x-axis) normal to
the grating lines, it is first necessary to obtain a record, in this case in digital
form, of the distance on the photo between at least two fiduciary marks. This
information is used to establish the overall magnification of the photographs
and to convert the strain results to actual specimen space. If optical
distortions might be present, then a network of fiduciary marks can be used
to create a correction grid to eliminate the resulting errors.

The first step, then, is to number the moire fringes and identify the various
fiduciary marks and the orientations of the specimens in the pictures. This
preparation is done for a data photo and for its matching baseline fringe
pattern. The counting of fringe orders can begin anywhere because absolute
fringe number is not important; it is very important, however, that numbers
of fringes are not repeated or skipped.

After initial setup and entry of identity and fiducial data, the locations of
fringe intersections with the axis under study are entered with the cursor. As
mentioned, the process is applied to both the at-load and the corresponding
baseline fringe patterns. These two data sets form a unit for computation of
the displacement component and strain component with proper correction
for grating pitch mismatch and magnification. The data sets are first plotted
(Fig. 12.10b) and then converted to displacement (Fig. 12.10c). Finally, the
displacements are obtained by differentiation of displacement, and the results
are plotted (Fig. 12.10d).

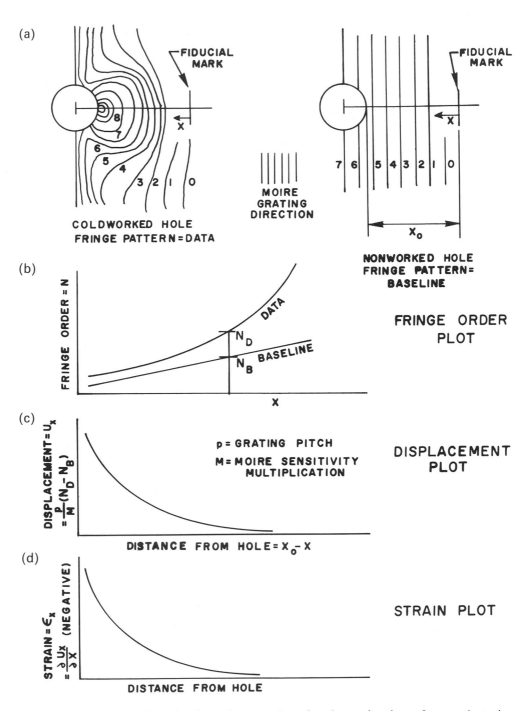

Figure 12.10. Steps in reduction of moire data for determination of normal strain parallel to the chosen axis.

12.7.2 *Digitizing fringes for normal strain perpendicular to the chosen axis*

Digital reduction of the moire fringe photographs to obtain a mapping of, say, normal strain ϵ_y along an x-axis is somewhat more complex. Some practitioners apparently do not realize that this measurement can be performed. To obtain ϵ_y the grating must be aligned with the x-axis. The plotting of fringe order and the differentiation must be done along the chosen y-axis, since $\epsilon_y = \partial u_y/\partial y$. A plot of fringe order with respect to x distance from a fiducial mark is of no use. Rather, it is necessary to obtain and analyze a series of plots of fringe order with respect to distance in the y-direction in the vicinity of the x-axis. This procedure is carried out for as many y-axes as needed to obtain a complete picture of ϵ_y along the x-axis. From the values of strain obtained at the intersection of the several y-axes and the x-axis, a plot of ϵ_y versus x can be developed. A pictorial summary of the process is shown in Figure 12.11.

After one catalogs the photographs and locates the fiducial marks, each of the data and baseline photographs that were taken with the gratings oriented for determining ϵ_y versus x has inscribed on it a series of y-axes at appropriate spacings (Fig. 12.11a). After this preparation, the photographs are digitized by the technique, apparatus, and program chosen, except that the locations of the moire fringe intersections with each y-axis are digitized, and the location of that y-axis along the x-axis is entered. The data from a given y-axis on a data photo are paired with the data from the corresponding y-axis on the appropriate baseline photo to form a set. The sets are plotted as shown in Figure 12.11b. Plots of displacement for each axis are constructed by subtracting baseline from data, multiplying by pitch, and dividing by multiplication factor (Fig. 12.11c). Strain for each axis is then found by differentiating the displacement plots (Fig. 12.11d).

12.7.3 *Computer data reduction*

There are several ways to set up software to reduce the digitized moire fringe data. Given the structure of the data sets just mentioned, a typical routine will perform the following tasks for a single data set to obtain the distribution of the normal strain component parallel to the chosen axis.

1. Read in data containing set designation (specimen number plus other identifiers), moire sensitivity multiplication factor from optical fringe data processing, moire grating spatial frequency, distance between fiduciary marks (for scaling purposes), the maximum fringe order, and the distance from a fiduciary mark to the intersection of each moire fringe with the chosen axis. The maximum fringe order and fringe locations for the corresponding baseline fringe pattern are also read.

(a)

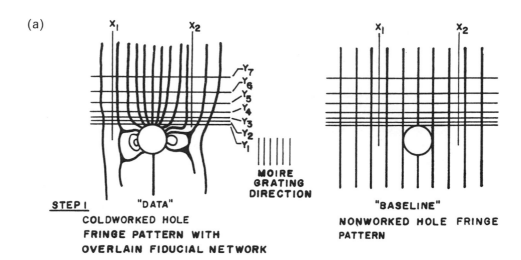

STEP I   "DATA"

COLDWORKED HOLE
FRINGE PATTERN WITH
OVERLAIN FIDUCIAL NETWORK

"BASELINE"

NONWORKED HOLE FRINGE
PATTERN

(b)

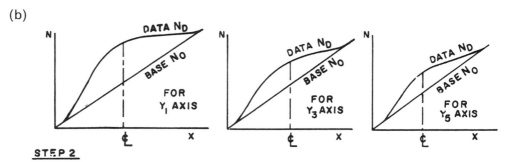

STEP 2

A PLOT OF FRINGE ORDER VERSUS DISTANCE Y FROM CONVENIENT
FIDUCIAL X-AXIS (E.G. X_1-AXIS) IS CONSTRUCTED FOR EACH Y AXIS.

(c)   (d)

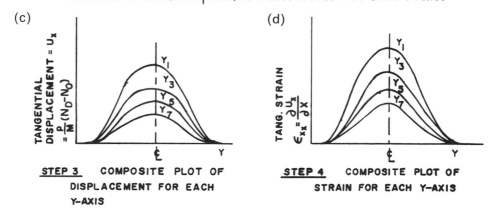

STEP 3 COMPOSITE PLOT OF
DISPLACEMENT FOR EACH
Y-AXIS

STEP 4 COMPOSITE PLOT OF
STRAIN FOR EACH Y-AXIS

Figure 12.11. Steps in reduction of moire data for determination of normal strain
perpendicular to the chosen axis.

2. Generate fringe order numbers to match each fringe location entered as data.

3. Fit the baseline and coldwork data with two continuous smooth curves by means of a cubic spline smoothing routine. The degree of smoothing should be specified by the user.

4. Interpolate the calculated curves to obtain fringe number as a function of distance from the fiducial mark at, say, 100 points on the data and baseline curves. The maximum range of the curves is sorted out and divided by 100 to establish the nodes, which must be common to both data and baseline curves.

5. Subtract the baseline fringe order from the data fringe orders for each of the 100 points; multiply this difference by the pitch (reciprocal of spatial frequency) of the grating on the specimen; and divide it by the sensitivity multiplication factor to obtain the displacement function u_x for the chosen x-axis.

6. Utilize the distance between fiduciary marks on specimen and photographs to calculate the magnification factor, which is used to convert all the data to specimen dimension space.

7. Compute by finite differences the first derivative of displacement with respect to distance; this results in the normal strain at each of the 100 nodes.

8. Print, if requested by the user, all values of input fringe orders, displacements, strain, and (if they are needed) higher derivatives of displacement with respect to distance.

9. Scale the data and generate a plot of the input data and baseline curves. This graph shows fringe data points and the smoothed curves.

10. Plot displacement as a function of distance from the chosen fiduciary mark.

11. Plot strain as a function of distance.

12. Start over with the next complete set of data and continue for the desired number of sets. The output plots should be available for individual data sets or as a composite of several data sets.

In addition to giving the strain distribution, one purpose of this analysis approach is to allow detailed study of each set of data and the results it produces. Input errors, such as skipping a fringe during digitizing, are immediately evident as discontinuities in the displacement plot.

The distribution of the normal strain component perpendicular to the chosen axis can be performed with the same routine, except that all the sets for the given array of parallel x-axes (recall section 12.7.2) are plotted on the same output. Then an additional step to pick off the strain for the given y-axis intersections is required. This last step is easy to perform manually.

Additional routines may be prepared to create composite or statistical summary plots, as required.

Two sample plots produced by the routines just outlined are shown in Figures 12.12 and 12.13.

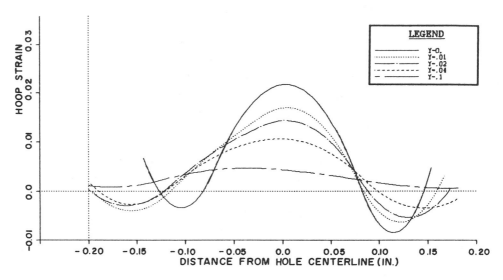

Figure 12.12. Typical computer plot of strain for various axes from moire fringe patterns.

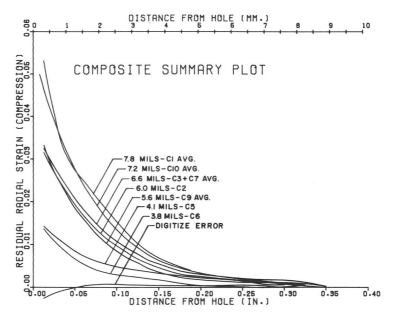

Figure 12.13. Typical composite statistical summary plot showing average radial strains for all coldworked levels and digitizing error analysis (Cloud 1980a).

12.8 Three-dimensional application

For many years, mechanicians have sought to extend their theoretical, computational, and experimental techniques to obtain dependable information about three-dimensional fields. Increased use of advanced materials and augmented demands for energy- and material-efficient designs have intensified the quest for three-dimensional methods and results.

A multiple embedded grid moire method has been developed to study the strain on the surfaces, quarter-planes, and midplane of thick specimens made of polycarbonate (Cloud and Paleebut 1984, 1992). The idea of using a single embedded moire grating for strain measurement on one interior plane has been utilized before (Durelli and Daniel 1961; Kerber and Whittier 1969; Sciammarrella and Chiang 1964), although there has been disagreement about the need to correct for errors resulting from stress-induced changes in the refractive index. Here the concept is extended to investigate multiple interior planes as well as the surface of the same specimen. The grids on the surface and in the interior are recorded for the unloaded condition and for various loaded states using high-resolution photographic techniques. The interior gratings are photographed through the out-of-focus intervening gratings. Moire fringes are extracted using optical Fourier processing. An empirically based correction procedure to eliminate the errors caused by the material refraction effect has been derived. The need for such a correction for errors caused by gradients of refractive index is not unique to embedded grid moire. The developed approach could be used to reduce errors when other optical methods, such as speckle, moire interferometry, or holointerferometry are used to measure strain or deformation in the interior of solids. It is also possible that the material refraction coefficient used in the derivation of the correction equations might be useful in stress analysis.

This improved three-dimensional moire technique has been utilized to determine the strain fields in the interior and on the surfaces of thick, compact tension specimens and on fastener specimens built up from layers of polycarbonate. Figure 12.14 is a schematic of a typical specimen.

In one such study, copper grids (two-way gratings) of 20 dots/mm were printed on some of the pieces using a stencil method. A grid side on the first piece was bonded to a clear side on the second piece. Nickel mesh with 20 lines/mm was placed between the second and third pieces (midplane). Finally, translucent copper grids were vacuum-deposited on the surface planes.

The surface and interior grids were photographed at zero load and at all other loads of interest. Each grid was photographed by focusing the camera on the grid plane to get maximum sharpness and contrast. Separate data plates were recorded by focusing on the front surface and quarter-plane

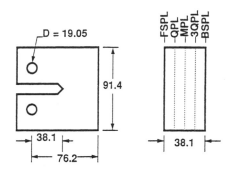

Figure 12.14. Multiple embedded grid moire specimen used for three-dimensional studies. All dimensions are in mm.

in turn. The midplane and back surface were photographed after reversing the arrangement. Each interior grid was photographed through at least one intervening grid. A large aperture giving small depth of focus is necessary in order to separate the gratings optically. Some loss of quality in the grid records was noted, but the results were adequate for successful optical processing. Each photographic replica of a grid was treated by optical Fourier filtering to obtain moire fringe patterns as described in the preceding sections.

The procedure is sufficient for determining surface strain fields. The raw interior strain results require correction for local changes in refractive index of the polycarbonate. Because these changes are not uniform, either through the thickness or in the lateral direction, they create an internal lens effect that changes the apparent pitch of the moire gratings in the recorded grating image and so result in a false strain, which is super-imposed on the actual strain.

Study of this problem showed that the false strain in the interior could be eliminated by photographing interior gratings from both sides of the specimen (Cloud and Paleebut 1992). Space does not permit presentation of the details. The relationships between the observed strains and the real strains are, for real strain on the quarter-plane,

$$\epsilon_{R1} = \frac{9\epsilon_{m1} - \epsilon_{m3}}{8} \tag{12.1}$$

and for real strain on the midplane,

$$\epsilon_{R2} = \frac{\epsilon_{m2} - (\epsilon_{m3} - \epsilon_{m1})}{2} \tag{12.2}$$

where　　ϵ_{m1} and ϵ_{m3} are the strains measured on a given quarter-plane by taking grid photographs through one-quarter thickness and three-quarter thickness; ϵ_{m2} is the strain observed on the midplane

These equations show that the actual strain on two interior planes can be estimated from three moire observations of the same planes, even though the moire data are contaminated by errors resulting from strain-induced gradients in refractive index.

This multiple embedded grid moire method is useful for studying three-dimensional problems. The technique is simple, and it requires no measurements other than the usual moire grid records. Grating records obtained from the interior, even though photographed through intervening gratings, are good enough to form useful moire fringe patterns when processed with an elementary optical Fourier analyzer.

There are some obvious shortcomings to the embedded grating approach. Any three-dimensional analysis is complex, and this method is no exception. The technique also is not actually three-dimensional because it gives strains only in the planes of the gratings.

References

Austin, S., and Stone, F. T. (1976). Fabrication of thin periodic structures in photoresist. *Applied Optics*, 15, 4: 1070–4.

Chiang, F. P. (1969). Production of high-density moire grids – discussion. *Experimental Mechanics*, 9, 6: 286.

Cloud, G. L. (1978) *Residual Surface Strain Distributions Near Fastener Holes Which Are Coldworked to Various Degrees*. Air Force Material Lab Report AFML-TR-78-1-53. Ohio: Wright Aeronautical Labs.

Cloud, G. (1979). Measurement of elasto-plastic strain fields using high-resolution moire photography, coherent optical processing and digital data reduction. *Proc. 8th Congress of International Measurement Confederation*, Moscow, USSR: S7, 13–26.

Cloud, G. (1980a). Measurement of strain near coldworked holes. *Experimental Mechanics*, 29, 1: 9–16.

Cloud, G. (1980b). Simple optical processing of moire grating photographs. *Experimental Mechanics*, 20, 8: 265–72.

Cloud G., and Bayer, M. (1988). Moire to 1370°C. *Experimental Techniques*, 12, 4: 24–7.

Cloud, G., and Paleebut, S. (1984). Surface and internal strain fields near coldworked holes obtained by a multiple embedded-grid moire method. *Engineering Fracture Mechanics*, 19, 2: 375–81.

Cloud, G., and Paleebut, S. (1992). Surface and interior strain fields measured by multiple embedded grid moire and strain gages. *Experimental Mechanics*, 32, 3: 273–81.

Cloud, G., Radke, R., and Peiffer, J. (1979). Moire gratings for high temperatures and long times. *Experimental Mechanics*, 19, 10: 19N–21N.

Cloud, G., and Sulaimana, R. (1981). *An Experimental Study of Large Compressive Loads Upon Strain Fields and the Interaction Between Surface Strain Fields Created by Coldworking Fastener Holes*. Air Force Technical Report AFWAL-TR-80-4206. Ohio: Wright Aeronautical Labs.

Durelli, A. J., and Daniel, I. M. (1961). A nondestructive three-dimensional strain analysis method. *Journal Applied Mechanics*, 28, 3: 83–6.

Holister, G. S., and Luxmoore, A. R. (1968). The production of high-density moire grids. *Experimental Mechanics*, 8: 210.

Kerber, R. C., and Whittier, J. S. (1969). Moire interferometry with embedded grids – effect of optical refraction. *Experimental Mechanics*, 9, 5: 203–9.

Luxmoore, A. R., and Hermann, R. (1970). An investigation of photoresists for use in optical strain analysis. *Journal Strain Analysis*, 5, 3: 162.

Luxmoore, A. R., and Hermann, R. (1971). The rapid deposition of moire grids. *Experimental Mechanics*, 11, 5: 375.

Sciammarella, C. A., and Chiang, F. P. (1964). The moire method applied to three-dimensional elastic problem. *Experimental Mechanics*, 4, 11: 313–19.

Straka, P., and Pindera, J. T. (1974). Application of moire grids for deformation studies in a wide temperature range. *Experimental Mechanics*, 14, 5: 214–16.

PART V

Moire interferometry
(with Pedro Herrera Franco)

13

Principles of moire interferometry

In this part, a moire method that combines the concepts of the moire effect, diffraction by a grating, and two-beam interference is described. The method is truly interferometric, and it is capable of high sensitivity. This chapter develops the theory.

13.1 Concept and approach

The preceding chapters have discussed two approaches for utilizing the moire effect in measurement of displacements, rotations, and strain. There is yet a third approach for performing moire measurements. It utilizes the fundamental concept of the moire effect, the concept of diffraction by a grating, and the phenomenon of two-beam interference to extend the capability and utility of moire measurement far beyond the limitations of geometric moire. It shares some basic ideas with intermediate sensitivity moire, which uses optical processing; but it bypasses the limitations imposed by the necessity to optically image the gratings. The result is a moire technique that is capable of truly interferometric sensitivities. That is, the wavelength of light is the metric, and displacements of fractions of 1 μm can be measured. This technique is finding increasing favor with experimentalists doing research in material characterization, fracture, and other areas for which high sensitivity is needed but other interferometric techniques are not suitable.

Moire interferometry can be modeled as a physical process in two distinct ways, and valuable insights are to be gained from each model. Before getting into the details of this powerful technique, which necessarily involve intricate geometric visualization, let us examine briefly these two physical models.

On the one hand, moire interferometry can be viewed strictly as a process involving two-beam interference and diffraction, and nothing needs to be said about the moire effect. In this model, the specimen carries a phase-type diffraction grating that alters the phase rather than the amplitude of incident

269

light in a regular, repetitive way. When the specimen is deformed, the grating
on its surface deforms with it. The specimen grating will change in frequency
and direction systematically from point to point. Consequently, plane wavefronts illuminating the specimen grating will be diffracted; but, because of the
localized changes in frequency and direction of the grating lines, the emergent
wavefronts will be slightly warped. If two such wavefronts are utilized
simultaneously, and if their angles of incidence are properly chosen, then their
diffraction angles are such that they interfere with one another. The resulting
interference fringe pattern of these diffracted warped wavefronts is a contour
map of the angular separation of the two diffracted orders. Because the
angular deviation of the diffracted light is a measure of the grating spatial
frequency, the contour map can be viewed as a map of distortion of the
specimen grating. Nothing has been said about the moire effect. But we find
that the interference pattern is the same as a moire pattern showing contours
of displacement. That it is the moire pattern that would result if the specimen
diffraction grating were directly superimposed with a similar but undeformed
grating, in the mode of geometric moire, is, perhaps, not immediately obvious.
This same question was answered in a different context in Chapter 11.

An alternative way of looking at the process is that the two beams incident
on the specimen create, by two-beam interference, a master grating. This
grating is sometimes called a virtual grating, although it is a real image that
can be viewed on a screen or used to expose a photoplate; we do not, however,
create a physical replica of it. This master grating is optically superimposed
upon the specimen grating to form the moire pattern. The pattern is viewed
by using the diffracted beams coming off the specimen to form an image in
a camera. This explanation submerges the fundamental diffraction and
interference aspects of the process, viewing it more as an extension of
traditional moire.

Several versions of the interferometric moire technique have been described
and put to use in recent years. Most of these utilize the ideas and designs
put forth in the early literature related to experimental mechanics applications
of the technique (McDonach, McKelvie, and Walker 1980; Post 1980; Post
and Baracat 1981; Walker and McKelvie 1978). A useful extensive review of
the historical development of moire interferometry has been published by
Walker (1994). A new book by Post, Han, and Ifju (1994) offers a detailed
analysis of high-sensitivity moire techniques with an insight derived from
many years of experience and study by an inventor of the technique. An
abbreviated treatment by Post (1987) is also valuable.

This chapter develops first the general equations for diffraction by a grating
under oblique incidence in three dimensions. This analysis is more general
than is customary in most moire literature, as a full three-dimensional
treatment is needed for understanding the operation of the 6-beam moire
interferometer that is described in subsequent chapters. The diffraction result

is combined with the knowledge of the interference of two plane waves that was developed in section 2.9, and with basic mechanics relations to derive the equations relating moire interference fringes to strain and rotation components. Some of the development is of necessity rather involved in geometry, but the final results are beautifully simple.

13.2 Geometrical treatment of diffraction by a grating

We have dealt carefully and quite rigorously with the important optical phenomena involving diffraction by apertures, including those that might contain a grating. The classic Huygens–Fresnel construction to explain the interference patterns produced by multiple slits was described but not pursued in favor of the more fundamentally refined, rigorous, and satisfying Kirchhoff integral approach.

At this point, it is a good idea to have a closer look at the simple geometric approach based on the Huygens principle. The reason is that this treatment, though it leaves out many important effects and is based on a drastic assumption, gives us an easy path to understanding where the diffracted beams end up when light is made to fall at oblique incidence upon a grating. The shortcomings of the Huygens construction are not relevant in this context, whereas they are critical in the context of Fourier optics. The same result can, in fact, be obtained by further manipulation of the results of the general diffraction theory in Chapter 10. But the basic moire interferometry relations are most easily developed using calculations of beam angles and path length difference, so it is just as well to start with the geometrical treatment for a single grating under oblique incidence.

When light is incident on a grating surface, it is somehow scattered or reflected from the grooves. A fundamental assumption, based on Huygens' concept and justified by many empirical observations, is that each grating groove acts as a very small, slit-shaped source of reflected or transmitted light. The usefulness of a grating depends on the fact that there exists a unique series of angles such that, for a given groove spacing and wavelength, the light from all the facets is in phase. This can be visualized from the geometry shown in Figure 13.1, which shows a plane wavefront incident at an angle α with respect to the grating normal, and diffracted at an angle β. The groove spacing is designated by d in this chapter.

It is easy to see that the geometrical path difference CD between light from successive grooves is simply $(d \sin \alpha - d \sin \beta)$. The principle of interference dictates that only when this difference equals the wavelength of light, or a simple integral multiple thereof, will the light from successive grooves be in phase and so give constructive interference. At all other angles there will be some degree of destructive interference among the wavelets originating at the

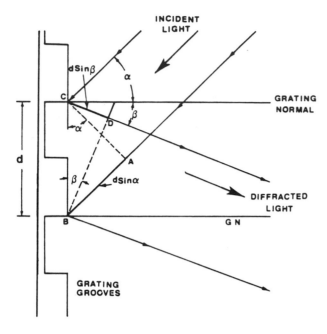

Figure 13.1. Path length difference between rays from two grating facets (Herrera Franco 1985).

groove facets. These relationships are expressed by equation 13.1, known as the grating equation since it governs the behavior of all gratings,

$$m\lambda = d(\sin \alpha \pm \sin \beta_m) \qquad (13.1)$$

where α is the angle of the incident rays with respect to the normal to the grating

β is the angle of diffraction with respect to the normal;

λ is the wavelength of the light

d is the grating pitch or distance between successive grooves

m is the diffracted order, usually a small integer

The minus sign in the preceding equation signifies that β is on the same side of the normal as α, whereas the plus sign signifies that β is on the opposite side of the normal. By convention, positive angles are measured counterclockwise from the grating normal. As they are limited by the grating surface and the diffraction equation, diffraction angles are in the range $-90° \leq \beta_m \leq +90°$. Also by convention, the positive diffraction orders are numbered increasing in the counterclockwise direction, beginning with the zeroth order. For the zeroth order one has $m = 0$ and $\sin \alpha = \pm \sin \beta$; this is Snell's law of reflection, which defines the path of the direct or undiffracted light. In other words, the zeroth order always emerges at an angle that corresponds to the specular reflection angle and that is equal to the angle of incidence. The numbering of diffraction orders aids in following the diffraction sequence of a beam, particularly when two moire gratings are involved.

Equation 13.1 can be rearranged as

$$\sin \beta_m = m\lambda f + \sin \alpha \qquad (13.2)$$

The $\pm$ sign can be omitted according to the convention just mentioned, and $f = 1/d$ is the grating spatial frequency. This equation is the two-dimensional grating equation.

Interest will be centered in the special case of diffraction referred to as "symmetrical diffraction." This situation occurs when, for every order that emerges from the grating at an angle $+\beta$, there is another diffraction order that emerges at angle $-\beta$. If the diffraction order $m = -1$ is chosen to emerge symmetrically opposed to the zero order, then $\beta_{-1} = -\alpha$ and the diffraction equation 13.2 reduces to

$$\sin \alpha = \frac{\lambda}{2d} \qquad \text{or} \qquad \sin \alpha = \frac{\lambda f}{2} \qquad (13.3)$$

13.3 General grating equation in oblique incidence

13.3.1 Directional relations

In the description of the behavior of the grating that was given, it was assumed that the incident and emergent beams both lie in the same plane. More general relations between the directions of incidence and emergence are needed for moire interferometry, and these will now be developed. Some of the fine explanations presented by Guild (1956) are utilized in this presentation.

Figure 13.2 shows the general geometry for incident and diffracted rays in three dimensions. In this figure, the following definitions are important:

A *principal plane* is defined as a plane normal to the surface of the grating that intersects it in a line perpendicular to the grating lines.

A *secondary plane* is a plane normal to the surface of the grating that intersects it in a line parallel to the grating lines.

An *incidence plane* is a plane formed by the incident ray and the grating normal (e.g., the plane Oac in Fig. 13.2).

A plane that is formed by an emergent ray of any diffraction order and the grating normal is a *plane of emergence* for that order (e.g., the plane Oec in Fig. 13.2).

To specify the direction of a ray of light in relation to the grating, we use two parameters. First, the direction of the incident ray is partially specified by the angle α between the ray and the normal to the grating surface. This angle is usually referred to as the angle of incidence. The second parameter is the angle ψ_i between the plane of incidence and the principal plane. Similarly, the direction of emergence of rays of the mth order can be specified first by the angle of emergence, β_m, between the rays of order m and the normal to the grating. Also needed is the angle ψ_m between the plane of emergence and the principal plane.

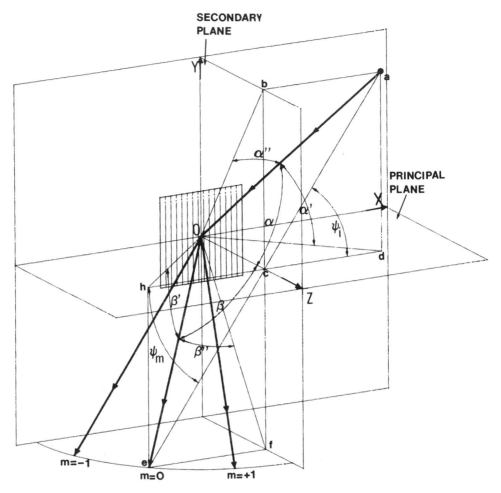

Figure 13.2. Three-dimensional geometry of the incident and diffracted rays of a grating (Herrera Franco 1985).

The direction of a ray can also be specified by its inclination to any two planes, an obvious choice being the principal and secondary planes of the grating. These two inclinations for the incident ray are α', which is the angle between the incident ray and its projection on the principal plane, and α'', which is the angle between the incident ray and its projection on the secondary plane. The corresponding specifications for the emergent ray of given order m are β', which is the angle between the emergent ray and its projection on the principal plane, and β'', which is the angle between the emergent ray and its projection on the secondary plane.

The geometrical relationships among these parameters are very simple and can be obtained from Figure 13.2. First, $\sin \alpha' = ad/ao$, $\sin \alpha = ac/ao$, and

$\sin \psi_i = ad/ac$, which combined yield

$$\sin \alpha \sin \psi_i = \sin \alpha' \tag{13.4}$$

Also, $\sin \alpha'' = ab/ao$, $\sin \alpha = ac/ao$, and $\cos \psi_i = cd/ac = ab/ac$, to yield

$$\sin \alpha \cos \psi_i = \sin \alpha'' \tag{13.5}$$

For the mth emergent order, $\sin \beta' = eh/oe$, $\sin \beta_m = ce/oe$, and $\sin \psi_m = eh/ce$, with the result after combining,

$$\sin \beta_m \sin \psi_m = \sin \beta'_m \tag{13.6}$$

Finally, $\sin \beta''_m = ef/oe$, $\sin \beta_m = ce/oe$, $\cos \psi_m = ch/ce = ef/ce$, together yield

$$\sin \beta_m \cos \psi_m = \sin \beta''_m \tag{13.7}$$

13.3.2 Derivation of the general equation

The following exposition draws heavily from the fine explanations of Stroke (1967) and James and Sternberg (1969). Consider a plane grating as shown in Figure 13.3, where the grating surface lies on the plane $z = 0$ of a rectangular coordinate system. Also, consider a ray incident on the grating at the origin of the coordinate system, with direction cosines L_1, L_2, L_3. Let there be another ray parallel to the first, meeting the grating at point $C(x, y, 0)$. Let them both be diffracted, with direction cosines L'_1, L'_2, and L'_3. As shown in Figure 13.3, lines CA and CB are drawn from C to the first incident and diffracted rays, meeting them perpendicularly. Then CA lies on the incident wavefront; and if the outgoing rays are in the direction of a principal maximum, CB lies on a diffracted wavefront. Therefore, for constructive interference, ACB must be an integral number of wavelengths in length. Let the incident waves at O and at C be specified by

$$L_1 x + L_2 y + L_3 z = 0$$

and

$$L_1 x + L_2 y + L_3 z = CA \tag{13.8}$$

respectively. Let the diffracted rays at O and at C be

$$L_1 x + L_2 y + L_3 z = 0$$

and

$$L_1 x + L_2 y + L_3 z = CB \tag{13.9}$$

Let the incident wave normal be, in Cartesian coordinates with the usual unit vectors,

$$\mathbf{N} = n_x \mathbf{i} + n_y \mathbf{j} + n_z \mathbf{k}$$

or

$$\mathbf{N} = L_1 \mathbf{i} + L_2 \mathbf{j} + L_3 \mathbf{k} \tag{13.10}$$

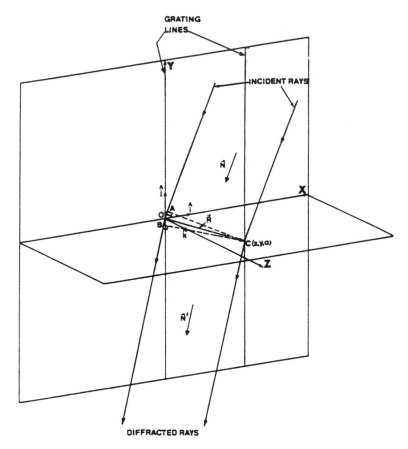

Figure 13.3. Light obliquely incident on a plane grating (Herrera Franco 1985).

and let the diffracted wave normal be

$$\mathbf{N'} = L_1'\mathbf{i} + L_2'\mathbf{j} + L_3'\mathbf{k} \tag{13.11}$$

On the first groove, at point C,

$$\mathbf{R} = d\mathbf{i} + y\mathbf{j} \tag{13.12}$$

Then, for the incident beam, the magnitudes of the vectors are

$$OA = N \qquad \text{and} \qquad R = L_1 d + L_2 y \tag{13.13}$$

and, for the diffracted wave, the magnitudes are

$$OB = N' \qquad \text{and} \qquad R = L_1' d + L_2' y \tag{13.14}$$

For constructive interference to occur,

$$OA + OB = m\lambda \tag{13.15}$$

or

$$(L_1 + L_1')d + (L_2 + L_2')y = m\lambda \tag{13.16}$$

This equation must hold for any value of y, and so the first condition for OB to be an outgoing ray is

$$L_2 = -L_2'$$ (13.17)

and so

$$L_1 + L_1' = \frac{m\lambda}{d}$$ (13.18)

Referring to Figure 13.2, it is easy to see that

$$L_1 = \cos\left(\frac{\pi}{2} - \alpha''\right) = \sin \alpha''$$

$$L_1' = \cos\left(\frac{\pi}{2} + \beta''\right) = -\sin \beta_m''$$

$$L_2 = \cos\left(\frac{\pi}{2} - \alpha'\right) = \sin \alpha'$$ (13.19)

$$L_2' = \cos\left(\frac{\pi}{2} + \beta'\right) = -\sin \beta_m'$$

Thus, equations 13.17 and 13.18 can be expressed as

$$\sin \beta_m'' = \frac{m\lambda}{d} + \sin \alpha''$$ (13.20)

$$\sin \alpha' = \sin \beta_m'$$ (13.21)

Combining equations 13.4 and 13.5 with the preceding results gives

$$\sin \beta_m'' = \frac{m\lambda}{d} + \sin \alpha \cos \psi_i$$ (13.22)

$$\sin \beta_m' = \sin \alpha \sin \psi_i$$ (13.23)

These are the three-dimensional diffraction grating equations; many of the subsequent geometrical derivations are based on them, so they should be well understood.

Note from Figure 13.2 that for $\psi_i = 0$, $\sin \alpha'' = ad/ao = cd/ad$ and $\sin \beta_m'' = ef/oe = ch/oh$. Use of these results causes equation 13.22 to reduce to

$$\sin \beta_m'' = m\lambda f + \sin \alpha$$

which is the two-dimensional case solved the hard way.

A very important idea is demonstrated by equation 13.23. The angle β_m', which is the angle between the emergent ray and its projection on the principal plane, is a function only of the angle of incidence α and the angle ψ_i between the plane of incidence and the principal plane. It is not a function of the diffraction order. It is constant for any given arrangement of apparatus. This result means that the angle between the projection of any diffracted ray and the principal plane is the same. For $m = 0$, equations 13.22 and 13.23 reduce to Snell's law, that is, $\alpha = \beta_{m=0}$ and $\psi_i = \psi_{m=0}$. Also, the sine of the angle between any diffraction order and its projection on the secondary plane will always be given by the summation of $\sin \beta''_{m=0}$ plus a quantity that depends on a multiple of the grating frequency.

It is the inclinations of the incident and emergent beams to the secondary plane that are relevant to the operation of the grating, not their inclinations to the normal. The inclination of the incident beam to the principal plane is merely carried through to the emergent beams without modification. This fact seems surprising at first look. It is also important to understanding the three-dimensional diffraction of moire interferometry.

For the analysis of the formation of moire fringes, a sign convention for the diffracted rays and the angles of incidence and emergence must be established. Note in Figure 13.2 that a diffracted order will be positive when it propagates with a component in the $+z$ direction and when it lies in a counterclockwise direction with respect to the zero order when viewed from the $+y$ direction. For the incident and emergent rays, the following convention applies: the angle between the ray and the z-axis is positive when its projection on the xz-plane is rotated counterclockwise from the z-axis, and it travels in the $+z$ direction when viewed from the $+y$ direction. Its projection on the yz-plane follows the same convention when viewed from the $+x$ direction. In the case of a two-direction grating, the same convention applies if x and y are interchanged.

13.4 Review of two-beam interference in oblique incidence

Section 2.9 dealt with this problem in detail, but the subject deserves a very short review within the context of moire phenomena.

When two coherent beams of collimated light, as from a laser, intersect at an angle 2ϕ, a volume of interference fringe planes is created. Figure 13.4 shows in cross section how the two incident wavefronts combine to form a stationary system of parallel interference bands in space. The bands are not merely lines, as shown; they are planes lying perpendicular to the bisector of wavefronts, and in the ideal case they exhibit a sinusoidal intensity distribution. Such an intensity distribution can be recorded if the two beams fall simultaneously on a screen. This phenomenon can be used to create very fine diffraction gratings. It also explains the formation of moire fringes as two diffracted beams meet in space.

The distance D between the adjacent walls of interference, that is, the fringe

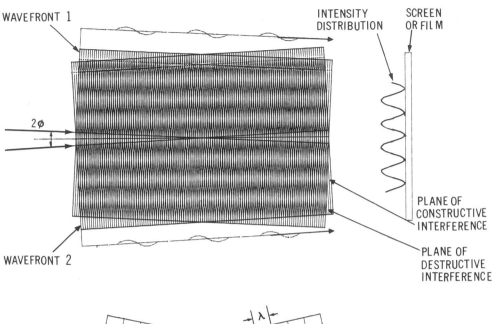

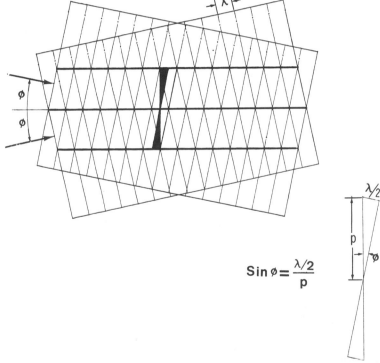

$$\text{Sin}\,\phi = \frac{\lambda/2}{p}$$

Figure 13.4. Interference fringes produced by the combination of two plane wavefronts (Herrera Franco 1985).

spacing, is calculated to be

$$D = \frac{\lambda}{2} \sin \phi \qquad (13.24)$$

where λ is the wavelength of the laser light, and ϕ is half the angle between the propagation axes of the two beams of light. The equation can be rewritten as

$$\sin \phi = \frac{f\lambda}{2} \qquad (13.25)$$

where f is the number of fringes per unit of length, commonly called just the fringe density or spatial frequency. It is the reciprocal of the fringe spacing p.

13.5 Geometry of moire interferometer

The basic concept of moire interferometry is that diffraction of two plane waves from a specimen grating will cause the waves to be redirected and their wavefronts to be distorted. When these two wavefronts interfere, the resulting fringe pattern will be a moire fringe pattern that is identical to the pattern that would be created by geometric moire using the same gratings.

The basic concept can be realized in the laboratory in several ways. The emphasis here is on the design of an instrument that can be used to obtain whole-field measurement of three strain components, thereby giving all information needed to completely characterize the strain field. Czarnek and Post (1984) suggested that specimen gratings with rulings oriented at ± 45 degrees (such a two-way grating is often called a grid) to the y-axis can be used to determine the u_x and u_y displacement fields, that is, the displacement fields in the x and y directions in the plane of the specimen surface. This approach can be utilized and extended to give the desired three measurements, those being the three displacement maps of u_x, u_y, and u_{45}. These displacements can then be differentiated to obtain the strains for the same axes. The strain rosette equations can then be called upon to obtain principal strains and principal angle, as was explained in a previous chapter.

Figure 13.5 illustrates a "six-beam arrangement" of a moire interferometer with $\pm 45°$ specimen gratings, as developed by Herrera Franco (1985), to obtain the complete full-field strain characterization just mentioned. Actually, only three beams are used, but each is used twice to create three two-beam devices having different measurement axes. By combining diffracted orders from beams AO and BO, deformation fields in the horizontal direction x can be measured; the combination of beams BO and CO provides the information for the vertical y direction, and combining beams AO and CO provides information in a third direction located at $45°$ with respect to the x-axis.

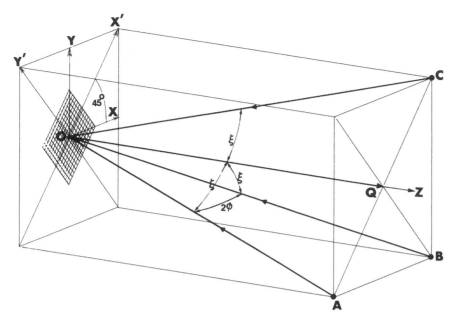

Figure 13.5. Six-beam moire three-axis interferometer (Herrera Franco 1985).

When the specimen grating is illuminated with beams AO and BO, an interference pattern between the two beams is formed. The frequency of this interference system is F, and it is calculated according to equation 13.25. This interference pattern will be referred to as "the reference grating," and its frequency is not the same as the frequency f of the specimen grating because of the 45° tilt of one grating with respect to the other. Now, the angle between these two beams of light is

$$\text{AOB} = 2\phi \qquad (13.26)$$

The first diffraction order (-1) of beam AO, incident on the specimen grating x', is desired to be perpendicular to the specimen grating surface, since for an oblique diffracted order the view is foreshortened and distorted. Also, a fixed distance from the observer or camera would make focusing easier. Therefore, the required angle of incidence is ξ, the angle AOQ. By calling upon the results of the previous section, we see that the angle ξ is related to the angle ϕ by

$$\sin \xi = \sqrt{2} \sin \phi \qquad (13.27)$$

Similarly, the angle of incidence for beam BO should be $-\xi$ for the first diffraction order $(+1)$ to be perpendicular to the specimen grating surface.

The frequency of the specimen grating f can be determined through the grating equation by letting $m = -1$, $\alpha = -\xi$, and $\beta_{-1} = 0$; this gives $f = \sin \xi / \lambda$. Recalling that $F = (2/\lambda) \sin \phi$, and using equation 13.27, one obtains for the relationship between the frequency f of the specimen grating and the frequency F of the reference grating created by interference of AO and BO,

$$f = \frac{F}{\sqrt{2}} \tag{13.28}$$

This result is consistent with physical expectations if one realizes that the nodes of the rotated specimen grating form the grating (actually an array of dots) that interacts with the reference grating having vertical lines. The horizontal separation of the nodes must equal F.

The first diffraction orders (-1) and $(+1)$ of beams AO and BO will be referred to as AO′ and BO′, respectively, and they will be perpendicular to the specimen grating surface if and only if the frequencies of both gratings x' and y' are related to specimen grating frequency f as in equation 13.28.

13.6 Deformation of the specimen grating

Because a grating that lies at $45°$ to the xy specimen coordinate system is being used, the relationships between the strain components and the resulting deformations of the two-way specimen grating must be studied. Consider one square formed by the intersection of two pairs of grating lines, originally parallel to the x'- and y'-axes. Geometrical representations of the changes in length and the rotations of the sides of the square are shown in Figure 13.6. The original positions of the lines are indicated by dashed lines.

The specimen on which the grating is mounted is being deformed by the action of externally applied loads. Assume that a state of plane strain parallel to the xy-plane exists at the point under consideration. In tensor form, the strain components can be written as

$$\bar{E} = \begin{bmatrix} \epsilon_x & -\epsilon_{xy} \\ \epsilon_{xy} & \epsilon_y \end{bmatrix} \tag{13.29}$$

where ϵ_x and ϵ_y represent normal strains along the x- and y-axes, respectively, and $\epsilon_{xy} = \epsilon_{yx} = \gamma_{xy}/2$ are the small rotations of the sides that were initially parallel to the x- and y-axes, respectively. The strain along the sides of the square whose sides are parallel to the x'- and y'-axes can be calculated simply by applying the following transformation to the strain in the specimen since it is a second-order tensor,

$$E_{ij} = A_{ip} A_{jq} E_{pq} \qquad \text{or} \qquad E' = A \ E \ A^{\mathrm{T}} \tag{13.30}$$

where A is the matrix of direction cosines and A^{T} is its transpose. Then

$$\bar{E}' = A \ E \ A^{\mathrm{T}} = \begin{bmatrix} m & n \\ -n & m \end{bmatrix} \begin{bmatrix} \epsilon_x & \frac{1}{2}\gamma_{xy} \\ \frac{1}{2}\gamma_{xy} & \epsilon_y \end{bmatrix} \begin{bmatrix} m & -n \\ n & m \end{bmatrix} \tag{13.31}$$

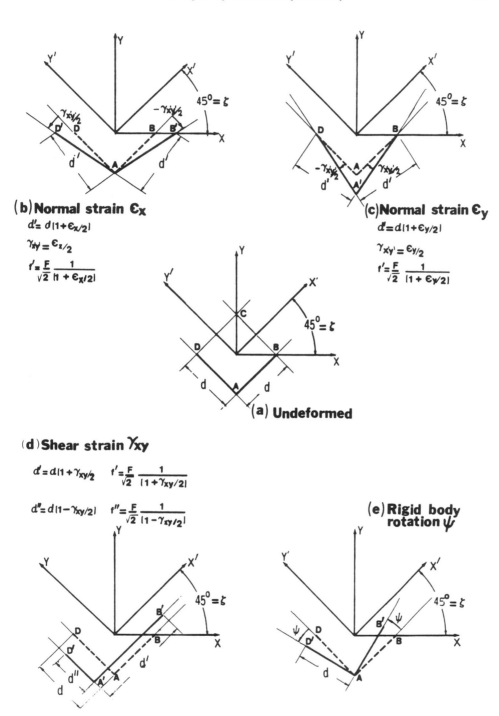

(b) Normal strain ϵ_x

$d' = d(1 + \epsilon_{x/2})$

$\gamma_{xy'} = \epsilon_{x/2}$

$f' = \dfrac{F}{\sqrt{2}} \dfrac{1}{(1 + \epsilon_{x/2})}$

(c) Normal strain ϵ_y

$d' = d(1 + \epsilon_{y/2})$

$\gamma_{xy'} = \epsilon_{y/2}$

$f' = \dfrac{F}{\sqrt{2}} \dfrac{1}{(1 + \epsilon_{y/2})}$

(a) Undeformed

(d) Shear strain γ_{xy}

$d' = d(1 + \gamma_{xy/2}) \qquad f' = \dfrac{F}{\sqrt{2}} \dfrac{1}{(1 + \gamma_{xy/2})}$

$d'' = d(1 - \gamma_{xy/2}) \qquad f'' = \dfrac{F}{\sqrt{2}} \dfrac{1}{(1 - \gamma_{xy/2})}$

(e) Rigid body rotation ψ

Figure 13.6. Geometrical representations of deformation of specimen grating by various strain and rotation components (Herrera Franco 1985).

where $m = \cos \zeta$ and $n = \sin \zeta$. The result of the multiplication is

$$\bar{E}' = \begin{bmatrix} \epsilon_x(m^2) + \epsilon_y(n^2) + \gamma_{xy}(mn) & (\epsilon_y - \epsilon_x)mn + \frac{1}{2}\gamma_{xy}(m^2 - n^2) \\ -(\epsilon_y - \epsilon_x)mn + \frac{1}{2}\gamma_{xy}(m^2 - n^2) & \epsilon_x(n^2) + \epsilon_y(m^2) - \gamma_{xy}(mn) \end{bmatrix}$$

$$(13.32)$$

For small strains, the principle of superposition can be used to write:

$$\bar{E}' = \begin{bmatrix} \epsilon_x(m^2) & -\epsilon_x(mn) \\ -\epsilon_x(mn) & \epsilon_x(n^2) \end{bmatrix} + \begin{bmatrix} \epsilon_y(n^2) & \epsilon_y(mn) \\ \epsilon_y(mn) & \epsilon_y(m^2) \end{bmatrix}$$

$$+ \begin{bmatrix} \gamma_{xy}(mn) & \frac{1}{2}\gamma_{xy}(m^2 - n^2) \\ \frac{1}{2}\gamma_{xy}(m^2 - n^2) & -\gamma_{xy}(mn) \end{bmatrix} \qquad (13.33)$$

These matrices represent the effect of ϵ_x, ϵ_y, and γ_{xy}, respectively, on the deformations of the grating lines. For $\zeta = 45°$, this equation reduces to

$$\bar{E}' = \begin{bmatrix} \frac{1}{2}\epsilon_x & -\frac{1}{2}\epsilon_x \\ -\frac{1}{2}\epsilon_x & \frac{1}{2}\epsilon_x \end{bmatrix} + \begin{bmatrix} \frac{1}{2}\epsilon_y & \frac{1}{2}\epsilon_y \\ \frac{1}{2}\epsilon_y & \frac{1}{2}\epsilon_y \end{bmatrix} + \begin{bmatrix} \frac{1}{2}\gamma_{xy} & 0 \\ 0 & -\frac{1}{2}\gamma_{xy} \end{bmatrix} \qquad (13.34)$$

These results show that ϵ_x and ϵ_y will produce both extension and rotation of the grating lines, and γ_{xy} will produce extensions and contractions but not rotations. The results shown in equation 13.34 apply to two sides of the square originally oriented along the positive direction of the x'- and y'-axes. From the first matrix, corresponding to the effect of ϵ_x, the two sides of the square are stretched by $\epsilon_x/2$ and the final angle between them is $(\pi/2 + \epsilon_x)$. Similarly, the second matrix, which corresponds to the effect of ϵ_y, indicates that the two sides are stretched by $\epsilon_y/2$ and that the final angle between them is $(\pi/2 - \epsilon_y)$. From the third matrix, it can be seen that γ_{xy} will produce a tensile strain $\gamma_{xy}/2$ in the line along the x'-axis and a compressive strain $-\gamma_{xy}/2$ in the line along the y'-axis, but they will remain perpendicular to each other.

If the frequency of the grating before deformation is $f = 1/D$, then the new frequency is given by $f' = 1/D'$.

13.7 Moire fringes of x-displacement

The problem now is to predict the orientations of the beams that are diffracted by the deformed and rotated specimen grating. This must be done carefully for all components of strain and for all possible rotations because they affect each of the three pairs of beams in the interferometer. It is tedious, but it is necessary for full and accurate understanding of the final moire patterns. The relationships will be constructed in detail for the pair of beams AO and BO, which are sensitive to the x component of displacement induced by the various strains and rotations. Then, permutation will be used to advantage in attending to the other cases. Also, considerable detail will be left out in the subsequent developments. It is here that full understanding of the workings of moire interferometry is gained.

13.7.1 Fringes produced by normal strain ϵ_x

As can be seen in Figure 13.7, the specimen grating is interrogated with beams AO (Fig. 13.7a) and BO (Fig. 13.7b). Figure 13.7c shows the combined result. Remember that the beams will be diffracted back along the z-axis if the specimen grating is not deformed; the grating and angles were designed to make this true.

Now, think of what happens when the grating pitch is changed or it is rotated. Part of the incident beam AO will be diffracted by the x' grating, and its -1 diffraction order will be referred to as OA'. Similarly, OB' will be the $+1$ diffraction order of the incident beam BO, diffracted by the y' grating.

A normal strain ϵ_x produces a change in frequency from $1/d$ to $1/d'$ (see Fig. 13.6b), and it also produces a rotation of the grating lines through an angle $\gamma_{x'y'}/2$ (see Fig. 13.6b). As a result the incident beams will not lie in the principal planes of their corresponding gratings. To determine the orientations of the emergent beams, it is necessary to use the three-dimensional grating equations. For beam AO, using equations 13.22 and 13.23, it can be seen that

$$\sin \beta''_m(x'z) = m\lambda f + \sin \alpha \cos \psi_i$$
$$\sin \beta'_m(y'z) = \sin \alpha \sin \psi_i \tag{13.35}$$

Similarly for beam BO,

$$\sin \beta''_m(y'z) = m\lambda f + \sin \alpha \cos \psi_i$$
$$\sin \beta'_m(x'z) = \sin \alpha \sin \psi_i \tag{13.36}$$

Note that the direction of each diffracted ray is given by the direction of its projection in the $x'z$- and $y'z$-planes, as indicated in parentheses. For the case of infinitesimal normal strains, the ψ_i will be small, and these equations reduce to the following forms. For beam AO,

$$\sin \beta''_m(x'z) = m\lambda f + \sin \alpha$$
$$\sin \beta'_m(y'z) = \psi_i \sin \alpha \tag{13.37}$$

and for beam BO,

$$\sin \beta''_m(y'z) = m\lambda f + \sin \alpha$$
$$\sin \beta'_m(x'z) = \psi_i \sin \alpha \tag{13.38}$$

(a)

(b)

(c)

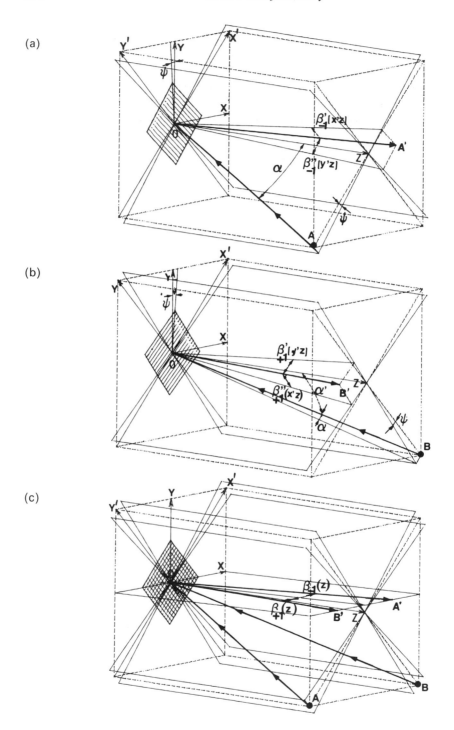

Figure 13.7. Geometry of the incident and diffracted rays from a grating deformed by ϵ_x when analyzed by (a) beam AO, (b) beam BO, and (c) both beams (Herrera Franco 1985).

It is worth taking time to understand fully the development of equations 13.37 and 13.38. They are used often as the starting point in subsequent manipulations to avoid the tediousness of going back each time to the more general three-dimensional grating equations (eqs. 13.20–13.23).

Figure 13.6 and the related equations show that it is appropriate to use the same ψ_i for beams AO and BO. Then, for beam AO, let $m = -1$, $\sin \alpha = \sin \xi = \sqrt{2} \sin \phi = \sqrt{2}(\lambda/2)F = \lambda F/\sqrt{2}$; also let $\psi_i = \gamma_{x'y'}/2 = \epsilon_x/2$. Then let $f' = (F/\sqrt{2})[1/(1 + \epsilon_x/2)]$, which can be expressed as a continued fraction $f' = (F/\sqrt{2})(1 - \epsilon_x/2 + \epsilon_x^2/4 - \cdots)$. Equations 13.37 yield (neglecting the high-order terms):

$$\sin \beta''_{-1}(x'z) = \frac{\lambda F \epsilon_x}{2\sqrt{2}}$$

$$\sin \beta'_{-1}(y'z) = \frac{\lambda F \epsilon_x}{2\sqrt{2}} \tag{13.39}$$

Similarly, for incident beam BO, letting $m = +1$, $\sin \alpha = \sin \xi = -\lambda F/\sqrt{2}$, and also letting $\psi_i = -\gamma_{x'y'}/2 = -\epsilon_x/2$, with f' as given before, equation 13.38 yields

$$\sin \beta''_{+1}(y'z) = \frac{-\lambda F \epsilon_x}{2\sqrt{2}}$$

$$\sin \beta'_{+1}(x'z) = \frac{-\lambda F \epsilon_x}{2\sqrt{2}} \tag{13.40}$$

It is not difficult to show that the diffracted rays AO′ and BO′ lie in the xz-plane (see Fig. 13.7) and that the sine of the angle between each of them and the z-axis is

$$\sin \beta_{-1}(z) = -\sin \beta_{+1}(z) = \frac{\lambda F \epsilon_x}{2} \tag{13.41}$$

At this point, we can see exactly how the fringe patterns can be formed by two-beam interference. Figure 13.8 shows the scheme. In an actual optical system, where a field lens decollimates beams AO′ and BO′, they converge to two bright points in the focal plane of the decollimating lens. Because of their angular separation, these two bright spots will focus a small distance apart. If they are close enough to overlap, then an interference pattern is produced. A more useful procedure is to use another lens and screen (that is, a camera) to construct images of the specimen grating x' and y' fields with the light contained in the two diffracted wavefronts OA′ and OB′. Essentially, the camera forms two images that lie atop one another. The degree of

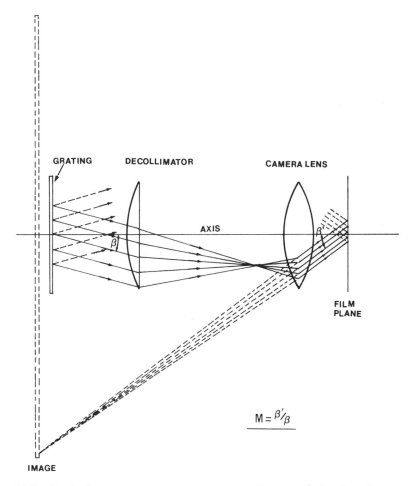

Figure 13.8. Optical arrangement to create an image of the interference pattern produced by two diffracted beams (M = magnification) (Herrera Franco 1985).

interference in the image depends mainly on the relative displacement of the two focal spots in the back focal plane of the decollimator, which, it must be recalled, depends on the relative inclinations of the diffracted beams. The fringe pattern is not affected by magnification of the lens system. The ratio β'/β, which governs the fringe spacing, is exactly equal to the magnification factor M. That is, any magnification of the specimen is carried into the fringe pattern, so the ratio of fringe spacing to specimen dimension remains constant.

The image in the camera, then, displays a pattern of interference fringes that are indicative of the local spatial frequency and orientation differences between the two gratings. The resulting interference system of *vertical* fringes, parallel to the y-axis (because we are allowing only u_x at this point), has a spatial frequency $P_{\epsilon x}$, which can be calculated using equation 13.25 to be

$$P_{\epsilon x} = \frac{2}{\lambda} \sin \beta(z) = \epsilon_x F \tag{13.42}$$

Equation 13.42 is important because it relates the strain ϵ_x in the specimen to the known frequency F of the interference system produced by beams AO and BO, as well as the observed moire fringe density $P_{\epsilon x}$. Therefore, the quantity ϵ_x can be determined in terms of $P_{\epsilon x}$ and F.

The relative displacement u_x (perpendicular to the direction of the lines of the reference grating) between a pair of points a distance p apart ($p = 1/P_{\epsilon x}$) on two adjacent fringes of orders, say, $N = 0$ and $N = 1$ will then be

$$u_x = \epsilon_x p = \frac{1}{F} \tag{13.43}$$

In general, the in-plane displacement in the direction normal to the grating lines is, in terms of the fringe order N_x,

$$u_x = \frac{(N_x)_{\epsilon x}}{F} \tag{13.44}$$

13.7.2 Fringes produced by normal strain ϵ_y

As shown in Figure 13.6c, a normal strain ϵ_y would also change the frequency of the specimen grating, as well as cause the grating lines to rotate through an angle $\gamma_{x'y'}/2$.

Again, from equations 13.37 and 13.38, it is possible to calculate the directions of the diffracted rays with respect to the $x'z$- and $y'z$-planes. With the directions known, the way in which the fringe patterns are affected by this displacement component can be understood. Figure 13.9 will aid in understanding this development.

For the incident beam AO (Fig. 13.9a), and letting $m = -1$, $\sin \alpha = \sin \xi = \lambda F/\sqrt{2}$, $\psi_i = \gamma_{x'y'}/2 = -\epsilon_y/2$, and $f' = (F/2)(1 - \epsilon_y/2)$, equation 13.37 yields for small angles and rotations

$$\sin \beta''_{-1}(x'z) = \frac{\lambda F \epsilon_y}{2\sqrt{2}}$$
$$\sin \beta'_{-1}(y'z) = \frac{-\lambda F \epsilon_y}{2\sqrt{2}} \tag{13.45}$$

(a)

(b)

(c)

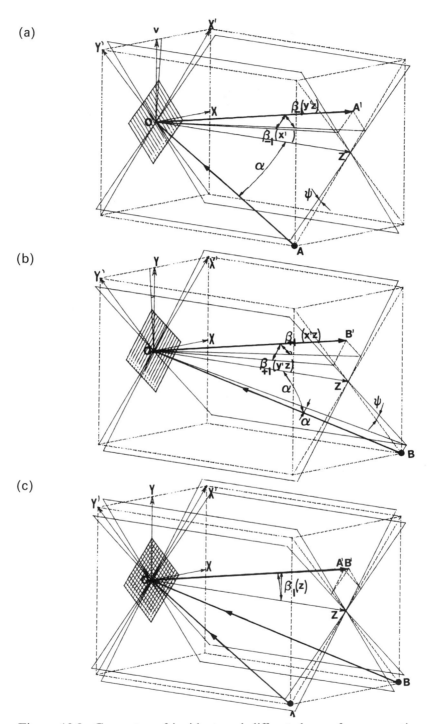

Figure 13.9. Geometry of incident and diffracted rays from a grating deformed by ϵ_y when analyzed with (a) beam AO, (b) beam BO, and (c) both beams (Herrera Franco 1985).

In a similar way for incident beam OB (Fig. 13.9b), letting $m = +1$, $\sin \alpha = \sin \xi = -\lambda F/\sqrt{2}$, $\psi_i = \gamma_{x'y'}/2 = \epsilon_y/2$, and $f' = (F/\sqrt{2})(1 - \epsilon_y/2)$, equation 13.38 yields for small angles and deformations

$$\sin \beta''_{+1}(y'z) = \frac{-\lambda F \epsilon_y}{2\sqrt{2}}$$

$$\sin \beta'_{+1}(x'z) = \frac{\lambda F \epsilon_y}{2\sqrt{2}}$$

(13.46)

In this case, the two diffracted rays lie in a plane parallel to the yz-plane (Fig. 13.9), and the sine of the angle between each of them and the z-axis is, for small ψ_i,

$$\sin \beta_{-1}(z) = \sin \beta_{+1} = \frac{\lambda F \epsilon_y}{2}$$

(13.47)

Because the diffracted rays AO' and BO' in this case lie parallel to one another, equation 13.25 yields

$$P_{\epsilon y} = 0$$

(13.48)

This finding means that when the specimen grating is interrogated by both incident beams AO and BO (Fig. 13.9c), the normal strain ϵ_y will not contribute to the interference system. This decoupling of the normal strains is expected and useful.

13.7.3 *Fringes produced by shear strain* γ_{xy}

As shown in Figure 3.6d, a shear strain γ_{xy} would induce normal strains $\epsilon_{x'}$ and $\epsilon_{y'}$ in the x' and y' directions, respectively. Because of $\epsilon_{x'}$ the pitch of grating x' would increase, consequently decreasing its frequency. Similarly, because of $\epsilon_{y'}$ the pitch of grating y' would decrease and its frequency would increase. Figure 13.10 aids in predicting the directions of the diffracted orders for this case.

Because the shear strain does not produce any rotation of the $x'y'$ grating, the two-dimensional grating equation can be used to determine the orientation of each diffracted ray. For incident ray AO, with $m = -1$, $\sin \alpha = \sin \xi = \lambda F/\sqrt{2}$, $\psi_i = 0$, $f = (F/\sqrt{2})(1 - \gamma_{xy}/2)$; and for incident beam OB, with $m = +1$, $\sin \alpha = \sin \xi = -\lambda F/\sqrt{2}$, $\psi_i = 0$, and $f'' = (F/\sqrt{2})(1 + \gamma_{xy}/2)$; we obtain with equation 13.2,

$$\sin \beta''_{-1}(z) = \sin \beta''_{+1}(z) = \frac{\lambda F \gamma_{xy}}{2\sqrt{2}}$$

(13.49)

(a)

(b)

(c)

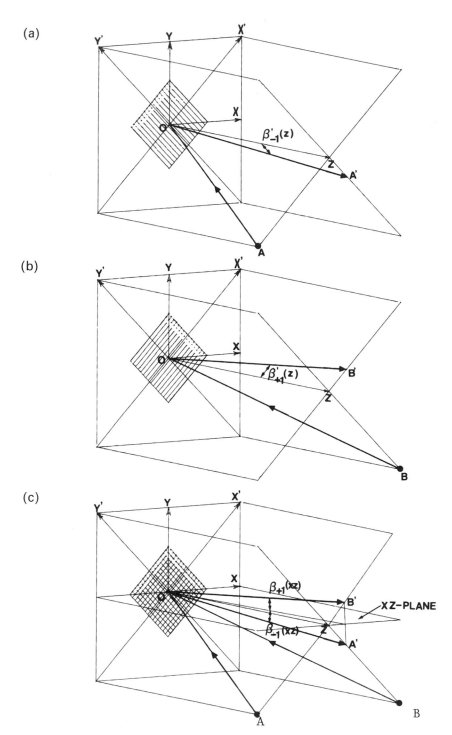

Figure 13.10. Geometry of incident and diffracted rays from a grating deformed by γ_{xy} when analyzed with (a) beam AO, (b) beam BO, and (c) both beams (Herrera Franco 1985).

From simple geometry (see Fig. 13.10), the sine of the angle between each of the diffracted orders and the xz-plane is found to be

$$\sin \beta = \frac{\lambda F \gamma_{xy}}{4} \qquad (13.50)$$

The resulting interference fringe pattern is composed of *horizontal* fringes parallel to the xz-plane whose frequency can be calculated using equation 13.25 to be

$$P_{\gamma xy} = \frac{\gamma_{xy} F}{2} \qquad (13.51)$$

13.7.4 Fringes produced by in-plane rotation ψ

A rotation of the x' and y' gratings through an angle ψ, considered positive in the counterclockwise direction, causes the two diffracted rays OA′ and OB′ to diverge from their respective planes $x'z$ and $y'z$, but their pitch and frequency will not be disturbed. Figure 13.11 will aid in predicting the orientations of the diffracted beams.

Through use of equations 13.37 and 13.38 (the second of each pair), the orientations of the two diffracted rays are found to be

$$\sin \beta'_{-1}(y'z) = \frac{\lambda F \psi}{\sqrt{2}}$$

$$\sin \beta'_{+1}(x'z) = \frac{-\lambda F \psi}{\sqrt{2}} \qquad (13.52)$$

and the sine of the angle between each of the two diffracted orders and the xz-plane is (see Fig. 13.11)

$$\sin \beta_\psi = \frac{\lambda F \psi}{2} \qquad (13.53)$$

Again, equation 13.25 gives the frequency of the resulting interference system to be

$$P_\psi = \psi F \qquad (13.54)$$

13.7.5 Fringes produced by out-of-plane rotations

Any small out-of-plane rotation about an in-plane axis will introduce equal deviations of beams AO′ and BO′. It will not contribute to the interference pattern, and so it can be ignored.

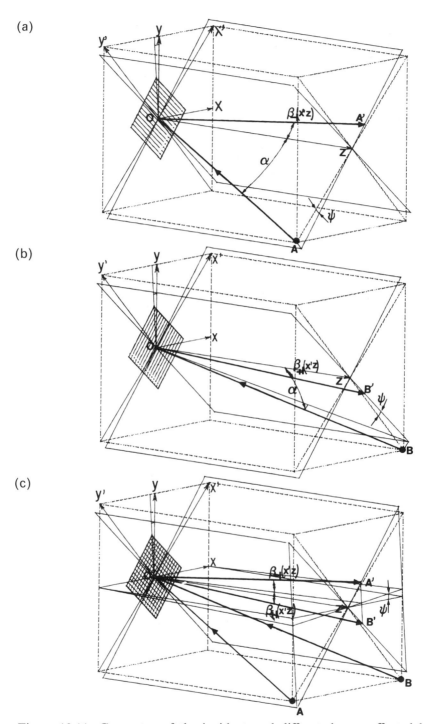

Figure 13.11. Geometry of the incident and diffracted rays affected by rigid body in-plane rotation of the gratings when analyzed by (a) beam AO, (b) beam BO, and (c) both beams (Herrera Franco 1985).

The preceding analysis has shown that by interrogating the specimen grating with beams AO and BO, moire patterns of displacement, u_x, can be obtained; and these fringe patterns will contain contributions from the normal strain ϵ_x, the shear strain γ_{xy}, and the in-plane rigid body rotation ψ.

13.8 Moire fringes of *y*-displacement

To obtain the moire fringes of displacements u_y, the specimen grating is interrogated with beams BO and CO. Owing to symmetry, the analysis of the deformations of the grating lines follows the same pattern as was pursued for ϵ_x, and there is no need to repeat it.

13.9 Moire fringes of 45°-displacement

The displacements u_{45} are obtained by interrogating the specimen grating with coherent beams AO and CO. The basic geometry of the moire interferometer needed for this operation is established in Figure 13.12.

The angle AOC is equal to 2ξ, so the spatial frequency of the reference grating produced by the interference of beams AO and CO is given by equation 13.25 as

$$F_{45} = (2/\lambda) \sin \xi \qquad (13.55)$$

Through use of equation 13.26, the frequency can be rewritten as

$$F_{45} = (2/\lambda)(\sqrt{2} \sin \alpha) = 2f \qquad (13.56)$$

The frequency of the reference grating is found to be exactly twice the frequency f of the specimen grating.

Now, an analysis similar to that followed in section 13.7 is undertaken to determine the effects of the various strain and rotation components on the 45° displacement and, thereby, on the moire fringe pattern.

13.9.1 Fringes produced by normal strain $\epsilon_{x'}$

Figure 13.13 shows the geometry of the diffracted rays for this case. The action of normal strain $\epsilon_{x'}$ is to change the pitch of the x' grating lines by a factor of $(1 + \epsilon_{x'})$, so its frequency will decrease to $f' = (F_{45}/2)(1 - \epsilon_{x'})$. Because both incident beams AO and CO lie in the $x'z$-plane, the angle of emergence of the diffracted rays can be determined using the two-dimensional grating equation 13.2,

$$\sin \beta_m(x'z) = m\lambda f + \sin \alpha \qquad (13.57)$$

To see how beam AO is diffracted, let $m = -1$, $\sin \alpha = \sin \xi = \lambda F_{45}/2$, and

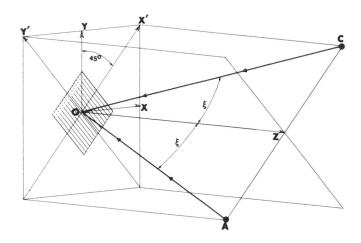

Figure 13.12. Angles of incidence of the moire interferometer when analyzing the 45° displacements with beams AO and CO (Herrera Franco 1985).

let f' be as given before. The grating equation yields

$$\sin \beta''_{-1}(x'z) = \frac{\lambda F_{45}\epsilon_{x'}}{2} \tag{13.58}$$

Similarly, for beam CO, let $m = +1$, $\sin \alpha = \sin \xi = \lambda F_{45}/2$, and f' as before, to get

$$\sin \beta''_{+1}(x'z) = \frac{-\lambda F_{45}\epsilon_{x'}}{2} \tag{13.59}$$

Both diffracted rays AO' and CO' lie in the $x'z$-plane (Fig. 13.13), and they are located at the same distance on opposite sides of the z-axis. The resulting interference pattern has the frequency

$$P_{\epsilon x'} = F_{45}\epsilon_{x'} \tag{13.60}$$

and the fringes are perpendicular to the $x'z$-plane for this strain component. Again, it is seen that the frequency of the interference pattern is a function of the strain $\epsilon_{x'}$ and the frequency F_{45} of the reference grating. The displacement along the x' direction is given by

$$u_{45} = \frac{N_{x'}}{F_{45}} \tag{13.61}$$

13.9.2 Fringes produced by in-plane rotation ψ

When the specimen grating is unstrained but undergoes a rigid-body rotation in its own plane by a small counterclockwise angle ψ, the three-dimensional

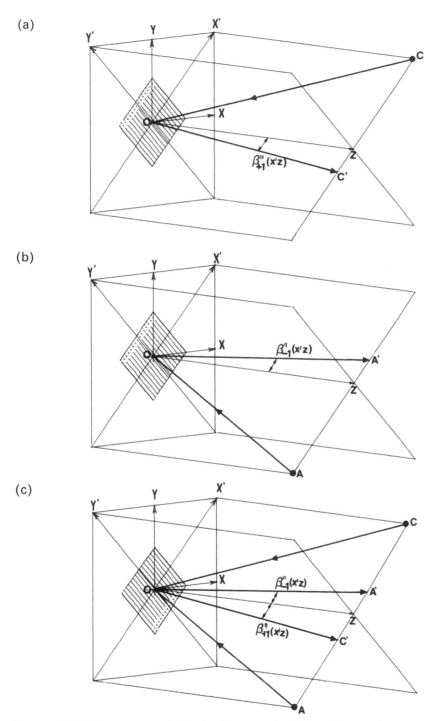

Figure 13.13. Geometry of the incident and diffracted rays from an x' grating deformed by $\epsilon_{x'}$ when analyzed with (a) beam CO, (b) beam AO, and (c) both beams (Herrera Franco 1985).

grating relations in the form of equations 13.37 and 13.38 can again be used to define the directions of the diffracted rays. Refer to Figure 13.14 for the schematic.

For AO,

$$\sin \beta''_{-1}(x'z) = 0$$

$$\sin \beta'_{-1}(y'z) = \frac{\lambda F_{45} \psi}{2}$$

(13.62)

and for CO,

$$\sin \beta''_{+1}(x'z) = 0$$

$$\sin \beta'_{+1}(y'z) = \frac{-\lambda F_{45} \psi}{2}$$

(13.63)

Because both rays AO' and CO' lie in the $y'z$-plane, at the same distance on opposite sides of the z-axis, the fringes of the resulting interference pattern are perpendicular to the $y'z$-plane, and their frequency is

$$P_\psi = \psi F_{45}$$

(13.64)

13.9.3 Fringes produced by shear strain $\gamma_{x'y'}$

Figure 13.15 shows the geometry for this situation. A positive shear strain $\gamma_{x'y'}$ will cause the x' grating lines to rotate clockwise through an angle $-\gamma_{x'y'}/2$. The three-dimensional grating equation is used as for in-plane rotation to get, for AO,

$$\sin \beta'_{-1}(y'z) = \frac{-\lambda F_{45} \gamma_{x'y'}}{4}$$

and for CO,

(13.65)

$$\sin \beta'_{+1}(y'z) = \frac{\lambda F_{45} \gamma_{x'y'}}{4}$$

Again, both diffracted rays are located at the same distance on opposite sides of the z-axis (see Fig. 13.15), and the frequency of the resulting interference pattern is

$$P_{\gamma x'y'} = \frac{F_{45} \gamma_{x'y'}}{2}$$

(13.66)

13.9.4 Fringes produced by normal strain $\epsilon_{y'}$

When the specimen grating is deformed in the y' direction by the action of a normal strain $\epsilon_{y'}$, the lines of the x' grating are stretched but the spacing does not change. Both diffracted orders leave the grating surface along its

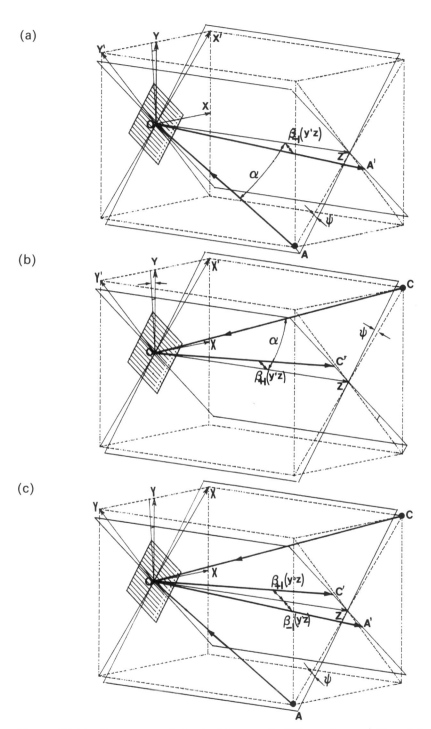

Figure 13.14. Geometry of the incident and diffracted rays affected by rigid-body rotation of the grating when analyzed by (a) beam AO, (b) beam CO, and (c) both beams (Herrera Franco 1985).

(a)

(b)

(c)

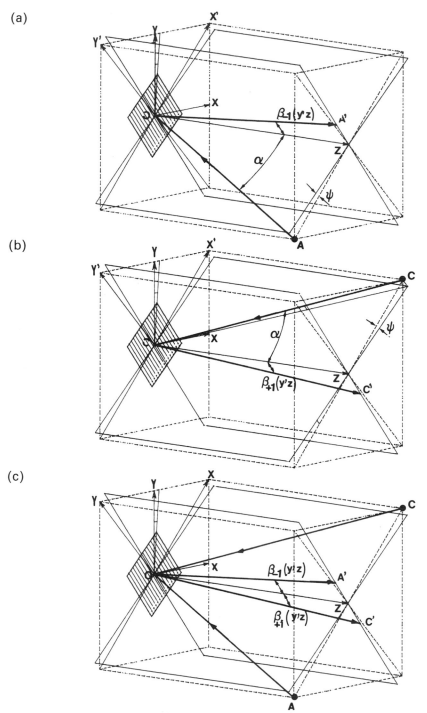

Figure 13.15. Geometry of the incident and diffracted rays for grating deformed by shear strain $\gamma_{x'y'}$ when analyzed by (a) beam AO, (b) beam CO, and (c) both beams (Herrera Franco 1985).

normal and, because they are parallel, the frequency of the interference pattern is equal to zero.

13.9.5 Fringes produced by out-of-plane rotation

It has been shown (Basehore and Post 1981) that a rotation about the x'-axis causes identical angular deviations in the diffracted rays AO' and CO', but the angle between them remains unchanged. Also, Walker and McKelvie (1978) showed that a rotation about an axis perpendicular to the plane under examination will produce some fringes, but for the case of small rotations, they are negligible.

13.10 Complete analysis of two-dimensional strain field

The theory presented in the previous sections demonstrates that the displacement component in a given direction can be measured by interrogation of the specimen grating with different pairs of beams. That is, in order to measure u_x, the specimen grating is interrogated by beams AO and BO. Similarly, u_y is obtained by combining beams BO and CO, and u_{45} is observed by combining beams AO and CO.

The state of strain throughout a general two-dimensional strain field can now be determined. Recognize that there are three unknown strain components ϵ_x, ϵ_y, and γ_{xy} at every point in "plane" elasticity problems. We will not distinguish between plane stress, plane strain, and generalized plane stress at this stage, since we are dealing with surface strain in the specimen.

Differentiation of the displacement components with respect to the appropriate space variables yields ϵ_x, ϵ_y, and ϵ_{45}, the three strain components that define the complete state of surface strain throughout the field of view. As with the other moire processes that have been presented, and remembering that F is now the spatial frequency of the reference grating,

$$\epsilon_x = \frac{\partial N_x / \partial x}{F}$$

$$\epsilon_y = \frac{\partial N_y / \partial y}{F} \tag{13.67}$$

$$\epsilon_{45} = \frac{\partial N_{45} / \partial x'}{F}$$

We find, as for geometric moire, that rotational effects are decoupled from normal strain, so they do not cause errors in normal strain. That is, when one interrogates the specimen grating with beams AO and BO, the resulting interference moire pattern is affected by three factors. Fringes of extension are produced by normal strain ϵ_x, and fringes of rotation are produced by shear strain γ_{xy} and the rigid-body rotation ψ. It was also shown, and this is important, that the fringes of extension are perpendicular to the xz-plane, whereas those caused by the small rotations are nearly parallel to it. Additionally, the interference fringes created by the intersection of beams AO and BO are perpendicular to the xz-plane.

Because the quantity to be measured is a normal strain (e.g., ϵ_x), the needed final result is the gradient or spacing of the moire fringes in the x direction. The fringes induced by shear and rotation run roughly parallel to the x-axis and, therefore, have small gradient in that direction. In algebraic form, the total fringe order is expressed as the sum of a part resulting from extension and another part caused by rotation:

$$N_x = (N_x)_\epsilon + (N_x)_R \tag{13.68}$$

Use the partial derivatives to get apparent strains from the extension and rotation parts,

$$\epsilon_x = \frac{\partial N_x / \partial x}{F} = \frac{\partial (N_x)_\epsilon / \partial x + \partial (N_x)_R / \partial x}{F} \tag{13.69}$$

But $\partial (N_x)_\epsilon / \partial x \ggg \partial (N_x)_R / \partial x$ because of the directions in which the fringes run for the two types of contributions. The conclusion is that rotation does not affect the measurement of normal strain. This finding is consistent with that of Chapter 8 for geometric moire.

In practice, the moire fringe order is not separated into two parts. The rotation element merely causes the moire strain fringes to deviate from the usual orientations parallel to the grill. This effect is demonstrated in Figure 13.16.

It would seem that the shear strain γ_{xy} could be evaluated by cross derivatives of displacements as

$$\gamma_{xy} = \frac{\partial u_x}{\partial y} + \frac{\partial u_y}{\partial x} \tag{13.70}$$

That this method of finding shear strain can be subject to large error was demonstrated in section 8.6. The reason is that the partial derivatives in the shear equation are gradients in the direction parallel to the analyzer grating lines. The gradient of the rotation-induced fringes in that direction can be at least as large as the gradient of extension-induced fringes, so the potential error is great. This situation is worsened by the lack of any direct way of estimating or eliminating the error. The rectangular strain rosette method involving three measurements of normal strain is one of two effective methods for eliminating rotation-induced errors when evaluating shear strain. That technique was explored in Chapter 8, and it need not be repeated. This approach also circumvents the experimental difficulties associated with rigid-body motion and accidental misalignment of the reference gratings.

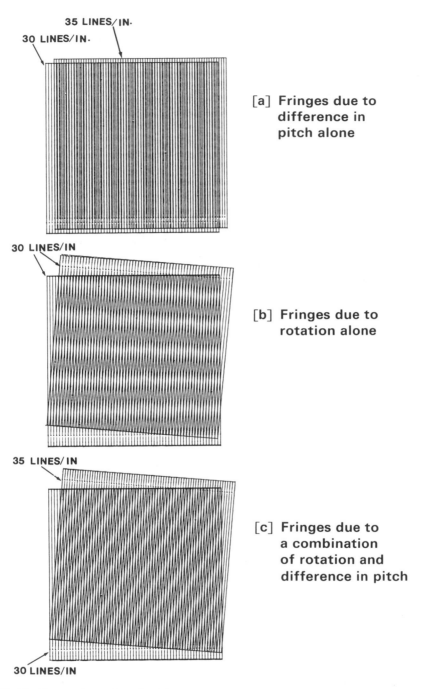

35 LINES/IN.

30 LINES/IN.

[a] **Fringes due to difference in pitch alone**

30 LINES/IN

[b] **Fringes due to rotation alone**

35 LINES/IN

[c] **Fringes due to a combination of rotation and difference in pitch**

30 LINES/IN

Figure 13.16. Formation of moire fringes caused by pitch and rotational mismatch. Fringes caused by (a) difference in pitch alone, (b) rotation alone, and (c) combination of rotation and pitch difference (Herrera Franco 1985).

The second method for eliminating rotation errors from shear strain is by assuring that the reference gratings for determining the orthogonal displacement components are exactly perpendicular to one another. This approach has been used by Post (Post, Han, and Ifju 1994; Post 1965) and others. Given the required precision of the moire interferometry apparatus, the reference gratings (termed virtual gratings by some practitioners) should be found to be orthogonal, and with precise data reduction, the potential rotation errors will not be a concern.

13.11 Use of pitch mismatch in moire interferometry

The accuracy of the in-plane moire method depends largely on how accurately the displacement curve is plotted, whether graphically or numerically; and the accuracy of the displacement curve, in turn, depends on the number of available points. For a given displacement field, the number of fringes depends on the gratings used. The finer the grating, the greater the number of resulting fringes. With the interferometric technique, the frequency of the reference grating and the specimen gratings can be adjusted easily to suit the problem. However, the finer the grating, the more severe are the requirements in terms of handling, optical bench stability, and so forth. For practical purposes, it is convenient to use frequencies in the range of 15,000 to 60,000 lines/in. (600–2,400 lines/mm).

So far, it has been assumed that the reference grating and the specimen grating are related by $F = \sqrt{2}f$ and that the angle of incidence of the laser light on the specimen grating is exactly ξ. If these conditions are not satisfied to within a small tolerance, pitch and rotational mismatches will be present. These initial differences are called "linear" and "rotational" mismatch, respectively; and, of course, they are measured against the reference grating. The resulting fringes are called "linear mismatch fringes" and "rotational mismatch fringes." An illustration of the fringes produced by the mismatch is shown in Figure 13.16.

13.11.1 Analysis of undeformed specimen

To illustrate the use of the mismatch, we will consider an undeformed specimen grating that is being interrogated with beams BO and CO to obtain the u_y displacement field. The relationship between the specimen grating and the reference grating frequencies is

$$\delta F = F - \sqrt{2}f \qquad (13.71)$$

where δF is frequency mismatch;
            F is reference grating frequency;
            f is specimen grating frequency

This equation can be rewritten as

$$f = \left(\frac{F}{\sqrt{2}}\right)\left(\frac{1 - \delta F}{F}\right) \tag{13.72}$$

It should be noted that the $\delta F/F$ term is a component of the frequency of the reference grating with its lines parallel to the xz-plane. To obtain the components along the $\pm 45°$ orientation, a rotation through $45°$ must also be imposed. This will yield a component of extension equal to $\delta F/2F$ and a component of rotation of the same magnitude. The first component will produce the same effect as a fictitious tensile strain, and the second is equivalent to a fictitious rotation of the specimen grating lines. By using the three-dimensional grating equation, one can determine the frequency of the resulting interference pattern with the aid of Figure 13.17.

For beam BO, let $m = +1$, $\sin \alpha = \sin \xi = -\lambda F/\sqrt{2}$, $f' = (F/\sqrt{2})(1 - \delta F/2F)$, and $\psi_i = \delta F/2F$. The grating equations used so often before give

$$\sin \beta''_{+1}(y'z) = \frac{-\lambda(\delta F)}{2\sqrt{2}}$$

$$\sin \beta'_{+1}(x'z) = \frac{\lambda(\delta F)}{2\sqrt{2}} \tag{13.73}$$

Similarly, for beam CO, let $m = +1$, $\sin \alpha = \sin \xi = -\lambda F/\sqrt{2}$, $f' = (F/\sqrt{2})(1 - \delta F/2F)$, and $\psi_i = -\delta F/2F$. The grating equation gives

$$\sin \beta''_{+1}(x'z) = \frac{-\lambda(\delta F)}{2\sqrt{2}}$$

$$\sin \beta'_{+1}(y'z) = \frac{\lambda(\delta F)}{2\sqrt{2}} \tag{13.74}$$

It can be shown that the diffracted rays OB' and OC' lie in the yz-plane (Fig. 13.17), and the sine of the angle between them and the z-axis is

$$\sin \beta_{\delta F}(z) = \frac{\lambda(\delta F)}{2} \tag{13.75}$$

The frequency of the resulting initial fringe pattern with its fringes parallel to the xz-plane is

$$P_{\delta F} = \delta F \tag{13.76}$$

(a)

(b)

(c)

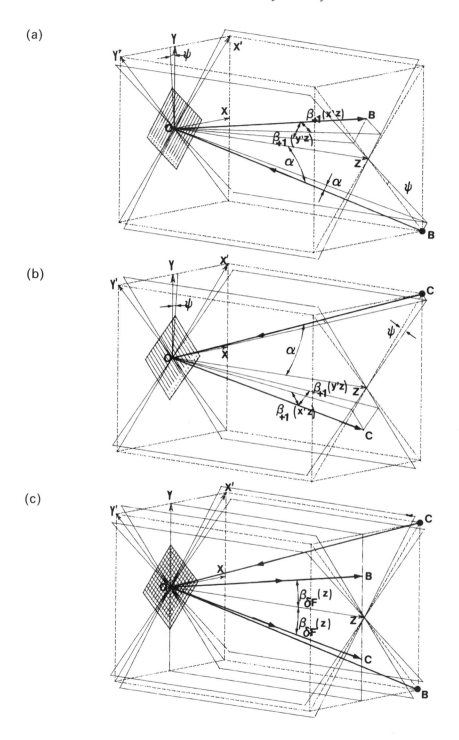

Figure 13.17. Geometry of the incident and diffracted rays affected solely by frequency and rotational mismatch when analyzed with (a) beam BO, (b) beam CO, and (c) both beams (Herrera Franco 1985).

and the fictitious displacement is given by

$$(u_y)_{\delta F} = \frac{(N_y)_{\delta F}}{F} \tag{13.77}$$

Thus, the fictitious strain produced by the mismatch is

$$(\epsilon_y)_{\delta F} = \frac{\partial (u_y)_{\delta F}}{\partial y} = \frac{\partial (N_y)_{\delta F}/\partial y}{F} \tag{13.78}$$

13.11.2 Analysis of the deformed grating (strain plus pitch mismatch)

Now, the combined effect of strain plus pitch mismatch will be considered. As usual, we refer to a schematic of the geometry of the system, in this case Figure 13.18.

To begin, take $f' = (F/\sqrt{2})[1 - (\epsilon_y)_\epsilon - \delta F/2F]$. Then for beam BO, $\sin \alpha = \sin \xi = -\lambda F/\sqrt{2}$, $\psi_i = (\epsilon_y)_\epsilon/2 + \delta F/2F$, and $m = 1$. The three-dimensional grating equation yields

$$\sin \beta''_{+1}(y'z) = \frac{-\lambda F}{\sqrt{2}} \left[\frac{(\epsilon_y)_\epsilon}{2} + \frac{\delta F}{2F} \right]$$

$$\sin \beta'_{+1}(x'z) = \frac{\lambda F}{\sqrt{2}} \left[\frac{(\epsilon_y)_\epsilon}{2} + \frac{\delta F}{2F} \right] \tag{13.79}$$

For beam CO, with $m = +1$, $\sin \alpha = \sin \xi = -\lambda F/\sqrt{2}$, and f' and ψ_i as before,

$$\sin \beta''_{+1}(x'z) = \frac{-\lambda F}{\sqrt{2}} \left[\frac{(\epsilon_y)_\epsilon}{2} + \frac{\delta F}{2F} \right]$$

$$\sin \beta'_{+1}(y'z) = \frac{\lambda F}{\sqrt{2}} \left[\frac{(\epsilon_y)_\epsilon}{2} + \frac{\delta F}{2F} \right] \tag{13.80}$$

Again, the diffracted rays BO' and CO' lie in the yz-plane (Fig. 13.18), and the sine of the angle between each of them and the z-axis is

$$\sin \beta_{(\epsilon+\delta F)} = \lambda F \left[\frac{(\epsilon_y)_\epsilon}{2} + \frac{\delta F}{2F} \right] \tag{13.81}$$

The frequency of the resulting interference pattern whose lines are parallel to the xz-plane is

$$P_{(\epsilon+\delta F)} = F[(\epsilon_y)_\epsilon + \delta F] \tag{13.82}$$

(a)

(b)

(c)

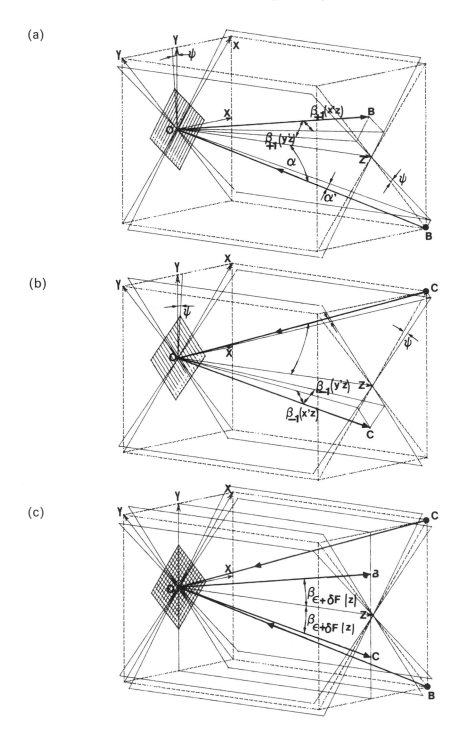

Figure 13.18. Geometry of the incident and diffracted rays from a grating affected by strain plus pitch mismatch when analyzed with (a) beam BO, (b) beam CO, and (c) both beams (Herrera Franco 1985).

The displacement is

$$(u_y)_{(\epsilon + \delta F)} = \frac{(N_y)_{(\epsilon + \delta F)}}{F} \tag{13.83}$$

and the strain is

$$(\epsilon_y)_{(\epsilon + \delta F)} = \frac{\partial (N_y)_{(\epsilon + \delta F)} / \partial y}{F} \tag{13.84}$$

Computation of the true displacement and strain at any point follows directly from prior derivations:

$$(u_y)_t = (u_y)_{(\epsilon + \delta F)} - (u_y)_{\delta F}$$
$$(u_y)_t = \frac{(N_y)_{(\epsilon + \delta F)} - (N_y)_{\delta F}}{F} \tag{13.85}$$

where $(U_y)_t$ is the true displacement. The true strain ϵ_t is computed as follows,

$$\epsilon_t = (\epsilon_y)_{(\epsilon + \delta F)} - (\epsilon_y)_{\delta F} \tag{13.86}$$

$$\epsilon_t = \frac{1}{F} \left\{ \frac{\partial [(N_y)_{(\epsilon + \delta F)} - (N_y)_{\delta F}]}{\partial y} \right\} \tag{13.87}$$

These results correspond to those developed in Chapter 8 for geometric moire, with minor differences. Notice that if the use of pitch mismatch is to be beneficial, the sign of the fictitious strain should be of the same sign as that of the true strain, otherwise a reduction in the number of fringes will result. Mismatch of the opposite sign can be used efficiently if its magnitude is at least twice the magnitude of the strain to be measured. Nevertheless, caution should be exercised, because the use of too much mismatch or wrong mismatch will hinder rather than help.

Note also that it is not absolutely necessary to know the exact size of the mismatch in order to calculate and eliminate the pitch difference and initial rotation effect. Before-strain and after-strain fringe photographs can be used directly. A detailed procedure for the elimination of the fictitious strain was given in earlier sections and will be discussed briefly again in Chapter 15.

Further, when a high level of rotational or pitch mismatch is used, the resulting high numbers of fringes can be thought of as "carrier fringes" (e.g., Post 1987). The displacement is then considered to be information that modulates these carrier fringes; that is, the mismatch fringes carry the strain intelligence. If the frequency of the carrier fringes is made high, then

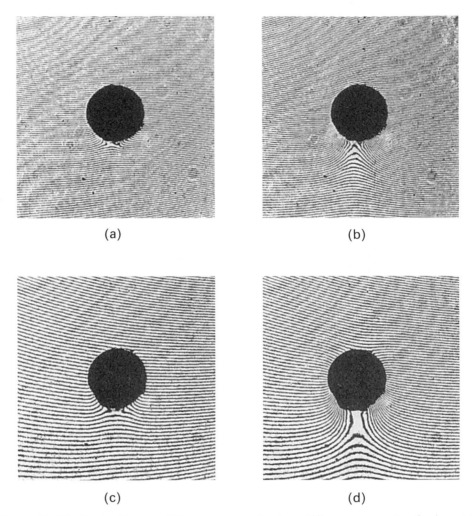

(a)   (b)

(c)   (d)

Figure 13.19. Moire fringes of displacements for two different amounts of mismatch and identical load conditions: (a) large mismatch at small load, (b) large mismatch at large load, (c) moderate mismatch at small load, (d) moderate mismatch at large load (Herrera Franco 1985).

superimposition of no-load and at-load patterns (moire of moire; see sec. 7.5) can be employed to determine the displacement fields. Otherwise, optical spatial filtering (discussed in Chapter 11) will serve to demodulate the carrier fringes and retrieve the displacement information.

Figure 13.19 shows photographs of two different amounts of mismatch for the same loading condition on a pin-loaded hole. Figure 13.20 shows the strain plots obtained for these photographs; notice that the strain plot obtained for the case of greater mismatch gives a more detailed strain contour than the other case.

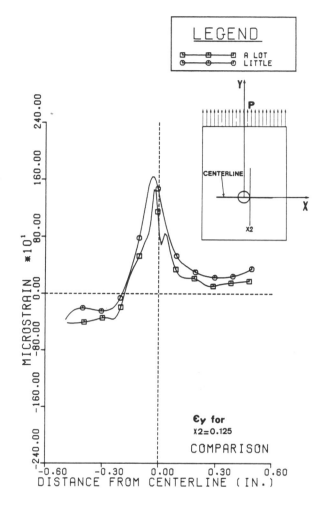

Figure 13.20. Strain ϵ_y obtained from two sets of moire fringe patterns for the same specimen with two different levels of pitch mismatch (Herrera Franco 1985). *Note*: The curves were drawn using 100 computer-generated points. The symbols are for identification only.

References

Basehore, M. L., and Post, D. (1981). Moire method for in-plane and out-of-plane displacement measurements. *Experimental Mechanics*, 21, 9: 321–8.

Czarnek, R., and Post, D. (1984). Moire interferometry with $\pm$45-deg. gratings. *Experimental Mechanics*, 24, 1: 68–74.

Guild, G. (1956). *Interference Systems of Crossed Diffraction Gratings*. Oxford: Clarendon Press.

Herrera Franco, P. J. (1985). *A Study of Mechanically Fastened Composite Using High Sensitivity Interferometric Moire Technique*. Ph.D. dissertation. East Lansing, MI: Michigan State University.

James, J. F., and Sternberg, R. S. (1969). *The Design of Optical Spectrometers*. London: Chapman and Hall.

McDonach, A., McKelvie, J., and Walker, C. A. (1980). Stress analysis of fibrous composites using moire interferometry. *Optics and Lasers in Engineering*, 1: 85–105.

Post, D. (1965). The moire grid-analyzer method for strain analysis. *Experimental Mechanics*, 5, 11: 368–77.

Post, D. (1980). Optical interference for deformation measurements – classical, holographic and moire interferometry. In *Mechanics of Nondestructive Testing*, Proceedings, Ed. W. W. Stinchcomb, pp. 1–53. New York: Plenum Publishing.

Post, D. (1987). Moire interferometry. In *Handbook on Experimental Mechanics*, Ed. A. S. Kobayashi, Ch. 7. Englewood Cliffs: Prentice Hall.

Post, D., and Baracat, W. A. (1981). High sensitivity moire interferometry – a simplified approach. *Experimental Mechanics*, 21, 3: 100–4.

Post, D., Han, B., and Ifju, P. G. (1994). *High Sensitivity Moire: Experimental Analysis for Mechanics and Materials*. New York: Springer-Verlag.

Stroke, G. W. (1967). Diffraction Gratings. In *Handbuch der Physik*, Ed. S. Flügge, Vol. 29, pp. 426–754. Berlin: Springer-Verlag.

Walker, C. A. (1994). A historical review of moire interferometry. *Experimental Mechanics*, 34, 4:281–99.

Walker, C. A., and McKelvie, J. (1978). A practical multiplied moire system. *Experimental Mechanics*, 8, 8: 316–20.

14

A moire interferometer

General considerations relevant to the design of moire interferometry apparatus are presented, followed by a description of the construction and adjustment of a particular 6-beam 3-axis device.

14.1 Approaches to design

There are three general design steps in the physical realization of a device for performing interferometry, including the moire variety. First, the needed special capabilities and tolerable limitations must be listed. Second, an optical arrangement that will satisfy the requirements must be designed. Finally, optical components must be obtained and arranged to perform the desired tasks.

Various optical arrangements for moire interferometry have been employed (Cloud and Herrera Franco 1986; Herrera Franco 1985; McDonach, McKelvie, and Walker 1980; Post 1980; Post 1987; Post and Baracat 1981; Post, Han, and Ifju 1994). Discussion of all the possibilities would be somewhat tedious and could be counterproductive in that the reader's attention might be distracted from the development process. Once the process is understood, the practitioner can review literature on the designs that have been employed and can then modify these designs or create new ones to fit the problems and resources at hand. For that reason, attention here is focused on the design and construction of a general-purpose 6-beam instrument with useful capabilities for experiments that require full knowledge of the surface strain fields in different types of specimens (Cloud, et al. 1987; Cloud, Herrera Franco, and Bayer 1989; Herrera Franco 1985).

The main objectives pursued in the construction of this moire interferometer were the following:

1. To avoid rigid connections between the optical elements and the specimen being tested in order to gain flexibility and allow the instrument to be used in several different experimental setups.

2. To incorporate capability to perform measurements in three different directions over the whole area of interest, so that the complete surface strain field could be measured.

3. To obtain good efficiency of light utilization so that a low-power laser would suffice.

4. To make the experimental setup suitable for performing measurements in environments not so ideal as an optics laboratory.

5. To incorporate some sort of unitized construction so that critical adjustments would be simple to perform and, once completed, would be fixed until circumstances dictated a controlled change.

6. To facilitate the creation of positive and negative pitch mismatch at desired levels to fit the problem under study.

7. To be compatible with photographic and video fringe recording and digital data reduction.

8. To assure that the fundamental grating frequency could be easily changed to suit the measurement problem.

14.2 Construction of interferometer

A system that uses three pairs of beams for measurement of strain in three directions is described. This device is properly called either a 3-axis interferometer or a 6-beam interferometer, even though only three physical beams are needed to form the desired three pairs of interfering beams. The instrument meets all the criteria just listed.

Figure 14.1 is a photograph showing an overall view of the experimental apparatus, including a small loading frame with a load transducer. Only part of the imaging system is included so that the optical elements will not be obscured.

Figure 14.2 is a conceptual sketch of the optical system that illustrates how the required three beams are created and directed to the specimen. Recall that this device utilizes three pairs of two beams each to obtain the strain measurements along three directions. The sketch is distorted to clarify the idea, and the fringe imaging system is not included in this rendition.

Finally, Figure 14.3 shows a two-dimensional schematic that establishes the basic plan and function of the optical system. A device built according to this plan could be used to measure one strain component, which might be adequate for demonstration purposes or for certain types of measurement applications.

Referring to Figure 14.3, one can follow the basic functions of the device. A 20-MW helium–neon laser (1) produces a beam of coherent light having a wavelength of 632.8 nm (24.913×10^{-6} in.). The two front-surface mirrors

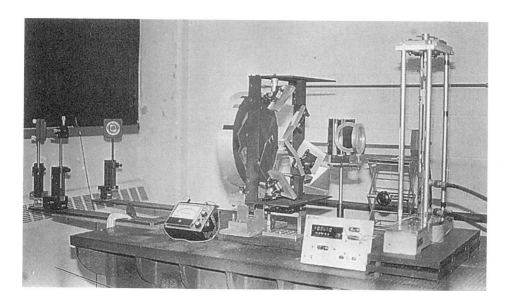

Figure 14.1. Overall view of the 3-axis 6-beam moire interferometer (Herrera Franco 1985).

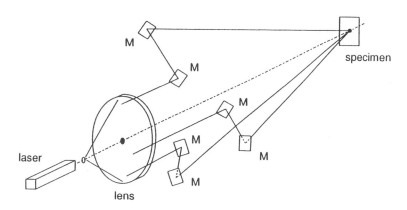

Figure 14.2. Conceptual optical scheme of the 3-axis 6-beam moire interferometer, not including fringe imaging optics.

(2 and 3) direct the beam to the spatial filter, where it is filtered and expanded with a 40 × microscope objective and a pinhole (4). A collimating lens (5) (a 13-in. diameter plano-convex lens, focal length = 1.0 m) changes the expanding beam into a parallel beam. This collimated beam reaches two first-surface flat mirrors (6 and 7), oriented at 45° with respect to the optical axis of the lens. Each of these two mirrors directs portions of the parallel rays to another

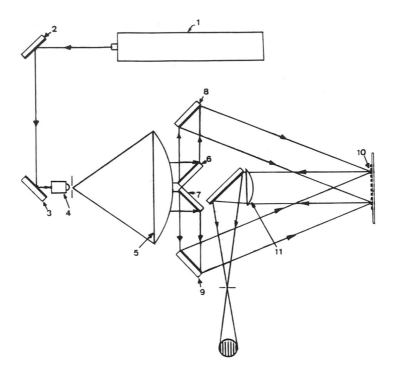

Figure 14.3. Two-dimensional schematic of the basic plan of each of the two-beam interferometers contained in the 3-axis 6-beam moire interferometer (Cloud, Herrera Franco, and Bayer 1989).

two mirrors (8 and 9), which can easily be oriented to direct the light toward the specimen grating (10) and to obtain the desired angle of interference between the two beams. These two beams, incident on the specimen grating, are diffracted. The diffracted rays pass through a converging (decollimating) lens (11) (focal length, 75 cm) and are directed by a front-surface mirror, oriented at 45° with respect to the normal of the grating, to the observing system or photographic camera. The beam narrows and passes through an aperture before reaching the camera.

Figure 14.3 and the preceding description do not adequately describe how the mirrors must be oriented in three dimensions to create the required three pairs of beams for measurement of the three strain components. For that purpose, consider Figure 14.4, which shows a plan view (looking along the main optic axis) of the arrangement of the mirrors in their mounting frame. This is the heart of the system. This sketch also shows how the mirror frame is mounted on a base that has micrometer adjustments for tilt, rotation, and position. The individual mirrors also are mounted on micrometer adjustable bases, but these are not shown.

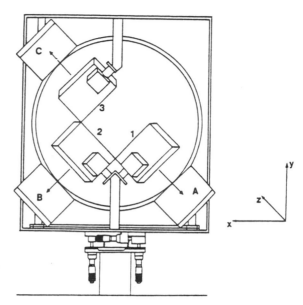

Figure 14.4. Plan view of arrangement of mirrors in the frame of the moire interferometer (Cloud, Herrera Franco, and Bayer 1989).

To describe the orienting of the flat mirrors, refer to Figure 14.4 and consider, as a reference system, a set of axes with the positive z-axis going into the page and the x- and y-axes located on the plane of the paper.

The three flat mirrors on which the plane wavefront from the collimator is incident are denoted by 1, 2, and 3. The other three mirrors, used to direct the beams of light toward the specimen grating, are denoted by A, B, and C.

To direct a portion of the light to mirror A, mirror 1 is rotated 45° from the xy-plane in the clockwise direction as viewed from the positive y-axis, and also rotated 45° counterclockwise from the xy-plane as viewed from the positive z-axis. To direct light to mirror B, mirror 2 is rotated 45° from the xy-plane counterclockwise as viewed from the positive y-axis, and 45° from the xy-plane in the clockwise direction as viewed from the positive z-axis. Similarly, to direct a portion of the light to mirror C, mirror 3 is rotated 45° counterclockwise from the xy-plane as viewed from the positive y axis, and 45° from the xy-plane counterclockwise as viewed from the positive z-axis.

Mirrors A, B, and C are located at three corners of a square. The distances between their centers can be adjusted easily to aid in obtaining the required grating spatial frequency and specimen coverage. A commonly used spacing is 11 in. These three mirrors provide the three beams of light that converge on the surface of the specimen and form three edges of a square-based pyramid. The angle of interference between the beams can be adjusted by changing the tilt and spacing of mirrors A, B, and C, and also by adjusting the distance between the interferometer and the specimen grating according to equation 13.25, which relates the angle of incidence and the desired interference frequency.

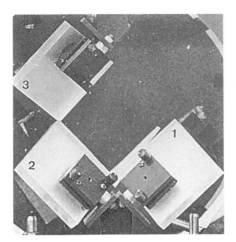

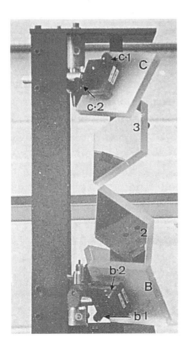
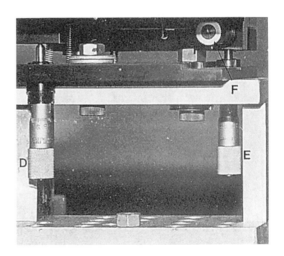

Figure 14.5. Mechanical elements used in positioning optical components of the 6-beam moire interferometer (Herrera Franco 1985).

Figure 14.5 shows several closeup views of the mechanical components that have been utilized in building an interferometer according to the plan described. All the mirrors and the mechanical components used to position and mount the mirrors can be purchased from suppliers of optical apparatus, particularly those that market holographic and optical processing equipment.

Only front-surface mirrors should be used. Their flatness should be one quarter of a wavelength, and their dimensions 100×100 mm for the system discussed here. All the mirrors can be simply glued onto the mirror mounts. The tilting element of the mirror mounts provides a kinematic orthogonal adjustment using balls that are spring-loaded between a conical recess and a flat. Orientation is controlled by stainless steel drive screws having 80 threads per inch, which push the alignment mechanism by means of hardened steel balls. The angular range is $\pm 4.45°$ about each of the orthogonal axes. Mirror mounts 1, 2, and 3 are fixed permanently to the supporting aluminum frame of the interferometer. Mirrors A, B, and C are connected to the aluminum frame by support posts that are held by post holders. The posts are held together using universal clamps. These universal clamps allow free positioning and orienting of mirrors A, B, and C. The post holders of mirrors A and B are mounted on optical carriers, which allow horizontal adjustment up to 0.25 in. by an 80-pitch screw. These carriers are mounted on an optical rail. The post holder for mirror C is fixed directly to the aluminum frame using a screw that can slide horizontally. The aluminum frame holding all the optical components is mounted on a tilt platform, which offers three axis adjustments: two tilts and one in-plane rotation. The tilt range in each direction is $-6°$ to $+8°$. The body of the platform is hard anodized aluminum and has hardened steel inserts to interface with the micrometer spherical tips. The platform is oriented to provide tilt about the x- and z-axes and in-plane rotation (xz-plane) about the y-axis.

14.3 Adjustment of moire interferometer

Setting up and adjusting an optical interferometer can be frustrating to the novice because seemingly minute changes cause such large changes in the output. This 6-beam moire apparatus can be especially tricky because each beam is used in two different two-beam interferometers. The mirror adjustments are all interdependent. The following sequence of steps is an efficient way to adjust the positions of mirrors A, B, and C:

1. Using equation 13.25, calculate the distance from the interferometer to the specimen grating that will give the desired angle of interference between the two incident beams.
2. Using this result, move the specimen together with the loading frame to the calculated distance. Loosen the universal clamps and manually orient the mirrors to direct the light toward the specimen grating. Care should be taken to avoid touching the surface of the mirrors, because hand grease cannot be removed from the mirrors without damaging the reflective surfaces. This positioning will give a rough adjustment of the orientation of the mirrors. As an aid in describing the fine adjustment of the orientation of the mirrors, we denote the knobs of the mirror

mount adjustment screws as follows: a_1 and a_2, b_1 and b_2, and c_1 and c_2 for mirrors A, B, and C, respectively (see Fig. 14.5a).

3. On the focal plane of the decollimating lens, notice three bright spots. They correspond to the first diffraction orders of the three incident beams. Adjust mirrors B and C first. Block the light from mirror A, then only two bright spots will remain on the focal plane of the decollimator. These bright spots should be vertical, that is, one above the other. To get the correct sign of the pitch mismatch – so that an increase in the number of fringes per unit of space will result from a tensile strain, and a decrease in the number of fringes per unit of space will result from a compressive strain – the bright spot corresponding to the diffracted order OB′ should be located above the bright spot corresponding to the diffracted order OC′. Then, both spots should be identified before continuing. The position and orientation of the bright spot OB′ can be adjusted using knobs b_1 and b_2. Knob b_1 will control the separation between the two bright spots. A clockwise rotation of this knob will move the bright spot OB′ upward. The angular orientation of the bright spot can be adjusted with knob b_2. A clockwise rotation of this knob will rotate the interference fringes clockwise as viewed from the camera. The function of the first knob is to produce extensional pitch mismatch, and rotational pitch mismatch is produced by the second, as in traditional moire. Once the two bright spots have been positioned correctly, the resulting interference pattern should be in the form of horizontal fringes.

4. Now, to adjust mirrors A and B, block the light from mirror C and uncover mirror A. In this step, mirror B should not be touched at all because that would produce a misalignment of the horizontal fringes. Next, block either of the two mirrors to identify their corresponding diffraction orders. Again, to get the correct sign of the pitch mismatch, spot OA′ should be to the right of spot OB′, that is, in a horizontal position. Knob a_2 can be used to produce extensional pitch mismatch, and a clockwise rotation will move spot OA′ farther to the right of spot OB′. Knob a_1 will produce a rotational pitch mismatch, and a clockwise rotation of it will produce a counterclockwise rotation of the interference pattern as viewed from the camera. When these two bright spots have been positioned correctly, the resulting interference pattern should be formed of vertical fringes.

5. After adjusting mirrors A and B, and B and C, for vertical and horizontal fringes, respectively, a third set of interference fringes can be obtained at 45°. To obtain this set of fringes, simply block the light reaching mirror B and uncover mirror C. While doing this, extreme care should be taken to avoid disturbing the adjustments for the other two orientations.

6. Also, for further alignment of the fringe pattern, the micrometers of the tilt table on which the frame is mounted can be used. These micrometers add three degrees of freedom and are denoted by D, E, and F. Micrometer D provides tilting about the z-axis and will rotate the interference pattern in the xy-plane, that is in the surface of the specimen as viewed from the camera. Micrometer E tilts the interferometer about the x-axis, and it will produce both extensional and rotational

pitch mismatch. Micrometer E will rotate the interferometer in the *xz*-plane. This rotation is used to position the surface of the specimen grating perpendicular to the optical axis of the collimator. For the no-load stage, the diffracted orders should retrace their path to the pinhole of the spatial filter.

7. A 35-mm camera and a 70–210 mm zoom lens with a 2 × focal length converter can be used to record the fringe patterns. A cable shutter release should be used to avoid any vibration of the camera. Alternatively, a video camera can be used, with output directed to a video monitor for real-time viewing or to a computer-based image acquisition system.

The specimen can be loaded using a small loading frame such as that manufactured by Scott Engineering Sciences (1400 S.W. 8th Street, Pompano Beach, Florida, 33060). Tension is applied using a hydraulic (double-acting) cylinder ram with the hydraulic pressure supplied by a hand pump. To measure the applied load, a force transducer can be made or purchased.

The virtues of a system of this type can be summarized as follows:

1. This system allows efficient light utilization.
2. No rigid connection is required between the specimen and the system.
3. With small improvements, the system can be utilized to perform measurements in environments not so ideal as an optics laboratory.
4. Grating frequencies are easily adapted to suit the problem.
5. Measurements can be performed in three different directions, giving a map of strains in the same number of directions and allowing calculations of maximum strains.
6. Pitch mismatch can be easily adjusted.
7. The system is compatible with various methods of data recording such as film and television camera.

References

Cloud, G., and Herrera Franco, P. J. (1986). Moire analysis of multiple hole arrays of composite material fasteners. *Proc. Spring 1986 Conference on Experimental Mechanics.* Bethel, CT: Society for Experimental Mechanics.

Cloud, G., Herrera Franco, P. J., and Bayer, M. H. (1989). Some strategies to reduce stress concentrations at bolted joints in FGRP. *Proc. 1989 SEM Spring Conference on Experimental Mechanics.* Bethel, CT: Society for Experimental Mechanics.

Cloud, G., Sikarskie, D., Vable, M., Herrera Franco, P., and Bayer, M. (1987). *Experimental and Analytical Investigation of Mechanically Fastened Composites,* U.S. Army Tank Automotive Command Report TACOM TR 12844, Warren, MI.

Herrera Franco, P. J. (1985). *A Study of Mechanically Fastened Composite Using High Sensitivity Interferometric Moire Technique.* Ph.D. dissertation. East Lansing. MI: Michigan State University.

McDonach, A., McKelvie, J., and Walker, C. A. (1980). Stress analysis of fibrous composites using moire interferometry. *Optics and Lasers in Engineering*, 1: 85–105.

Post, D. (1980). Optical interference for deformation measurements – classical, holographic and moire interferometry. In *Mechanics of Nondestructive Testing*, Proceedings, Ed. W. W. Stinchcomb, 1–53. New York: Plenum Publishing.

Post, D. (1987). Moire Interferometry. *Handbook on Experimental Mechanics*, Ed. A. Kobayashi, Ch. 7. Englewood Cliffs: Prentice Hall.

Post, D., and Baracat, W. A. (1981). High sensitivity moire interferometry – a simplified approach. *Experimental Mechanics*, 21: 100–4.

Post, D., Han, B., and Ifju, P. G. (1994). *High Sensitivity Moire: Experimental Analysis for Mechanics and Materials*. New York: Springer-Verlag.

15

Experimental methods in moire interferometry

Described here are details of procedures for analysis of displacement and strain in deformable bodies using moire interferometry. Some of these techniques have parallels in other optical methods, but the rest are specific to moire interferometry. The chapter closes with some sample results from various applications of the method.

15.1 Specimen gratings

Utilization of the moire effect in experimental measurements depends on the successful forming of line grids (or dots) on the surface of the specimen. The spatial frequencies of the gratings employed in moire interferometry fall in the range of a few hundred to several thousand lines/millimeter (roughly 5,000 to 50,000 lines/in.). If we recollect from the discussion of Fourier optics that the spatial frequency bandpass of a quality lens is limited to about 200 lines/mm, then we understand that no method of optical imaging is adequate for replicating gratings for moire interferometry, even supposing that a master grating is available to begin with.

The problem has three aspects. In the first place, some sort of master grating at the desired frequency must be created. Then, gratings suitable for transfer to the specimen must be manufactured. Finally, the gratings must, indeed, be somehow fastened to the specimen. Only after solving these problems can moire interferometry be undertaken. Walker (1994) outlines the several approaches that have been pursued through recent years. Post, Han, and Ifju (1994) also describe various techniques.

Of course, very fine and precise diffraction gratings have long been made by mechanical means. Such gratings tend to be very expensive and to exist only in small sizes. Since the development of the laser, an inexpensive source of nicely coherent light, gratings have increasingly been made by using the interference of two waves to create a fine fringe pattern. This fringe pattern can be directed onto a photographic material (silver halide emulsion or organic photoresist) to record the grating. Various grating geometries can be made in this way with little trouble, and they can be made in large sizes at relatively little cost. Gratings for many purposes are now made in this way. For moire experiments, the gratings must be transferred somehow to the specimen.

Recall that the moire interferometer itself creates a very fine grating structure in space. If this grating could be used to make a specimen grating, then the problems are solved, with the added advantage that any irregularities in the optical setup, such as distortions, will affect specimen and reference grating alike and so will probably not cause spurious fringes that could lead to measurement errors. To repeat, the idea is to use the moire interferometer to create the specimen grating and then to place that specimen grating back into the same interferometer for measurement.

The tasks, then, are to somehow record the grating created in the interferometer at the specimen plane and to get this grating attached to the specimen at proper orientation.

One obvious approach is to coat the specimen with a photoemulsion or photoresist. After exposure in the interferometer, the photo medium is developed and probably vacuum coated with metal to increase diffraction efficiency. This technique has been employed with various degrees of success. One problem is that the emulsions and resists available for this type of application tend to be "slow" and orthochromatic, so that heavy exposures of ultraviolet or deep visible blue light are needed. Consequently, rather a large laser is needed in the interferometer, and these are, typically, under-performers in interferometric applications for various reasons.

Other approaches center on techniques of transfer molding for grating replication (Walker and McKelvie 1978). A relief master grating (phase grating) is made using the interferometer or other technique. A thin layer of silicone cement or other replication medium is spread on the surface of the specimen. The master is pressed into this medium and left in place until the medium solidifies. The master is then pulled away, leaving its reverse replica behind with the specimen. The same idea can be used in reverse to replicate the deformed gratings on unwieldy structures for later study in the laboratory (Brown et al. 1979).

An important grid transfer method has been developed (Basehore and Post 1984; Post 1987; Post et al. 1994). In a way, this technique is a combination of the transfer molding idea and the old lithographic stripping film method for creating specimen gratings for geometric moire (see section 8.4). In brief, a phase-type relief grating is made by exposing a photoplate in the interferometer. The emulsion side is coated with metal and then glued to the specimen. The glass substrate with the emulsion is pulled away, leaving a metal-coated relief grating attached to the specimen. Although simple in

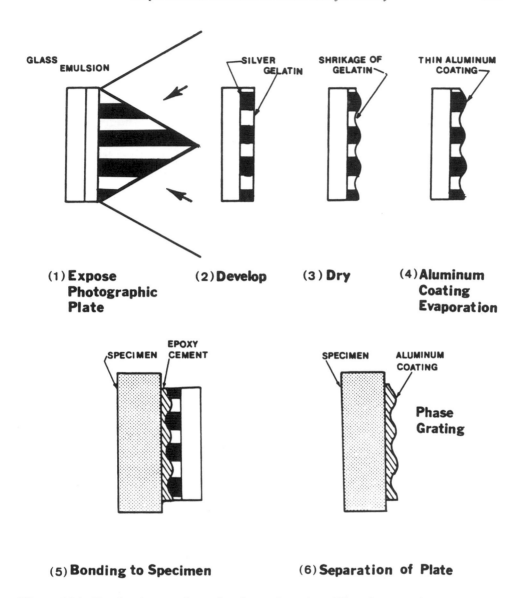

Figure 15.1. Production and replication of moire diffraction gratings (Herrera Franco 1985).

concept, the process is tricky to carry out at first, and much credit is due to the developers.

The entire process of creating a grating and transferring it to the specimen is summarized in Figure 15.1. This technique seems to have become the method of choice, so details of the steps are described in the next few paragraphs.

The master gratings or molds are produced as follows. A high-resolution silver halide photographic plate (e.g., holographic plate Agfa type 8E75) is exposed to two beams of collimated coherent laser light. The two beams intersect at the surface of the emulsion. These interfering beams generate a three-dimensional pattern of constructive and destructive interference. The frequency of this interference pattern can be calculated using equation 13.25. When the exposed plate is developed, silver compound remains in the areas exposed to constructive interference whereas the silver is washed out of those areas that are not exposed because of destructive interference. On drying, the gelatin of the emulsion shrinks, but the reinforcing effect of the residual silver causes the shrinkage to be less in the exposed areas. This phenomenon, which is termed the lenticulation effect, is well known and understood, and it has generally been considered a nuisance in the photographic industries. The effect causes a textured surface consisting of alternate furrows and raised terraces. In other words, it is a relief or phase grating that has the same spatial frequency as the interference system. If a cross-line grating is desired, as is usually the case, then the plate is rotated 90° after a first exposure, and a second exposure is made.

To produce these photoplates, the moire interferometer can be used effectively, but it must be set up to create only the two crossed gratings. The interferometer described in Chapter 14 can be adjusted to produce two horizontal beams using mirrors A and B, following the procedure in section 14.3. If a reference grating close to the desired known frequency is available, the angles are adjusted by observing the moire pattern created when the reference is placed in the interferometer. Usually, a traveling microscope with a $40 \times$ (or so) microscope objective, and a scale in millimeters, is used to count the number of fringes per millimeter in the interference system. Then, the mirrors can be fine-adjusted until the desired frequency is obtained. As a final verification step, one can place a small piece of the developed plate in a scanning electron microscope and take a photomicrograph with a precisely known magnification. Figure 15.2 shows a pair of such photographs at different magnifications. Note that the gratings have many irregularities, most probably caused by the grain of the photoemulsion. They are, however, eminently suitable for the purpose. Measurements from these particular photographs give a spatial frequency of 640 lines/mm (16,256 lines/in.).

The diffraction efficiency of a grating depends partly on the depth of the grooves. This characteristic can be controlled to a limited extent by the amount of exposure of the photoplate. The best time must be worked out by trial and error for the given interferometer. It has been reported (Basehore and Post 1984) that bleaching can be used to increase the depths of the corrugations of the grating. Others (Herrera Franco 1985) found that bleaching does augment the diffraction efficiency. But, at the same time, the resultant texture of the photoplate surface does not produce a highly reflective grating; instead it produces a diffusive surface.

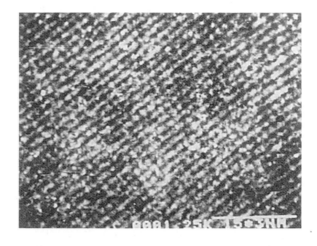

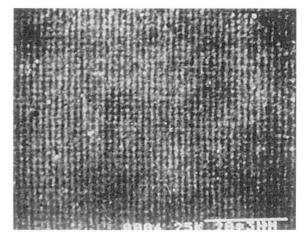

Figure 15.2.
Photomicrographs at two
magnifications of two-way
phase grating taken with
scanning electron
microscope (Herrera
Franco 1985).

After development and drying, a thin film of aluminum is deposited
on the whole surface of the photoplate. Aluminum is chosen because
it resists tarnishing, is highly reflective in thin films, and is low in cost.
An additional requirement that is satisfied by the use of aluminum is
that the metallic film should not contribute a significant reinforcing effect
to the specimen. The coating is best accomplished by vacuum deposition, although sputtering or other techniques would probably suit. Some
trial and error is involved in getting the thickness adjusted for maximum
efficiency of the grating and also to assure uniform coating. The overcoated
photoplate should be stored in a dust-free environment to await replication
on the specimen. Once the system for creating these gratings is set up
and working, it is a good idea to make several plates and store them until
needed.

15.2 Replication of gratings onto the specimen

Refer again to Figure 15.2, which summarizes the process of replicating the grating onto the surface of the specimen. The specimen is first positioned next to the two steel blocks that serve as a fixture to position the grating at the proper angle with respect to the specimen. For the 6-beam interferometer, the angle between the longitudinal axis of the specimen and the grating lines is 45°. A small amount of cement (such as PC-lOC photoelastic coatings cement, from Measurements Group, Photoelastic division, Rayleigh, NC) is poured onto the specimen. Next, the aluminum-coated plate is placed onto the specimen with the emulsion side inward. The excess cement is squeezed out by applying a pressure of approximately 10 psi to the glass-plate/specimen sandwich by means of weights. Curing time for the epoxy mentioned is about 4 hours, but the plates are removed after 12 hours. When the epoxy has hardened, the glass photoplate is separated from the specimen by a twisting-prying action. This step requires some practice, and one is advised to wear heavy gloves; sooner or later a photoplate will certainly break during the separation process, which requires forceful finesse. The thin film of aluminum adheres to the epoxy, which now carries a replica of the lenticulation grooves that were formed in the photoplate. This combination results in a highly reflective phase grating on the surface of the specimen.

Before the gluing operation, some important preparation steps are required. First, the specimen must be cleaned thoroughly using Freon degreaser or a nonpolluting equivalent. Necessary holes and cutouts in the specimen can be filled first with molding plastic or modeling clay. This material is given a concave shape on its upper surface to avoid spreading it out on the surface of the specimen during subsequent cleaning. Also, to avoid having some of the squeezed-out epoxy adhered permanently to the edges and bottom of the specimen, they can be covered with teflon tape, held in place with masking tape. Final cleaning takes place after these masking and filling operations.

After removing the photoplate, excess epoxy that squeezed out and adhered to the masking tape is removed easily. Because the epoxy is brittle, it can be broken free from the specimen edges.

At this point, the specimen is ready to be labeled and tested. Care must be taken to avoid contamination of the surface of the grating by either fingerprints or airborne dust. Figure 15.3 shows a flat, composite specimen with a reflective grating printed on its surface.

15.3 Recording fringe patterns

After all the adjustments of the interferometer are completed and the specimen is in place, the observing/recording camera is mounted with its lens or iris plane at the focus of the decollimating lens. The camera is focused in the

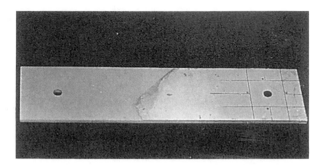

Figure 15.3. Specimen prepared with reflective diffraction gratings for moire interferometry (Herrera Franco 1985).

plane of the image of the surface of the specimen as seen through the decollimating (field) lens. The fiducial marks on the surface of the specimen serve as a focus aid. As usual in laser illuminated systems, focusing the imaging system is best accomplished by illuminating the specimen temporarily with white light, as from a desk lamp.

At this point, a final check for alignment is performed. A video camera and a monitor are very useful during this stage. Typically, the specimen is loaded lightly to keep it in a fixed position and to avoid any rigid-body rotation. Next the fringe patterns in all three directions are checked to verify that the initial loading does not introduce any rotational mismatch. Such rotation is common to all three fringe patterns, and it is corrected by rotating the tilt table about an axis perpendicular to the specimen surface. Final tweaking of the mirror adjustments is made at this time to get the initial mismatch fringes at optimum. These adjustments, if needed at all, should be minute and carefully done; otherwise it may be necessary to repeat the initial adjustment routine.

Recall that moire measurement, indeed most experiments, are best conducted in differential mode. The moire data reduction is best accomplished by recording-data in sets consisting of a baseline, no-load, or initial fringe pattern and an at-load fringe pattern. This approach allows automatic elimination of pitch mismatch fringes and reduction or total elimination of other errors resulting from film shrinkage, lens distortions, and the like.

From this point on, the recording of moire fringe data is routine. One of the beams in the interferometer, say beam CO to start, is blocked, and the initial fringe patterns for the AO–BO combination are photographed. Then a different beam is blocked, and the second interferometric beam pair is used for photography. Then the process repeats for the third interferometer. After recording all initial fringe patterns, the specimen load is changed to the desired level, and the fringes are recorded again for all three interferometers in turn. This continues for as many load stages as are desired.

Details such as film type and processing are the same as those discussed in Chapter 12, so they are not repeated here. Considerable variation is

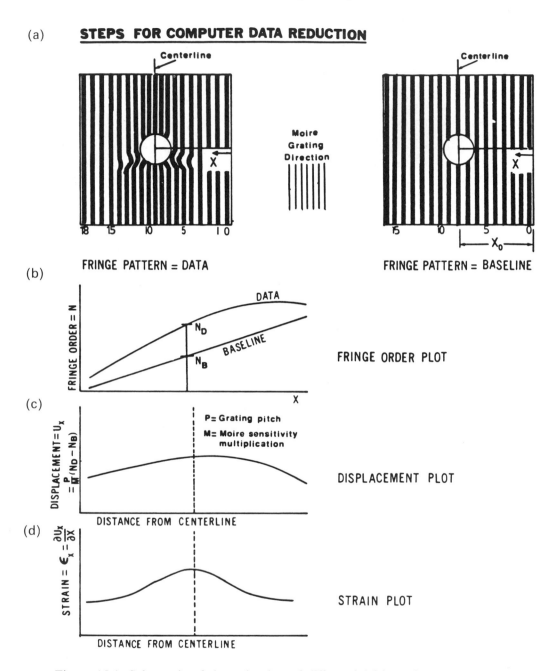

Figure 15.4. Schematic of the reduction of differential fringe data (Herrera Franco 1985).

possible, depending on available resources and results sought. The important thing is to get these fringe patterns stored in a photograph, in a computer, or on videotape.

15.4 Data reduction

We have seen in section 8.9 and elsewhere that the map of displacements can be determined more accurately by moire methods if a small amount of mismatch is used. It is possible to just measure or calculate the mismatch and its resulting false strain, and then subtract the false strain from subsequent apparent strain to get true strain. Such a procedure is not the best because it tends to propagate errors, especially those created by less-than-perfect optical systems.

A better procedure is to use a differential measurement scheme that utilizes two fringe patterns, as shown in Figure 15.4a. An initial moire fringe pattern, produced by using some mismatch, is obtained. Its initial fringe orders are measured, perhaps plotted (Fig. 15.4b), and subtracted from fringe orders recorded under all at-load conditions. The result, when multiplied by the appropriate factors, gives displacement (Fig. 15.4c). Of course, any two states of the specimen can be directly compared by using one as the initial state and the other as the at-load state. The subtraction and subsequent differentiation effectively define the strain produced by the deformation caused by the applied loading (Fig. 15.4d). Ideally, if the fictitious initial strain were homogeneous, the fringe density would be constant over the area viewed on the specimen by the moire setup. However, owing to imperfections in the lenses, the beam expanders, and the mirrors, the fringe density will rarely be constant over the length viewed. The subtraction of fringe densities between the at-load plus mismatch and the mismatch-only states (the former referred to as data and the latter as base or baseline) must be done locally.

We realize now that the reduction of moire interferometry data is identical to the reduction of other types of moire patterns. This process has been described in Chapter 12, and no more of it will be repeated here.

15.5 Sample results from moire interferometry

Here we present a set of moire interferometry fringe patterns and some plots of reduced data. The intent is not to give research results, although these samples are from a research application, but to offer the potential user of moire interferometry a vision of what to expect.

Figures 15.5, 15.6, and 15.7 are baseline and data fringe patterns showing 0°, 90°, and 45° displacement fields for a pin-loaded fastener hole in a composite plate. The hole is lined with an isotropic plastic insert (Herrera

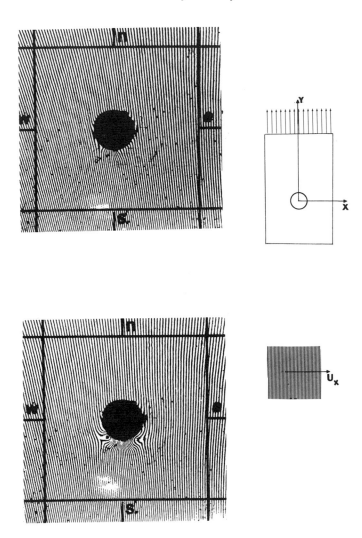

Figure 15.5. Moire interferometry fringes for displacements perpendicular to the load axis for a pin-loaded hole in composite with plastic insert (Herrera Franco 1985).

Franco 1985). These fringe patterns are very typical of those obtained in everyday application of the technique. They show some noise, much of which is caused by the roughness of the composite surface. Spurious fringes are seen; these are Newton's rings, caused by internal reflections during the making of the gratings, and so they cannot be eliminated. They are only of cosmetic interest since they do not mask the data fringes.

Figures 15.8, 15.9, and 15.10 are samples of the strain plots that are developed by computer data reduction of digitized fringe locations taken from the patterns just shown. These results, and the data obtained in similar studies, give a complete picture of the strain field. The shear strain γ_{xy} can be

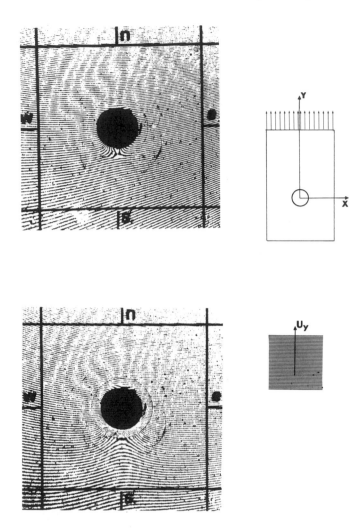

Figure 15.6. Moire interferometry fringes for displacements parallel to the load axis for a pin-loaded hole in composite with plastic insert (Herrera Franco 1985).

developed where needed by use of the rosette equations. In the end, a clear picture of the strain distributions can be derived for this and other problems.

Figure 15.11 shows a most interesting fringe pattern depicting the v-displacement field (the displacement in the y direction) and the resulting shear strain map from a study of the micromechanical strain field in a laminated composite (Guo, Post, and Han 1992). The ply sequence was $[90_2/0_2/+45_2/-45_2]_n$. The symbols in the figure represent plies with $90°$, $0°$, $+45°$, and $-45°$ fiber directions, respectively. The ply thickness was 0.19 mm (0.0075 in.). The reference grating frequency was $2,400$ lines/mm ($60,960$ lines/in.),

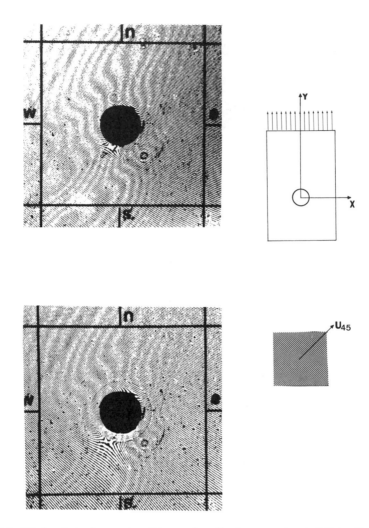

Figure 15.7. Moire interferometry fringes for displacements at 45° to the load axis for a pin-loaded hole in composite with plastic insert (Herrera Franco 1985).

which provided a sensitivity of 2.4 fringes per μm displacement. The axial strain, determined by $\epsilon_y = \partial v/\partial y$, is nearly constant at 2,700 μm/m. Very high shear strains were found near the junctions between $+45°$ and $-45°$ plies. The shear strain is evaluated by $\gamma_{xy} = \partial v/\partial x + \partial u/\partial y$. The final term of the shear strain is zero, however, for this case, so the shear strain is revealed completely by the v-field.

An interesting, final example of an application of moire interferometry is offered in Figure 15.12 (Hyzer and Walker 1989). The object of the study was to determine the residual strains in a thick-walled hollow cylinder and in an asymmetric lug, both of which were expanded by drawing an oversize mandrel

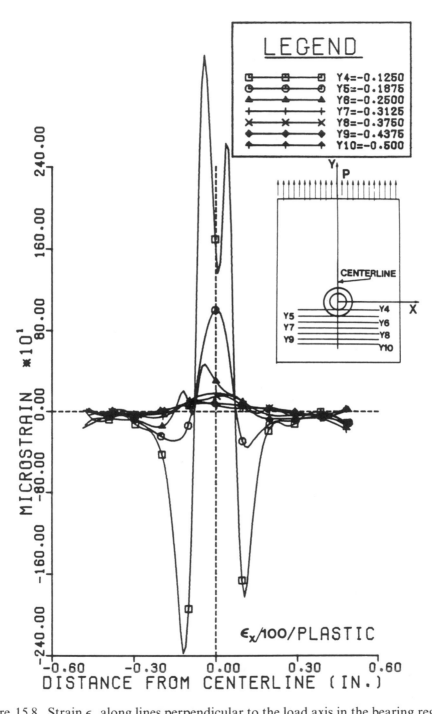

Figure 15.8. Strain ϵ_x along lines perpendicular to the load axis in the bearing region of a pin-loaded hole in composite with plastic insert (Herrera Franco 1985).

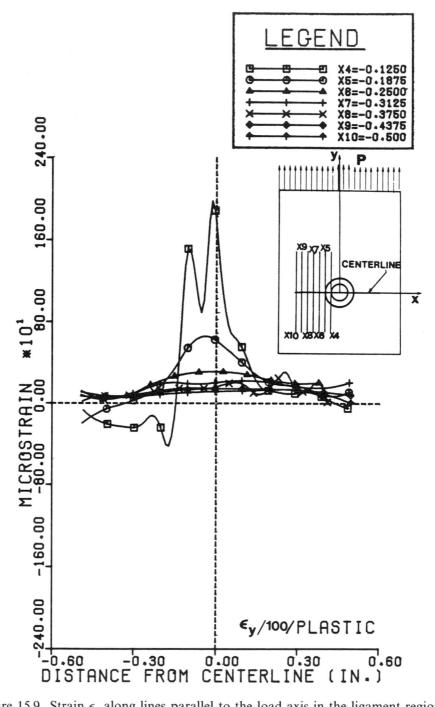

Figure 15.9. Strain ϵ_y along lines parallel to the load axis in the ligament region of a pin-loaded hole in composite with plastic insert (Herrera Franco 1985).

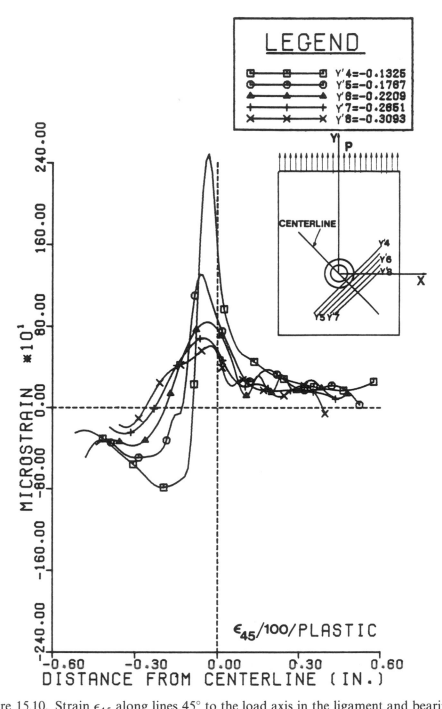

Figure 15.10. Strain ϵ_{45} along lines 45° to the load axis in the ligament and bearing regions of a pin-loaded hole in composite with plastic insert (Herrera Franco 1985).

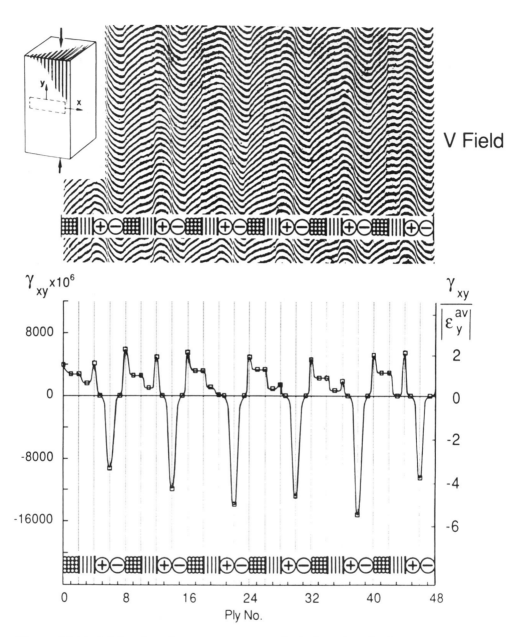

Figure 15.11. Moire interferometry pattern depicting the displacement in the *y* direction for a laminated composite in compression (top), and the resulting shear strains (bottom) (Guo, Post, and Han 1992).

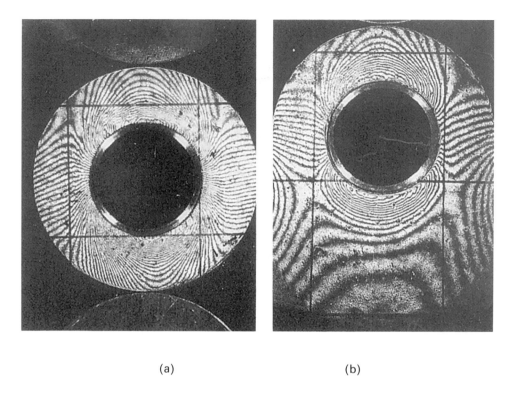

(a)   (b)

Figure 15.12. Moire interferometry fringe patterns showing vertical residual displacement fields caused by drawing an oversize mandrel through the hole in (a) a cylinder and (b) an asymmetric lug (Hyzer and Walker 1989).

through the hole. Very high strains and strain gradients are produced near the hole edge. The projection on one side of the lug affects the displacement field in a striking way.

References

Basehore, M. L., and Post, D. (1984). High frequency, high reflectance transferable moire gratings. *Experimental Techniques*, 8, 5: 29–31.

Brown, I. C., Mckelvie, J., Perry, S. H., and Walker, C. A. (1979). Site measurement of strain by a moire fringe technique. *Structural Engineer – Part B*, 57B, 3: 62–5.

Guo, Y., Post, D., and Han, B. (1992). Thick composites in compression: an experimental study of micromechanical behavior and smeared engineering properties. *Journal of Composite Materials*, 26, 13: 1930–44.

Herrera Franco, P. J. (1985). *A Study of Mechanically Fastened Composite Using High Sensitivity Interferometric Technique*, Ph.D. dissertation. East Lansing, MI: Michigan State University.

Hyzer, J. B., and Walker, C. A. (1989). New developments and applications of moire techniques to engineering stress analysis. In *Applied Solid Mechanics*, 3. Ed. I. M. Allison and C. Ruiz, pp. 244–57. Amsterdam: Elsevier Science Publishers.

Post, D. (1987). Moire Interferometry. In *Handbook on Experimental Mechanics*, Ed.
    A. Kobayashi, Ch. 7. Englewood Cliffs: Prentice Hall.
Post, D., Han, B., and Ifju, P. G. (1994). *High Sensitivity Moire: Experimental Analysis
    for Mechanics and Materials*. New York: Springer-Verlag.
Walker, C. A. (1994). A historical review of moire interferometry. *Experimental
    Mechanics*, 34, 4: 281–99.
Walker, C. A., and McKelvie, J. (1978). A practical multiplied moire system.
    Experimental Mechanics, 8, 8: 316–20.

PART VI
Holographic interferometry

16

Holographic interferometry theory

This chapter begins the study of a topic that is very interesting for many reasons, technical and otherwise. Holography is a method for creating three-dimensional images without a lens, and it offers many intriguing possibilities. Holographic images are precise enough to be used in interferometry, and this idea leads to several powerful techniques for measurement in a broad range of application areas.

16.1 Orientation

Holography is a unique method of storing and regenerating all the amplitude and phase information contained in the light that is scattered from an illuminated body. Because all the information is reproduced, the regenerated object beam is, in the ideal case, indistinguishable from the original. Here is a technology that offers the possibility of perfect three-dimensional photography. Since it is possible to record the exact shape and position of a body in two different states, then it is also possible to compare the two records to obtain a precise measure of the movement or deformation. This measurement technique is called holographic interferometry.

Holography was invented by Dennis Gabor in about 1948 as a method for improving the usefulness of microscopy. It did not work very well, partly because the two beams created in holographic reconstruction, which form what are commonly thought of as the real and virtual images, ended up in line with one another in the arrangement necessitated by the apparatus then available. His theory was not, apparently, limited in this way; but the potential beauty and utility of the technique were masked by the problem with the two images. Research was also greatly hampered by the lack of light sources with coherence lengths of more than a few millimeters; holograms could be made to reproduce only thin objects such as photographic transparencies.

The invention of the laser in 1960 gave researchers a light source with a great coherence length. Thus, interferometry could be carried out over path lengths of meters. This capability made possible the invention, by Leith and Upatnieks in the early 1960s (1962, 1964) of off-axis holography, which solved Gabor's problem by allowing the two reconstruction images to be separated in space. Now, it was possible to record and view holographic images of large objects and entire scenes. The reconstructed image was perfect in three-dimensional aspects, although contaminated by speckle effects and the monochromaticity of the laser. The intervening years have brought many improvements and simplifications to the technique. Holograms are now routinely seen on magazine covers and credit cards, in jewelry and fishing lures, and in many technical applications such as holographic lenses.

Holographic interferometry (HI) was demonstrated by Stetson and Powell (1965) for vibration study, and at about the same time for other applications by Haines and Hildebrand (1966), by Horman (1965), and by Brooks, Heflinger, and Wuerker (1965). This form of interferometry has many apparent advantages as a measuring technique, and some enthusiasts predicted that it would displace all other approaches. It is a noncontacting, nondestructive method with very high sensitivity. In addition, the information is presented as a three-dimensional image of the test object that is covered with a fringe pattern. From the appearance of this pattern one can quickly determine the overall deformation and the location of stress concentrations.

In spite of the obvious advantages, industry in general has been slow to accept this new technique. HI has been mostly confined to research laboratories, with some notable exceptions (e.g., Brown 1974). The reasons for this, apart from the usual distrust toward unfamiliar techniques, lie in two aspects. First, the stability requirement in holography is not readily compatible with industry environments unless pulsed lasers are used. Second, the photographic recording process and subsequent development introduce an annoying time delay, which prevents on-line inspection. Although self-developing and reusable recording media have been developed, a time delay between the recording and observation of the fringe pattern still remains, and direct computer data processing is not applicable.

Even so, holographic interferometry is an important tool. It alone will afford a solution to certain measurement problems. Some industrial applications, such as the inspection of aircraft tires, have yielded significant benefits in safety and economy. It is not really very difficult to do. Furthermore, study of holography gives insight and understanding of many related optical phenomena.

16.2 Fundamental basis of holography

The theory of holography is actually quite simple. It utilizes only the concepts of interference of two waves and diffraction by a grating. Before getting involved in the details, let us describe qualitatively how it works. Then the relevant optical phenomena need be reviewed only briefly, since they have all been discussed in previous sections of this text. That is to say, there is little here that is new except for the way in which known ideas are put together to make striking and useful three-dimensional pictures.

The requirement of a hologram is that it permanently record the complete structure of the light waves that are scattered or reflected from an illuminated object. Of course, none of our optical recording devices, most notably photographic emulsions, are sensitive to phase. So, we do what we always do in this situation. The phase data are converted to amplitude data by mixing the object beam with a reproducible reference beam. Making a hologram involves nothing more than the interference of two wavefronts. This subject was discussed in Chapter 2, and it was utilized in the treatment of other techniques, such as moire interferometry. The product of this interference of two beams is a diffraction grating.

Recovering the object wave, and thereby creating a reproduction of the image for a camera or our eyes, requires the illumination of the hologram with a replica of the original reference beam. This process involves diffraction by a grating. This phenomenon was treated in considerable depth in Chapter 10, and it was applied in several subsequent sections.

Holographic interferometry, in one of its forms, is a bare extension of these steps in that the hologram grating for two states of the object are recorded on the same photoplate. Essentially, two replica object waves are created on reconstruction, one for each state of the object. These two object waves interfere to produce a fringe pattern over the reconstructed image of the object. Notice that this is again nothing more than a process of interference between two beams.

In the following sections, the equations describing the making of a hologram and the reconstructing of a holographic image will be presented. Then, the use of holograms in interferometric measurement will be taken up. Attention will be confined to holograms of the transmission type, as these are the ones most often used for interferometric measurement. Many of the ideas discussed apply to other varieties of holography.

The trigonometric representations for light waves are not entirely suitable for developing the holography equations. The optical elements are usually scattered around the laboratory bench, and the trigonometry becomes cumbersome. The equations are shorter and easier to manage if they are put in complex variable form, so that representation will be used for a change. Review of section 2.8 is recommended to the extent necessary.

Given the interest in holography, it is not surprising that there exists a large body of literature on the subject. The basic concepts, procedures, and metrological applications, which formerly were to be found only in scientific papers, have been incorporated into several fine books. Those by Vest (1979),

Smith (1975), Waters (1974), Haskell (1971), Goodman (1968), Stroke (1969), Jones and Wykes (1983, 1989), and Hariharan (1984) are viewed as especially valuable. Taylor (1989) offers a useful brief treatment of the subject.

16.3 Producing a transmission hologram

Section 2.10 demonstrated that when two beams of coherent light having plane wavefronts are caused to intersect, the interference between the two beams causes the creation of a three-dimensional pattern of interference fringes filling the space where the two beams overlap. When a screen is placed within this overlap volume, one sees on the screen a grating of parallel lines. If the screen is a photoplate, the emulsion records the grating. This process is used, for example, in making gratings for moire interferometry. It is exactly how a hologram is made. In fact, recording the interference pattern produced by two plane waves creates a hologram that can be used to reproduce one of the incident beams. A moire interferometry grating, such as those described in the previous section, is actually a hologram.

The analysis presented in Chapter 2 can be extended intuitively to the case in which one or both of the interfering beams does not have a plane wavefront. The basic effects will be the same in that a stationary pattern of interference bands will be produced. The fringe pattern will not, however, be a simple system of parallel bands.

Figure 16.1 illustrates the case in which a plane wave interferes with the wave that is scattered and reflected by an illuminated object. Laboratory practicalities are discussed in the next chapter; for now, just concentrate on the interactions of the beams at the film.

It is this "object wave" that contains all the intelligence about size, shape, color, parallax, and so on. It has, of course, a very complex wavefront because every point in the surrounding space receives light simultaneously from every illuminated point in the surface of the object. The result is a wavefront that is complicated and that is different for each receiving point in the space near the object.

As the plane wave, hereafter called the reference beam, and the object beam merge, they interact to create a microscopically complicated interference pattern. This fringe pattern varies continuously in space, and a record of it may be made at any particular cross section by inserting a photographic emulsion on a plate or film. The optical system must be stable enough so that the film is exposed to a stationary pattern. It is important to recall that the interference "bands" have a spatial frequency on the order 2,000 lines/mm, and no element of the optical system must move more than a small fraction of the fringe spacing, which is to say that relative movements more than a fraction of the wavelength of light cannot be tolerated. The data-carrying

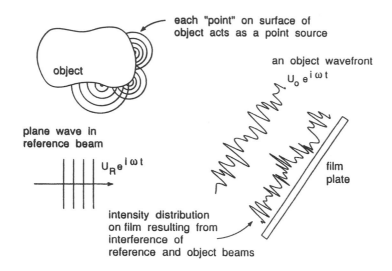

Figure 16.1. Interference of plane reference wave and object wave.

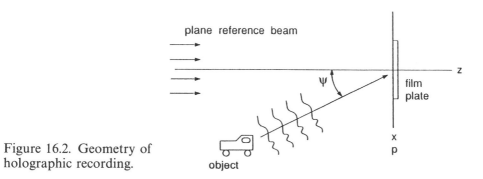

Figure 16.2. Geometry of holographic recording.

interference pattern in the film record cannot be examined except under high magnification. The photographic record of the wave interference pattern produced by two beams is called a hologram because it contains all available optical data about the object and reference beams at the film location. Detailed calculation of the complicated intensity distribution at the recording plate is not worthwhile; for one thing, it will be different for each point and for each object. It is sufficient to perform the calculation in terms of the general complex amplitudes of the reference and object beams.

In order to begin the calculation for the information stored in the hologram, a coordinate system is chosen arbitrarily. Consider the system shown in Figure 16.2, with the recording film in the xy-plane and the reference beam propagating along the z-axis. The object beam has a warped wavefront, and it meets the recording film more or less along inclination ψ to the z-axis. The implication of this restriction is that the object is small and far away from a small piece of film, a limitation that is not important in actual practice.

To ease the transition to complex notation, the first several equations are presented in both trigonometric and complex forms.

The equation for the object wave is

$$E_O = A_O(x, y, z) \cos\left[\omega t - \frac{2\pi}{\lambda}(x \sin \psi + z \cos \psi) + \phi_{OO}(x, y, z)\right] \quad (16.1)$$

At the film plate, $z = 0$, so, letting (p, q) be coordinates in the film plane,

$$E_O = A_O(p, q) \cos\left[\omega t - \frac{2\pi}{\lambda}(p \sin \psi) + \phi_{OO}(p, q)\right] \quad (16.2)$$

In complex notation,

$$E_O = \text{Re}(U_O e^{i\omega t})$$

where, at the film plane,

$$U_O = A_O(p, q) e^{-i[kp \sin \psi - \phi_{OO}(p,q)]} \quad (16.3)$$

where $k = 2\pi/\lambda$.

For the reference beam,

$$E_R = A_R(x, y, z) \cos\left(\omega t - \frac{2\pi z}{\lambda}\right) \quad (16.4)$$

where the initial phase is taken to be zero. At the film plate, $z = 0$ again, so

$$E_R = A_R(p, q) \cos(\omega t) \quad (16.5)$$

In complex form,

$$E_R = \text{Re}\left(U_R e^{i\omega t}\right)$$

where, remembering that $z = 0$ at the film,

$$U_R = A_R(p, q) \quad (16.6)$$

It is reasonable and convenient to assume that the intensity of the reference beam is uniform.

$$U_R = A_R = \text{const} \quad (16.7)$$

The light striking the film is the resultant of the object and reference beams. The equations are handled exactly as they were in section 2.9. As usual, the result can be obtained in several different forms. A direct approach with complex amplitudes yields a general form, which has the advantage of brevity,

$$U_S = U_O + U_R$$

Use equation 2.48 for intensity or irradiance,

$$I_S = U_S U_S^* = [U_O + U_R][U_O^* + U_R^*]$$
$$= U_O U_O^* + U_R U_R^* + U_O^* U_R + U_O U_R^* \quad (16.8)$$

Either form of the complex amplitude may be inserted in equation 16.8. Examination shows that the irradiance distribution at the film includes all amplitude and phase data in the incident beams as well as their complex conjugate beams.

Study of the recording process must now be undertaken. For a certain range of irradiance, the film acts as a linear detector; that is, the transmittance of the film, $T_a(p, q)$ is a linear function of the irradiance of light striking it. Such a response is also called "square law detection" since the transmittance (density of the negative) or output voltage in a photocell is proportional to the square of the amplitude. A constant is added to take care of normal residual fogging and so forth.

$$T(p, q) = K_0 + K_1 I_S(p, q) \qquad (16.9)$$

For the hologram, then

$$T(p, q) = K_0 + K_1(U_O U_O^* + U_R U_R^* + U_O^* U_R + U_O U_R^*) \qquad (16.10)$$

The exposed and developed film contains, therefore, a record of all data contained in the object and reference beams and their conjugates. The remaining problem is to recover this information in usable form. To retain more information about the detailed structure of the hologram, substitute the complex amplitudes of the image and object beams at the film plane into equation 16.10 to obtain

$$T(p, q) = K_0 + K_1 A_R^2 + K_1 A_O^2(p, q) + K_1 A_R A_O(p, q) e^{i[kp \sin \psi - \phi_{OO}(p,q)]}$$
$$+ K_1 A_R A_O(p, q) e^{-i[kp \sin \psi - \phi_{OO}(p,q)]} \qquad (16.11)$$

This wealth of information is recorded in the emulsion as a very fine and very complicated grating structure. The exposed and developed film will contain no recognizable image; in fact, the film will look as if the recording failed and the emulsion became fogged by stray light. Each small element of the emulsion contains all the data about the object as it would be observed from that point in the film. That is, each small element of the film is itself a complete hologram, as is easily demonstrated by cutting the film into small pieces and reconstructing an image from each piece.

In practice, the transmittance filter hologram is often converted to a phase filter hologram through bleaching of the exposed film. Economy of light in the data recovery process is achieved, with brighter images as the tangible reward. This aspect of holographic processing will not be discussed since it is a refinement that is not necessary to the basic holographic process.

16.4 Reconstruction of the holographic image

A hologram, according to the theory just developed, stores all information carried in the object beam. In addition, data about the reference beam and the complex conjugates of both beams are recorded. The problem is to recover the object beam in usable form. A visible image of the object is usually sought, and the image must be free from interference by spurious images or noise.

A special case of holographic reconstruction is most useful and sufficient for practical situations. This case calls for reconstruction by a replica, which is exact except for intensity, of the original reference beam. In this text, attention has been confined to the use of a plane wavefront reference. This restriction is not important in practice as long as the recording and reconstruction reference beams are identical.

The analysis of the reconstruction process is begun by rewriting the expression for the reference beam, giving it a different amplitude and calling it the reconstruction beam E_C,

$$E_C = \mathrm{Re}\; U_C e^{i\omega t}$$
$$U_C = A_C e^{-ikz}$$

(16.12)

This beam is simply used to illuminate the hologram within the same setup as established in Figures 16.1 and 16.2, with the reconstruction beam replacing the reference beam. The problem is to understand what is coming out the other side of the hologram, which is acting like a complicated diffraction grating.

At this point, one has a choice of how to proceed, and the path to be chosen depends on what is sought. The sophisticated choice is to try to predict the exact nature of the images by using diffraction theory. After all, this reconstruction process matches exactly the general diffraction model developed in the first few sections of Chapter 10. An aperture containing a transparency is illuminated by coherent light having a plane wavefront of uniform amplitude. The general theory showed that, at a distance from the aperture, the light field will involve the Fourier transform of the aperture function, that is, the hologram transmittance. This approach will be utilized to a limited extent after first gathering information by means of another line of thought. In fact, three different approaches will be utilized in order to develop maximum physical insight about the holographic image with relatively little labor.

16.4.1 *The hologram as transmittance filter*

For the moment, we adopt a path that is both simple and conventional, although somewhat troublesome of interpretation. The hologram is assumed to act like a simple transmittance filter to the reconstruction beam, with the understanding that the transmittance function refers to amplitude rather than to intensity. The light passing through the hologram will be termed E_I, which is the magnitude of the electric vector in the image beam. What we hope to get is an indication of what the light field is as it exits the hologram. We cannot press this analysis too far without Fourier transforms. If, however, the structure of the beams is recognizable, then we will have a valuable indication of how a hologram works without the complexities of diffraction theory.

With the limitations and the potential of this approach in mind, simply multiply the reconstruction beam amplitude by the hologram transmittance,

$$E_I = T(x, y)E_C \tag{16.13}$$

Equation 16.11 is combined with equations 16.12 and 16.13 to give the expression for the image beam. It is convenient to suppress the oscillation at optical frequency by working with only complex amplitudes. Rather than just use the symbols for complex amplitudes, the exponential form is maintained to enhance physical insight. After making the substitutions and carrying out the multiplications, we find the complex amplitude of the image beam to be

$$U_I = K_0 A_C e^{-ikz} + K_1 A_R^2 A_C e^{-ikz} + K_1 A_O^2(p, q) A_C e^{-ikz}$$
$$+ K_1 A_R A_C A_O(p, q) e^{i[k(p \sin \psi - z) - \phi_{OO}(p, q)]}$$
$$+ K_1 A_R A_C A_O(p, q) e^{-i[k(p \sin \psi + z) - \phi_{OO}(p, q)]} \tag{16.14}$$

Now, consider the physical interpretation of each of the terms in equation 16.14. The first is simply an attenuated reconstruction beam that propagates directly through the plate and keeps going along the z-axis without being deflected from its original path. The same is essentially true of the second and third terms. These three beams are not diffracted and are indistinguishable for the case under study. The third term does contain an amplitude modulation that depends on the original object wave, but it cannot be separated from the other two that have the same path.

Skip the fourth term of this equation for the moment. After lumping the constants, we rewrite the fifth term in the form,

$$K A_O(p, q) e^{-i[k(p \sin \psi + z) - \phi_{OO}(p, q)]} \tag{16.15}$$

Comparison with equation 16.1, after converting it to complex form (like equation 16.3 with the $z \cos \phi$ put back in), shows that the beam described is, except for the p and a missing $\cos \psi$ in the exponent, basically a constant times the original object wave. This beam, which is called the primary beam or primary wave, carries all the amplitude and phase information carried in the object wave. An eye or camera placed so as to receive a bit of this beam will not be able to tell whether it comes from the object or from the illuminated hologram. The missing $\cos \psi$ and the substitution of p for x in the exponent might be the price paid for temporarily excusing oneself from using diffraction theory. This question will be attended to in the next section.

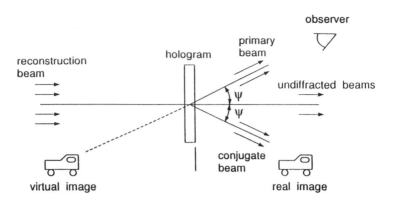

Figure 16.3. Reconstruction of holographic images.

At least, the reconstructed beam is the same as the original object beam at the film plane where $z = 0$. There is no x-term appearing in the exponential, so the direction of propagation of this beam is not clearly established by this approach. These modifications of the exponent can be said to induce a phase modification that changes the direction of the reference beam in the direction $\sin \psi$, but that is not readily apparent as yet.

Returning to the fourth term: it is seen to be the same as the one just discussed except that the sign of the exponent is partly changed. This beam, which is called the secondary or conjugate beam, contains all amplitude data $A_O(p, q)$ carried by the original object wave. The exponential has, however, a negative sign on the phase term $\phi_{OO}(p, q)$. The inference is that this beam carries image information that is correct in amplitude data and reversed in phase. If the primary image is diffracted to one side of the z-axis, then this image appears on the other side. Further physical implications are rather startling. The image, first of all, will be seen to be real; that is, it usually forms on the observer's side of the hologram, and in this setup, it can be cast directly on a screen or film plate for photography. In addition, the parallax is reversed. The side of the object that faced the viewer faces the film and is farthest from the viewer in the image. If one wants to look around the right side of the image, then the observer must move to the left! As the eye moves to the left, the image appears to rotate twice as much in that direction, resulting in more exposure of the right side. An image of this type is said to be pseudoscopic. It is interesting and entertaining, but it seems to have minimal practical use in experimental mechanics. Psychologists and researchers in physiological optics have shown some interest in it. This image is, incidentally, not easy to locate in space without some practice and/or the use of calculations or nomographs to predict where it should be.

A diagram of the reconstruction beams can now be drawn; one is shown in Figure 16.3. The locations and specific identifications of the beams are good only for the setup described. Qualitatively, however, the diagram serves for all transmission holography where this layout is approximated. Many other optical setups can be used for making and viewing transmission holograms. Then there are many ways to make other varieties of holograms, including reflection, volume, and transform types.

16.4.2 Diffraction theory applied to holography

A troubling deficiency in the conventional derivation of the nature of the reconstructed wavefronts, as described in the preceding section, is that it does not account for diffraction. The result is a lack of clear indication of the direction of travel of the beams that are exiting the hologram. Because light is being sent through a grating in a large (optically) aperture, the process is certainly one of diffraction.

General diffraction theory can be applied to the hologram reconstruction process to glean additional information about the location and structure of the various beams that exit the hologram. Considerable effort and space are needed if all the beams are to be treated in detail, and the benefits are not necessarily worth the price. For most applications of holography, only the primary beam is of use, so only that beam need be studied from the viewpoint of diffraction analysis. A very useful aspect of the crude transmittance filter approach undertaken earlier is that it tells us that the fifth term of the complex amplitudes coming through the hologram is the important one.

The holographic recording system pictured in Figure 16.2 was cleverly chosen to match the basic layout in the development of diffraction theory in Chapter 10. Rather than work with the complex amplitude of the reconstruction beam, the Fraunhofer diffraction integral, equation 10.33, will be employed. This transform equation contains the amplitude transmittance function for the diffracting aperture.

Select the fifth term from the hologram transmittance, equation 16.11. To save writing, the problem will be viewed in two dimensions only; so eliminate q and y from the transmittance. The constants can be lumped together. For simplicity of notation in the diffraction integral, replace the coordinates (x_2, z_2) of receiving point P by (x, z). Also switch the aperture variable from ξ to p to match current notation. The Fraunhofer integral for this case is found to be

$$U_p = Ke^{\frac{ikx}{2z}} \int A_O(p)e^{-i[kp \sin \psi - \phi_{OO}(p)]}e^{\frac{-ikxp}{z}} \, dp \qquad (16.16)$$

The equation looks unwieldy, and it cannot be integrated because the dependence of A_O and ϕ_{OO} on p is not known, in general. Regrouping the exponential terms does facilitate answering the question about image beam location that has been raised. The recast equation is

$$U_p = Ke^{\frac{ikz}{2z}} \int A_O(p)e^{i\phi oo(p)}e^{-ik\left[\frac{x}{z} + \sin\psi\right]p}\, dp \qquad (16.17)$$

This result is just the Fourier transform of the object beam phase and transmittance at the aperture (see eq. 16.3). But the spatial frequency metric $f = x/\lambda z$ in the transform, which represents deviation from the z-axis in transform space, has been augmented by $\sin \psi$. This situation is similar to that faced when analyzing the diffraction by a harmonic grating in section 10.5. It is another example of the frequency shifting property of transforms. At a given z the transform beam is shifted sideways by a certain amount. The integral can be reduced to normal form by taking $f' = f + (\sin \psi)/\lambda$, which amounts to a coordinate rotation in transform space. The primary image beam is now known to be the Fourier transform of the original object beam as propagated along the line given by $x/z = \sin \psi$. The small-angle assumption was invoked in developing the Fraunhofer equation. To within that limitation, the reconstructed object beam is shown to continue on the path the original object beam had in traveling from the object to the holographic film.

There is one more useful observation to be made. How do we know that if we put an eye or a camera downstream in the primary beam that an image of the object will be seen? We learned in Chapter 10 that the Fourier transform for this large-aperture system will not be seen on a screen unless it is several kilometers away from the hologram. Within only a meter or so, it would seem that the complex amplitude has not changed much from that obtained in equation 16.14. At longer distances, the object (image) will indeed seem to be very small. This idea can be developed further by rotating the coordinate system as mentioned before.

16.4.3 The hologram as a diffraction grating

A third and very fruitful line of thought about holograms is to consider them to be complicated diffraction gratings. To illustrate the approach and to gain two more increments of physical insight, an interesting special case is examined briefly. Suppose that the object beam in the holographic system has a plane wavefront. The interference of two such plane waves produces a volume grating in space, as was demonstrated in section 2.9. The planes of zero irradiance are separated by the distance $D_\perp = \lambda/2 \sin(\psi/2)$; these planes, in the holographic setup we are using, lie at an angle of $\psi/2$ with the z-axis. If a photographic film is placed in the xy-plane, it will record a grating with a separation between the dark lines of pitch $= D_\perp/\cos(\psi/2) = \lambda/\sin \psi$. The spatial frequency is $f = \sin \psi/\lambda$. If what we have taken as the original object beam impinges on the grating, it will be diffracted by an angle given by $\sin \theta = \lambda/\text{pitch} = \sin \psi$ (see section 10.5). The diffracted wavefront will be found to be plane. In other words, the object wave has indeed been regenerated. This is a special case of holography.

Think now what will happen if, for this same arrangement, the reference beam (the one along the z-axis) is switched off, and the original object beam is switched back on. Calculations similar to those just finished will show that the old reference beam will be recreated by the diffraction at the grating.

This reciprocal relationship between the beams exists for any set of wavefronts. If the object beam is a converging wavefront, for example, the hologram can be used to recreate a converging wave from a plane reference beam. Something that can convert a plane wavefront to a spherical one is usually called a lens. The conclusion is that it is possible to create a holographic lens that is just a flat piece of film. Other types of holographic optical components can be manufactured by extending this idea. One of the primary practical applications of holography is in the manufacture of inexpensive, compact, and unusual optical devices such as flat lenses.

The approach just outlined highlights the use of two-beam interference to record a hologram and diffraction to reconstruct the images. It is easy to extend the thinking to a more general case. If the object beam is more complicated, as might come from an oddly shaped object when it is illuminated by coherent light, the diffraction grating will become very complex. On the microscopic scale it is the same as the simple grating previously discussed. The grating spatial frequency and orientation change from one small area to another. But the diffraction of each piece will be the same as before.

In fact, the complicated diffraction grating can be seen as a perturbation of the simple grating created with two plane waves. Over the extent of the object wave, the angle of incidence of the object wave will not deviate greatly from the basic value of ψ that is dictated by the holographic setup (no matter if it does). The object phase term might vary wildly from 0 to 2π. Thus, the basic simple grating can be viewed as a set of carrier fringes that are modulated by the intelligence in the object beam. People who are familiar with communications theory find the carrier fringe model to be useful in understanding holography and other forms of interferometry.

16.5 Holographic interferometry

The basic idea in holographic interferometry is that the image formed in holography is so exact that it can be compared interferometrically with the object or with another holographic image of the object. If the object is disturbed by stress, vibration, or heat, by even a minute amount, then a pattern of interference fringes will be observed in the image. By counting fringes, one can determine the amount of displacement of the object. The sensitivity of measurement is on the order of the wavelength of light: less than a millionth of a millimeter.

The difficulty is in getting the image and object superimposed. The only way to do this accurately enough is to use exactly the same arrangement for making the hologram and for viewing the image. This stringent requirement has led to the development of three basic techniques for performing holographic interferometry. These approaches will be discussed in turn, with emphasis on concept and physical insight. In fact, most practical applications of holographic interferometry to date have been in qualitative observations of mechanical response. For such applications – which include, for example, holographic nondestructive testing – knowledge of the quantitative interpretation of holographic interferograms is not required. Following the qualitative discussion, some methods for quantitative interpretation are presented.

16.5.1 Double-exposure or frozen-fringe technique

A logical approach is to record holograms of the object in two or more different states on the same photoplate by using double exposure. Given the wealth of information contained in the microscopic structure of a hologram, it is difficult to believe that two entirely different such structures can be stored in the same thin emulsion. Actually, it is possible to store several holograms in the same photoplate.

All information to create the fringe pattern is permanently stored in the hologram, and the fringes will be superimposed on the image as it is reconstructed. To see this, the holographic equations developed previously are utilized with two changes. For simplicity of demonstration, let the object beam be the one traveling along the z-axis of Figure 16.2. For double-exposure interferometry, the hologram carries two complete and separate holograms. One of these records the object wave,

$$U_{O1} = A_1(x, y)e^{i\phi_1(x,y)} \tag{16.18}$$

The other object wave that is stored is

$$U_{O2} = A_2(x, y)e^{i\phi_2(x,y)} \tag{16.19}$$

Because the motion of the object between exposures is minute and nothing else changes, the amplitude distribution from the object does not change appreciably, so $A_1(x, y) = A_2(x, y) = A(x, y)$. The new phase may be expressed as the initial phase plus the phase change: $\phi_2(x, y) = \phi_1(x, y) + \Delta\phi(x, y)$. The motion of the object will be large relative to the wavelength of light, so the phase change will be large. Use the fifth term of equation 16.14 to obtain the complex amplitude of the primary reconstruction beam corresponding to the virtual image:

$$U_1 = KA(x, y)\{e^{i\phi(x, y)} + e^{i[\phi(x, y) + \Delta\phi(x, y)]}\} \qquad (16.20)$$

The irradiance of the beam is then found to be

$$I = U_1 U_1^* = K^2 A^2(x, y)[2 + e^{-i\Delta\phi(x, y)} + e^{i\Delta\phi(x, y)}] \qquad (16.21)$$

To extract physical meaning from this result, it may be converted back to the trigonometric form:

$$I = 2K^2 A^2(x, y)\{1 + \cos[\Delta\phi(x, y)]\} \qquad (16.22)$$

This result represents the irradiance distribution of the object multiplied by a phase term. That is, the object image is modulated, or has superimposed on it a system of light and dark fringes that are characteristic of simple two-beam interferometry. The dark fringes occur for

$$\cos \Delta\phi + 1 = 0 \qquad \Delta\phi = (2n - 1)\pi, \quad n = 1, 2, 3, \ldots \qquad (16.23)$$

The bright fringes are loci of points at which the phase shifts are even multiples of π.

There is much more to be said about quantitative fringe interpretation to obtain object motion since geometric factors and parallax problems (fringe localization) must be taken into account. Some of these matters will be taken up later.

Figure 16.4 is an example of the fringe patterns that can be recorded and reconstructed by the double-exposure method. The subject is an aircraft gear. This and several other results shown in this chapter were kindly provided by Dr. Karl Stetson, one of the inventors of hologram interferometry, and the author is most grateful.

The double-exposure technique is considered to be easier to carry out successfully than is the real-time method. The major difficulty is in predicting just how much disturbance of the object between exposures will give a good fringe pattern without loss of correlation between the two reconstructed images. This motion of the object must be predictable and controlled. Again, motions in the setup must be held to near zero through the recording phase of the experiment.

A disadvantage is that one cannot watch the fringes form as displacement is applied. The ability to correlate fringe changes with object state in real time is lost. The related advantage is that there is no requirement against disturbing the apparatus between making the hologram and creating the fringe patterns. The fringes are permanently stored, and they can be reconstructed at a later time and with a different setup. For some experiments, it is the only workable approach. For example, holograms of two stages of transient motion can be recorded using a pulsed laser. The resulting

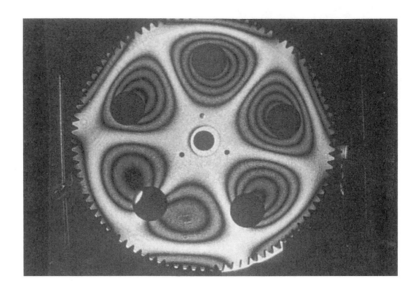

Figure 16.4. Fringe pattern created by double-exposure holographic interferometry on a helicopter gear (courtesy of Karl Stetson).

two-exposure hologram with its interference fringes can be reconstructed at leisure using a continuous-wave gas or ion laser.

16.5.2 Real-time or live-fringe technique

This approach requires that the photographic plate be physically developed in place without disturbing its position, or else one can use a plate holder that will mechanically return the plate to its original position after development.

After the single-exposure hologram is made in the normal way, it is illuminated by the reference beam while the object is lit by the original object beam. It might be necessary to change the amplitude ratio between the two beams in the observation stage. An observer will then see the holographic virtual image of the object superimposed on the object. If the object is disturbed in any way, or if the hologram is distorted or moved, interference fringes will be seen. As an illustration of the idea, Figure 16.5 shows how the fringes are formed in a simple case. The formation of fringes by the two object beams can be predicted in a manner similar to that used above for the double-exposure method. One exposure for the original state of the object is recorded and reconstructed. Its complex amplitude is

$$U_{OR} = -K_1 A(x, y)e^{i\phi(x, y)} \tag{16.24}$$

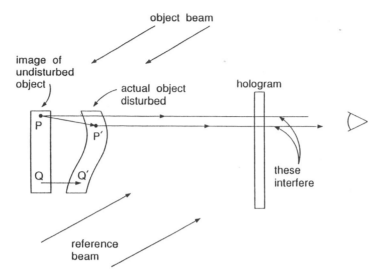

Figure 16.5. Fringe formation in real-time holographic interferometry.

A negative multiplier has been imposed because the hologram is actually a negative. This fact has been ignored as not important until this point. The various constants have again been lumped into a single parameter. The complex amplitude for the object in its displaced state has the same form, except for a phase change resulting from displacement and a different multiplicative constant. Again, the amplitude distribution has not changed appreciably because of the small displacements involved. The new object wave as received at the hologram plane is

$$U_O = K_2 A(x, y) e^{i[\phi(x, y) + \Delta\phi(x, y)]} \tag{16.25}$$

The observer at the viewing plane will receive the sum of these two beams. For convenience, it is assumed that the amplitude of the reconstruction wave and the object wave has been balanced with the transmittance and diffraction efficiency of the hologram so that the two constants K_1 and K_2 are equal. In the laboratory, this measure is a practical one in that it gives maximum fringe visibility. A variable beam splitter or attenuating filter is adjusted to give the most visible fringes, and this assumption is satisfied. The complex amplitude sum is

$$U_S = U_O + U_{OR} = KA(x, y)e^{i\phi}[-1 + e^{i\Delta\phi(x, y)}] \tag{16.26}$$

The irradiance is computed as usual, with the result

$$I_S = U_S U_S^* = K^2 A^2(x, y)[2 - e^{i\Delta\phi(x, y)} - e^{-i\Delta\phi(x, y)}] \tag{16.27}$$

$$I_S = 2KA^2(x, y)[1 - \cos \Delta\phi] \tag{16.28}$$

Once more, the problem has been reduced to one of two-beam interferometry. The amplitude function for the illuminated object is modulated by a series of fringes. Dark bands define the loci of points for which

$$\Delta\phi = 2(n - 1)\pi, \quad n = 1, 2, 3, \dots \tag{16.29}$$

As with the other forms of holographic interferometry, more must be said about how the fringes form in space. In general, they are not localized on the object.

The procedure just described, the real-time or live-fringe technique, allows immediate observation of the changes in the fringe pattern as the object is disturbed through various states. Figures 16.6 and 16.7 are samples of fringe patterns that can be obtained by this method. They illustrate two different but common applications. The contour fringes of Figure 16.6 were obtained by placing the laboratory mortar in an immersion tank. The refractive index of the immersion medium was changed between exposures. A very simple setup was used for this student exercise, so speckle effects are serious and fringe visibility is not good. Figure 16.7 shows fringes indicative of out-of-plane displacement of a circuit board as it is heated with an internal trace. The board was set with its edge toward the viewer and two mirrors were employed in order to make a hologram containing views of both sides of the board.

The major difficulty with the real-time approach lies in the requirement to return the hologram exactly to its original position after development. It must be within a few wavelengths or so for the object beams to be correlated. Even these small errors will generate an initial fringe pattern that might distort the fringes resulting from object motion. Development of the photoplate in place is one way to get around this problem, but that creates processing difficulties. Special plate holders and developing systems are available, and these facilitate use of the live-fringe method. A second source of trouble is in the necessity for absolute control of the displacement of the object during fringe formation. Any unintentional bumps will cause registration of the image with the object to be lost, and the experiment will have to be started over. A related concern is that the other elements of the optical system must not be disturbed in any way.

16.5.3 Time-average holographic interferometry

This approach is used for modal analysis of vibrating bodies. A hologram of the moving object is made with an exposure time lasting over several periods of the vibrational motion. Figure 16.8 is a time-average holographic interferogram of the vibrations of a plate.

At first glance, this method seems to violate the severe stability requirement

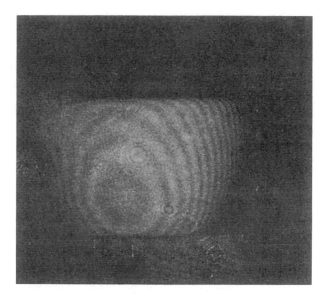

Figure 16.6. Fringes obtained by real-time holographic interferometry using a simple refractive index method on a laboratory mortar; the fringes are a contour map connecting points of equal distance from the observer.

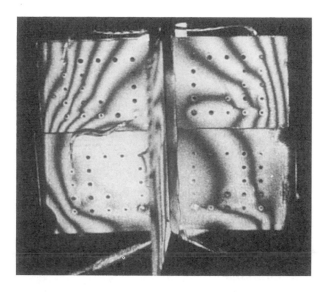

Figure 16.7. Displacement fringes obtained by real-time holographic interferometry simultaneously for both sides of a circuit board that is heated (courtesy of Karl Stetson).

Figure 16.8. Time-average holographic interferogram of a vibrational mode in an aircraft engine impeller fan (courtesy of Karl Stetson).

that is inherent in interferometry. In qualitative physical terms, the explanation lies in the fact that the motions are absolutely periodic. The complex amplitude from the object varies in a cyclic manner. The time average of the complex amplitude is translated in the hologram into an average of an oscillating grating. A harmonically vibrating object spends most of its vibrational period near its extreme positions, where it stops and changes direction. The average of the complex amplitude, and therefore of the holographic grating, over several periods will be biased toward the two extremes of the object motion. The result is a hologram with essentially two exposures corresponding to the two averages. Naturally, the two hologram gratings are not as clearly defined as they are for an object that is stationary in its two positions, especially for points on the body that undergo large motions. As a result, the holographic interference fringes show diminished contrast. The brightest and most visible interferometric fringes will correspond to the stationary vibrational nodes on the object.

The approach to mathematical description of time-average interferometry is similar to that of the double-exposure variety. Some details will be left out. Persons who anticipate heavy involvement should review the extensive literature on the subject (e.g., Stetson 1974; Vest 1979). Discussion is confined to the case where the time varying displacement is along the z-axis, which, in the holographic arrangement used, is along the line of sight between object and observer. The displacement is a periodic function of time, and the development is simplified if the displacement is allowed to vary only with x and time; take it to be $z(x) \sin \omega t$. If $\phi(x, y)$ is the resting phase distribution, then the object complex amplitude at the film plane is

$$U_O = A(x, y)e^{i\left[\phi(x, y) + \frac{4\pi}{\lambda} z(x) \sin \omega t\right]} \tag{16.30}$$

The time-average hologram is recorded with this object beam and the usual reference beam for a time T that is longer than several periods of the motion. The reconstructed object wave has a complex amplitude that is proportional to the time average of the U_O over the time T, which is

$$U_{O\,av} = A_{av}(x, y)\frac{1}{T}\int_0^T e^{i\left[\phi(x, y) + \frac{4\pi}{\lambda} z(x) \sin \omega t\right]} dt \tag{16.31}$$

The integral can be shown to be

$$J_0\left[\left(\frac{4\pi}{\lambda}\right)z(x)\right] \tag{16.32}$$

The J_0 is the zero-order Bessel function of the first kind. The irradiance is calculated as $U_{O\,av} U_{O\,av}^*$, which yields

$$I(x, y) = A^2(x, y)J_0^2\left[\left(\frac{4\pi}{\lambda}\right)z(x)\right] \tag{16.33}$$

In this case, the image has superimposed on it a system of fringes that correspond to the minima of the square of the zero-order Bessel function. These minima are not equally spaced, and they should not be interpreted like other fringes. The first order has much larger amplitude than subsequent orders, which is what makes the nodal areas the brightest in a time-average interferogram. A plot of the function and the location of the zeros is given in section 21.6.

16.6 Qualitative interpretation

Knowledge of the general form and meaning of the various hologram interferometry equations allows one to use these methods in a variety of applications where quantitative data about a displacement or strain field are not needed. With some background, it is possible to look at a fringe pattern and gain insight into the location of flaws, effect of stress concentrations, or vibrational mode.

As an example, one of the main engineering uses of holographic interferometry is in the detection and qualitative observation of hidden voids and flaws in objects. The subsurface discontinuity will affect the surface displacements enough so that, as the specimen is strained or heated, interference fringes will gather around the flaw area. One specific money-saving industrial application is in the inspection of honeycomb structures used in fabricating aircraft. Faulty bonds of honeycomb or honeycomb-to-skin are easily found by this nondestructive technique. Potential applications in design and production applications seem limitless, although the method has not yet been widely accepted.

A similar application uses one of these interferometric approaches to discover faulty contouring of complex machined or molded precision parts, such as lenses or mirrors, when they might still be in a semifinished state. As with many such uses, the goal is to discover only important anomalies in the fringe pattern. There is no fringe counting or calculation of displacements in applications of this qualitative type.

16.7 Measurement of out-of-plane motion

As an introduction to quantitative use of holographic interferometry, a special but very useful case is considered. Let the object be one that is displaced in only one direction, and arrange the optical setup so that displacement is along the line from object to hologram. The beam that illuminates the object is also close to the z-direction. The object need not be flat, as is often thought. Figure 16.9 illustrates the arrangement.

The path length difference, which creates the phase difference $\Delta\phi$ in the interferometry equations developed in section 16.5, is easily calculated and related to the observed fringe orders for this simple geometry. For the first position of point P on the object, the path traveled is SPQ. For the second position it is SP'Q. Subtraction gives the optical path length difference, which is 2d. The observer will see a pattern of fringes on the image as the two beams interfere along the common path. The fringe order will change by one increment between points that give a phase change of 2π or one wavelength of light. If the dark-fringe order can be established at one point, then the fringes can be assigned orders n starting from that point. The relationship between fringe order and displacement for this case is found to be

$$d = \frac{n\lambda}{2} \tag{16.34}$$

where n is an integer. This problem reduces to the same result as some forms of classical interferometry that involve amplitude division, such as Newton's rings.

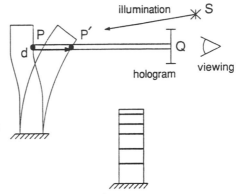

Figure 16.9. Holographic measurement of one-dimensional motion: (a) optical arrangement; (b) typical pattern of fringes for this case.

Many practical applications of holographic interferometry can be made to correspond well enough to the case just described. The bending of plates and beams is one such example. Contour mapping and determination of refractive index variations in a volume of fluid are two more; although more must be said about the latter case, because the factor of 2 will probably not be necessary.

16.8 Sensitivity vector

The computation of three-dimensional displacements for objects that might have complicated shapes is much more difficult than it is for one-dimensional motion. Geometrical factors affecting the path length difference must be accounted for. In general, three observations are required for the three unknown displacement components. To minimize uncertainty, it might be best to use two or three simultaneous holograms to obtain the required three pieces of data. This problem has been studied by several outstanding analysts. Their work is summarized and presented systematically by Vest (1979) and by Ennos (1974). Their presentations have evolved into a standard form, and it is well to utilize it with gratitude to the talented investigators who devised it. The latter treatment (Ennos 1974) is in the context of strain analysis, which is much more difficult than simply determining displacements because of the necessity to differentiate the primary data.

As mentioned, this derivation presupposes that the phase information coming from the observed specimen point P for the initial state can be combined with the phase data coming from the same point after the point has moved. This assumption is not exactly fulfilled in practice except for the one-dimensional problem or for vanishingly small displacements. It is met sufficiently well for real optical systems in which the object undergoes small

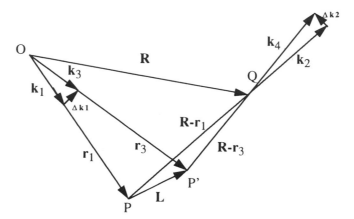

Figure 16.10. Position and propagation vectors in holographic and speckle interferometry.

displacements. Also, light from adjacent points is assumed to not interact with light from the point being examined.

The sensitivity of holographic and speckle interferometry measurement is governed by the orientation relation between three vectors: namely, the illumination vector, the displacement vector, and the viewing vector. Figure 16.10 shows these vectors in a general measurement situation.

The object is illuminated by a point source located at O. Light is scattered by an object point P to an observer or viewing plane at point Q. When the object is displaced, so that the point P is displaced by vector **L** to P′, there is a change in the optical path from the source O to a point in the viewing plane via a given point P in the object. The change in phase associated with this change in optical path is the basis of holographic and speckle correlation techniques for measuring surface displacements.

Let δ represent the phase shift as an angle, usually called $\Delta\phi$, of light scattered by point P on the object in a particular direction. In Figure 16.10, several vectors are defined for use in determining the relation between δ and **L**. Vectors **R** and $\mathbf{r}_1$ lie in the plane defined by points O, P, and Q; and $\mathbf{k}_1$ and $\mathbf{k}_2$ are the propagation vectors of the light illuminating P and the light scattered toward the viewing plane, respectively. Since the magnitude of a propagation vector is $2\pi/\lambda$, the phases of the two light rays that reach the viewing plane are as follows:

$$\phi_1 = \mathbf{k}_1\cdot\mathbf{r}_1 + \mathbf{k}_2\cdot(\mathbf{R} - \mathbf{r}_1) + \phi_0 \tag{16.35}$$

and

$$\phi_2 = \mathbf{k}_3\cdot\mathbf{r}_3 + \mathbf{k}_4\cdot(\mathbf{R} - \mathbf{r}_3) + \phi_0 \tag{16.36}$$

where ϕ_1 is the phase of the light scattered by P before displacement

ϕ_2 is the phase of the light scattered by P after displacement

ϕ_0 is the arbitrary phase assigned to these rays at the point source O

At the detector or viewing plane, the two beams are somehow combined in a way that is sensitive to phase. This may be by direct interferometric combination or by addition of two fields that have been converted to amplitude data by separate interference with reference beams. The latter case is important in speckle interferometry. For the moment, visualize the process as adding interferometrically two fields, one for each state of point P. The phase difference measured at the viewing plane would be

$$\delta\phi = \phi_2 - \phi_1 \qquad (16.37)$$

This quantity δ is already in radian units, so it should not be interpreted as a distance. After displacement of P to P′, the propagation vectors in the illumination and viewing directions are $\mathbf{k}_3$ and $\mathbf{k}_4$. Define the small changes, $\Delta\mathbf{k}_1$ and $\Delta\mathbf{k}_2$, in these propagation vectors by

$$\mathbf{k}_3 = \mathbf{k}_1 + \Delta\mathbf{k}_1 \qquad \mathbf{k}_4 = \mathbf{k}_2 + \Delta\mathbf{k}_2 \qquad (16.38)$$

Combining the preceding equations gives

$$\delta\phi = (\mathbf{k}_2 - \mathbf{k}_1) \cdot (\mathbf{r}_1 - \mathbf{r}_3) + \Delta\mathbf{k}_1 \cdot \mathbf{r}_3 + \Delta\mathbf{k}_2 \cdot (\mathbf{R} - \mathbf{r}_3) \qquad (16.39)$$

In practical situations, the magnitudes of $\mathbf{r}_1$ and $\mathbf{r}_2$ are much larger than $L = |\mathbf{r}_1 - \mathbf{r}_3|$; so for practical purposes, $\Delta\mathbf{k}_1 \perp \mathbf{r}_3$ and $\Delta\mathbf{k}_2 \perp (\mathbf{R} - \mathbf{r}_3)$. Because of these simplifications, the last two scalar products in equation (16.39) vanish, and the phase difference becomes

$$\delta\phi = (\mathbf{k}_2 - \mathbf{k}_1) \cdot \mathbf{L} \qquad (16.40)$$

This relation forms the basis of quantitative interpretation of the fringes of both holographic and speckle interferometry.

A sensitivity vector $\mathbf{K}$ may be defined as

$$\mathbf{K} = \mathbf{k}_2 - \mathbf{k}_1 \qquad (16.41)$$

so that

$$\delta\phi = \mathbf{K} \cdot \mathbf{L} \qquad (16.42)$$

This phase change is already in radians, as mentioned before.

The result just derived shows that if the object and illuminating beam directions are known, and if the phase change is measured, then the displacement vector can be determined. As expected, if the displacement has three unknown components, then three observations with different viewing directions are required, and there are three equations to solve simultaneously.

A useful observation is that the sensitivity vector defined by equation 16.41 has physical meaning. Let 2θ be the angle between the illumination and object beam directions. Then the magnitude of $\mathbf{K}$ is $(2\pi/\lambda) \cos \theta$, and it is directed along the bisector of the angle between viewing and object beam directions. This fact can be helpful in establishing the best directions to use for measuring a particular displacement component.

Consider for a moment the problem involving only out-of-plane displacement discussed in the previous section. For that situation, $\mathbf{k}_1 = -\mathbf{k}_2$, and $\mathbf{L}$ is parallel to both $\mathbf{k}$'s. The resulting relationship between fringe order and displacement is the same as obtained directly in producing equation 16.33. The equation is now easily modified to account for the fact that, in practical applications, the object beam directions cannot be made exactly along the line of sight. For that matter, the equation is easily solved for any given setup if it is known that there is only one displacement component to worry about, that is, if the direction of the unknown displacement is known.

Another special situation that is tractable is when the displacement is known to lie in the plane of the object and illuminating beams.

A serious problem arises in more general situations because of uncertainty and error propagation. Simply stated, the window size of a typical hologram is not large enough to take three observations from widely spaced viewing locations. Thus, there is not a great difference between the fringe orders viewed from the three locations. Also, the uncertainty in fringe order observations tends to be fairly large. The procedure requires determination of the small difference between large quantities that are not known with good precision. The errors are worsened when strains are sought, because they involve the spatial derivatives of the displacements.

Some clever approaches have been developed that give improved results. One method uses two or three separate but simultaneously made holograms to increase the effective size of the viewing window. The conditioning of the equations is thereby improved. Another idea is to use a telescope to count the fringes and to take four such observations instead of three.

The derivation of the sensitivity vector demonstrates that holographic interferometry is subject to the same limitation that affects other interferometric methods, from Newton's rings on up. The phase difference is a function of the viewing and illumination angles as well as of the displacement of the point being examined. It is not possible to calculate with practical ease the exact displacement in certain arrangements. A setup that uses spherical wavefronts at close range, so that the incidence and viewing angles vary over the field, is an example of a case in which the fringes cannot be easily interpreted to obtain exact displacements.

The use of holography for quantitative determination of general displacements will not be pursued further here. The educational value of further analysis seems marginal. Many important applications can be reduced to the case of one unknown by careful planning. Much of the practical use of

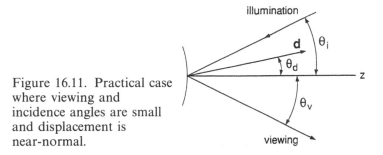

Figure 16.11. Practical case where viewing and incidence angles are small and displacement is near-normal.

holography in engineering involves only qualitative or semiquantitative analysis.

One further observation with regard to a practical situation will be useful in subsequent chapters. It also underlines the importance of the relationship between angles of viewing and incidence and phase. For most feasible holographic setups, the fringe pattern represents in-plane as well as out-of-plane displacement. If, however, the illumination and viewing angles are small, say less than 15° to the normal of the object surface, and if the displacement vector is known to lie between the observation and viewing directions, then the fringes give a good approximation to only the normal displacement component. To see this, consider the practical setup of Figure 16.11. The path length difference can be calculated directly as

$$\delta = \frac{2\pi}{\lambda} \left[d \cos(\theta_i - \theta_d) + d \cos(\theta_v + \theta_d) \right] \tag{16.43}$$

where
θ_i is the angle of illumination to surface-normal
θ_v is the angle of viewing to surface-normal
d is the magnitude of the displacement vector oriented somewhere within θ_i and θ_v

After the use of some identities,

$$\delta = \frac{2\pi d}{\lambda} \left[\cos \theta_i \cos \theta_d + \sin \theta_i \sin \theta_d + \cos \theta_v \cos \theta_d - \sin \theta_v \sin \theta_d \right] \tag{16.44}$$

If the angles are reasonably small, then equation 16.44 reduces to

$$\delta = \frac{2\pi}{\lambda} \left[\cos \theta_i + \cos \theta_d \right] d_z \tag{16.45}$$

where $d_z = d \cos \theta_d$ is the normal displacement. This approximate equation reduces easily to equation 16.34 for the one-dimensional problem.

To gain further insight into this angular dependence, let d_m be the displacement measured according to equation 16.45, and let d_z be the actual

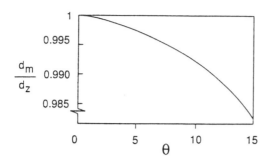

Figure 16.12. The variation of d_m/d_z resulting from changes of illuminating and viewing angles.

normal displacement (as would be measured if the angles were zero). The preceding equation can be used to obtain

$$\frac{d_m}{d_z} = \frac{\cos \theta_i + \cos \theta_v}{2} \tag{16.46}$$

The approximation relationship of d_m/d_z as a function of incidence angle is graphically illustrated in Figure 16.12.

16.9 Fringe localization

Section 16.8 incorporates an implicit assumption that there is parallax between the object image and the perceived fringes. That this assumption is true is easily demonstrated for most holographic interferograms. One only need view the fringe-covered image through different areas of the hologram (move the eye from side to side) to see that the fringes move with respect to the image. A similar effect is noticed if the eye is held stationary but the plane of focus of the eye is changed consciously. That means that the fringes are not localized on the surface of the object/image. They exist somewhere else in space, and the eye sees them as projected onto the virtual image. Indeed, this parallax is what allows one to utilize three observations from different viewpoints to obtain enough data to solve for the displacement vector in the general case.

A very simple model will illustrate the fundamental causes of fringe parallax in a way that is sufficient for our purposes. Consider the system shown in Figure 16.13 for viewing holographic fringes caused by object deformation or displacement. In this sketch, P and Q are points on the original object, and P′ and Q′ are the corresponding points on the deformed object. P″ and Q″ are the corresponding image points on the camera film or the retina of the eye. The aperture is assumed very small, so that the pencils of light from an object point that pass through the aperture to form the image are represented as lines. The eye or camera is first focused at the plane RS. The images of the object points might be defocused at the retina, but that does

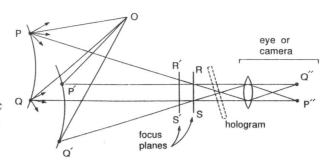

Figure 16.13. Fringe localization in holographic reconstruction when using the eye or camera to view interferometric fringes.

not affect the locations. It is helpful to assume that the setup is such that adjacent interferometric fringes are formed with the light waves that come from P and Q. That is, the path length difference OPRP″ − OP′RP″ is such that destructive interference will cause a dark fringe at P″. The same is true for OQSQ″ − OQ′SQ″. It is useful to realize that the fringe-forming interference does not occur until the waves meet at the detector, in this case the retina or the film. The focus of the eye does not affect the fringe sharpness.

Now, imagine that the eye is focused at plane R′S′. Redraw the ray diagram so that rays from P and P′ converge through a point on the new plane of focus and manage to get through the aperture of the eye lens and form the image at P″. (This is easier to imagine than it is to draw in a foreshortened sketch where the displacements PP′ and QQ′ are exaggerated.) To change the focus physically, either the lens-to-film distance must change (simple camera) or the focal length of the lens must change (eye). All the path lengths are therefore changed from their previous values. There is no reason to think that the new path length differences are such that dark fringes should be reproduced at the image points on film or retina. If the change of focus plane is caused to occur slowly, it will look as if the fringes are sliding away from their initial positions with respect to the image of the object.

A more complete analysis of parallax, or fringe localization, augments understanding of fringe formation in holographic interferometry, and the results are helpful in interpreting the details of fringe patterns. Given the typical uses to which holography is put, and in consideration of the purpose and scope of this book, the analysis is not undertaken. Readers who anticipate using holographic interferometry for measurement of strain and deformation in three-dimensional problems should study the special literature on the topic (e.g., Vest 1979).

The analysis of fringe parallax does lead to some observations that are useful in even elementary applications of interferometry. One is that a small viewing aperture is required in order to see fringes at all. Such an aperture limits the solid angle of the pencil or cone of light used by the imaging optics to form the image. If the aperture is increased, the fringes become less visible and eventually vanish. If the aperture is made too small, then speckle effects (see Chapter 18) begin to interfere with fringe visibility. One can verify these opposing effects by using a camera to look at the fringes.

The more precise treatment, in which the cone or pencil of light from an object point that passes through the aperture to form the image is used, also shows that the fringes will appear to be localized at a curved surface in space that does not, in general, coincide with the specimen surface.

Another useful fact is that there are cases for which the fringes are located at the object surface, and no parallax will be seen. It is no surprise that the special one-dimensional problem discussed in section 16.7 is in this category. If that setup is modified by the very practical measure of having the illumination beam at 45°, the parallax is again zero. A related finding is that there are combinations of displacement and optical arrangement for which no fringes will be seen, regardless of viewing aperture. Many cases have been tabulated (e.g., Vest 1979).

References

Brooks, R. E., Heflinger, L. O., and Wuerker, R. F. (1965). Interferometry with a holographically reconstructed comparison beam. *Applied Physics Letters*, 7: 248–9.

Brown, G. W. (1974). Pneumatic tire inspection. In *Holographic Nondestructive Testing*, Ed. R. K. Erf, Sec. 8.11. New York: Academic Press.

Ennos, A. E. (1974). Strain measurement. In *Holographic Nondestructive Testing*, Ed. R. K. Erf, Sec. 8.5. New York: Academic Press.

Goodman, J. W. (1968). *Introduction to Fourier Optics*. New York: McGraw-Hill.

Haines, K. A., and Hildebrand, B. P. (1966). Surface deformation measurement using the wavefront reconstruction technique. *Applied Optics*, 5: 595–602.

Hariharan, P. (1984). *Optical Holography*. Cambridge University Press.

Haskell, R. E. (1971). *Introduction to Coherent Optics*. Rochester MI: Oakland University.

Horman, M. H. (1965). An application of wavefront reconstruction to interferometry. *Applied Optics*, 4: 333–6.

Jones, R., and Wykes, C. (1983). *Holographic and Speckle Interferometry*. Cambridge University Press.

Jones, R., and Wykes, C. (1989). *Holographic and Speckle Interferometry*, 2nd ed. Cambridge University Press.

Leith, E. N., and Upatnieks, J. (1962). Reconstructed wavefronts and communication theory. *Journal Optical Society of America*, 52: 1123–1130.

Leith, E. N., and Upatnieks, J. (1964). Wavefront reconstruction with diffused illumination and three-dimensional objects. *Journal Optical Society of America*, 54: 1295–1301.

Smith, H. M. (1975). *Principles of Holography*, 2nd ed. New York: John Wiley and Sons.

Stetson, K. A. (1974). Holographic vibration analysis. In *Holographic Nondestructive Testing*, Ed. R. K. Erf, Ch. 7. New York: Academic Press.

Stetson, K. A., and Powell, R. L. (1965). Interferometric hologram evaluation and real-time vibration analysis of diffuse objects. *Journal Optical Society of America*, 55: 1694–5.

Stroke, G. W. (1969). *An Introduction to Coherent Optics and Holography*, 2nd ed. New York: Academic Press.

Taylor, C. E. (1989). Holography. In *Manual on Experimental Stress Analysis*, 5th ed., Ed. J. F. Doyle and J. W. Phillips, pp. 136–49. Bethel, CT: Society for Experimental Mechanics.

Vest, C. M. (1979). *Holographic Interferometry*. New York: John Wiley and Sons.

Waters, J. P. (1974). Holography (Ch. 2) and Interferometric Holography (Ch. 4). In *Holographic Nondestructive Testing*, Ed. R. K. Erf. New York: Academic Press.

17

Holographic interferometry methods

This chapter deals with the apparatus, materials, and experimental details involved in making simple holograms and in performing holographic interferometry to obtain data indicative of shape, deformation, stress, or other phenomena. The practitioner is reminded that, except for stability requirements and restrictions on maximum path length difference, holography is very forgiving. There are some fundamental rules, but a multitude of setups and materials can be made to work. There is much room for inventiveness and resourcefulness. Also, there are many details, and experimenters who want to go beyond the basics should review at least some of the standard literature (Jones and Wykes 1989; Ranson, Sutton, and Peters 1987; Smith 1975; Vest 1979; Waters 1974).

17.1 Some basic rules

Before getting into laboratory details, some fundamental requirements for successful holography should be summarized. These are derived from theoretical considerations and from experience.

1. The apparatus must be stable for the duration of the exposure. In interferometry, this stability requirement extends through the viewing of the real-time fringes or for the period of recording both exposures in the frozen-fringe technique. Recall that the process involves recording a grating structure caused by two-beam interference. Motions attaining a fraction of a wavelength of light between any of the optical components will cause the grating to move in space so it cannot be recorded. The setup must be isolated from floor vibrations, air-coupled sound waves that might cause resonance of one of the optical components, and thermal transients.

2. The optical path length differences must not be so large that interference cannot occur. The difference between object beam path (from beam divider to film plate) and the reference beam path must be within the coherence length of the laser. For

374

practical purposes, this difference limit can be taken as approximately the length of the laser used. A hank of string is useful for comparing these lengths. The actual path lengths are not important, the path length difference is.

3. Adverse polarization changes are not allowed. Only beams having the same polarizations can interfere. Polarization of a beam can be changed by reflection, and sometimes one is inadvertently trying to make a hologram with object and reference beams that have differing polarization states. This problem usually does not appear with setups in which the beams are all close to one plane. Polarization must be considered if one of the beams is caused to deviate from the common plane.

4. The photographic material must be capable of resolving the grating. Reference to the calculations of grating spatial frequency indicates that the film should resolve on the order of 2,000 lines/mm.

5. This is not a requirement but rather a useful tip. Do not open the entire box of holographic plate or film in the holography laboratory. It is altogether too easy to forget to close it before turning on the lights. Transfer a few plates to another box while in a darkroom and use this box in the laboratory.

17.2 Holographic platform

The requirement for interferometric stability has caused holographers to make a fetish of their optical tables. In fact, there are a great many solutions to this stability problem.

The major challenge is to isolate the optical components from room and air vibrations. Motions transmitted through the structure are the chief cause of difficulty. Every holographer has experienced a failure because a truck thundered by the laboratory at the wrong time, or somebody used the elevator, or the custodian slammed a door down the hall. Adding to the problem is that the mirrors and other components in the system are often mounted on rods. The resulting spring-mass combination will have a resonant frequency that may be close to the excitation frequency provided by an air conditioner, fan, or boiler feed pump. Tracking down these vibration sources can be a major headache.

A platform that isolates the optics from sources of disturbance greatly reduces the chance of failure caused by the environment. Usually, the platform consists of some sort of rigid and heavy table that rests on a support that is soft and well damped. Some combinations that have been used include the following:

A sheet (or two or more) of heavy plywood resting atop a twin-size bed mattress

A sandbox, with the optical components mounted on rods that are simply pushed into the sand

A slab of concrete resting on partially inflated tire innertubes.

A tombstone sitting on a box of old auto springs or a few old tires

A surplus cast-iron surface plate resting on partially inflated truck air bag overload helper springs

A surplus granite surface plate sitting on machine-tool mounting pads or on a table rolling on rubber tires

A piece of aluminum tooling plate sitting on a slab of foam rubber

Holographic tables designed and sold specifically for the purpose

The last-mentioned item is the best solution if the price is agreeable. Holographic tables typically are of honeycomb construction with a stainless steel top that is drilled and tapped for mounting components. They are not nearly as heavy as other types of tables. With them come matching legs that contain air bladders or pistons to provide air suspension for the table. A strongly favorable feature is that these systems are tuned so that they nearly totally cancel vibrations above a few hertz.

The platform should be located in a room that can be darkened, although some stray light is allowable if necessary. The room should be free of great thermal oscillations, and air currents should be minimized, if necessary by building a box around the table. Film processing facilities should be close by.

17.3 Optical setups for holography

Here are described some arrangements of apparatus for making holograms. These range from inexpensive and simple to somewhat sophisticated. Many others are possible (e.g., Smith 1975).

A system for holography of opaque objects reduced to bare essentials is shown in Figure 17.1. In this arrangement, the beam from laser L is first expanded by a simple lens or a microscope objective, BE. Only one lens is required. The beam splitter BS divides the expanding beam into two parts for object and reference. If a partial mirror is not available, then this arrangement can be made to work with just a piece of glass for the splitter. The ratio of front surface reflection to transmission will provide a reasonable ratio of object beam to reference beam illumination. The mirrors M redirect the beams to object and holographic film. Front-surface mirrors are best, but ordinary back-surface cosmetic mirrors can give reasonable results. The screen S is optional, but it often helps to eliminate spurious light from the laser tube or the scattered light from the expander or splitter. It need be nothing more than a piece of cardboard. H is the holographic film in a holder. For elementary work, this holder need be nothing more than a metal frame with grooves or spring clamps to hold the film or plate. Because there is no spatial filter (pinhole) at the beam expander, the holographic image will probably contain various spurious interference fringes. Quite acceptable holograms can be made with this simple apparatus. If care is taken to assure a stable setup, and if the motion or position of the object can be carefully

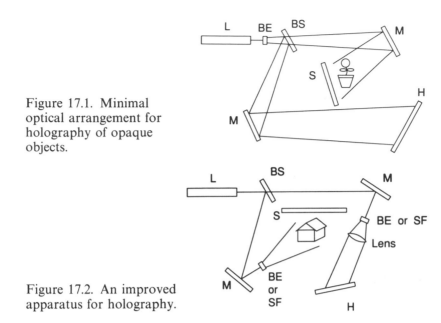

Figure 17.1. Minimal optical arrangement for holography of opaque objects.

Figure 17.2. An improved apparatus for holography.

controlled, then holographic interference fringes can also be created and observed with no more equipment than this. Even simpler setups can be devised, but they are usually intended for making reflection holograms, which have not been discussed in this text because of their limited use in measurement. Otherwise, the simpler setups for transmittance holograms suffer the disadvantages that there is a beam crossing in the vicinity of the hologram film plane, and getting the optical path lengths equalized is difficult or impossible.

The arrangement shown in Figure 17.2 is slightly more sophisticated but still very easy to implement with limited resources. The identifications of the components are the same as for Figure 17.1. Two beam expanders are used. The beam divider is likely a partially aluminized mirror, and it is best if it has a gradient of reflectance/transmittance so that the optimum beam ratios can be set. Extraneous fringes in the image can be eliminated if the beam expanders incorporate pinholes, so that now they are called spatial filters, SF. Section 10.8 explained the functioning and the technique for adjusting such a device. Front-surface mirrors should be used. With this setup, again, no beams are crossing in inconvenient places. Additional mirrors can be used to set the laser at an angle and so make the setup more compact.

If desired, collimating lenses can be placed in the reference beam and/or object beam, and one such is shown in this setup at location L. One advantage of using collimated radiation is that it is easy to reproduce if the hologram is used again at a later date. If the holograms are being made for quantitative interferometric purposes, then use of collimated light simplifies fringe interpretation because the incidence and viewing angles do not vary over the field.

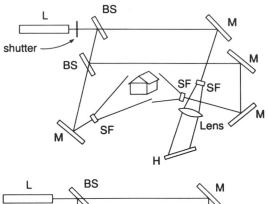

Figure 17.3. Holography with multiple reference beams.

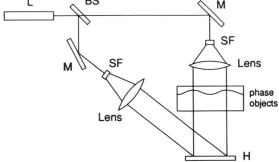

Figure 17.4. A layout for holography and holographic interferometry of phase objects.

A potential disadvantage of these first two layouts is that only one object beam is used, and it must come from one side of the hologram location. One side of the object will be in partial shadow, which might mask important data in holointerferometry. There is no reason why more than one object illumination beam cannot be used, provided the path length condition is met by all the beams. Figure 17.3 shows such an arrangement. The object beams come from opposite sides of the recording film, so if the beam ratios are well adjusted, there are no masking shadows. Some effort is required to get the path lengths equal, especially if the table is small. The diagram shows a collimated reference beam, which is optional but desirable. For pictorial or entertainment holography, the use of more than one object-illuminating beam will give greater control over the aesthetic quality of the image. Highlights, sunsets, spotlights, and the like can be simulated by controlling these illuminating beams. Notice that an optional shutter of the electromechanical or mechanical type has also been introduced into this setup.

Most of the development in this book has centered on the holography of opaque objects. Interferometry is often conducted with phase objects, and the path length difference is then a function of refractive index. A setup for recording holograms, viewing them, and performing holointerferometry with phase objects is shown in Figure 17.4. The identities of the components are as listed for Figure 17.1.

17.4 Recording and viewing a hologram

The following is a detailed list of steps for making and reconstructing a hologram using a setup that approximates the one in Figure 17.2.

1. Arrange all the optical elements on the optical table except for the beam expander–spatial filters. Arrange elements so that the reference beam comes to the center of the film/plate holder. The object beam should intersect the object on the front side of center. Check the path lengths with string to see that they are within 10 cm or so of being equal. If necessary, rearrange the mirrors to get the path lengths and angles right. If the mirrors are adjustable, some time spent in getting the beam path parallel to the table will pay dividends in time later on. A combination square with a steel rule is helpful in adjusting beam height.

2. Place the beam expander lenses with their holders in front of the appropriate mirrors. Center and align these by eye so that the laser beams hit the center of the lens. Use a white card to see that they are aligned so as to illuminate the film and the object properly. Adjust them as required to attain proper coverage and centering. At this point, the setup can be used to record a hologram if collimating lenses and/or pinholes are not used.

3. If collimating lenses are used, introduce them into the layout at this time. Again, check for alignment and centering of illlumination at the film plane and on the object.

4. Install the pinholes into the spatial filters. The technique will depend to some extent on the design of the holders. The ones that have the pinholes mounted on magnetic wafers are easiest to use. They cling to the micrometer adjusters and slide on a front surface plate. Use the technique outlined in section 10.8 to adjust the pinholes. Any other approach will be a waste of time. Again, check for proper illumination of film and object.

5. Take the box of holographic film or plate into a dark place and remove a few plates. Put them into a separate film box or a dark envelope as your working supply. Put the covers back on the main supply. Do not work in the laboratory from the full box, as sooner or later they will be ruined because someone will switch on the lights when the cover is off the box. Better to lose only one or two plates in the event. Return to the holographic setup and turn off the laboratory lighting. It is a good idea to keep a penlight in hand.

6. The next step is to adjust the beam ratios, if such facility is incorporated in the setup. This step is easiest if a variable beam splitter is being used. If not, then partial mirrors or attenuating (gray) filters can be introduced. Also, the adjustment is facilitated if a sensitive light meter or photometer is at hand. It need not be calibrated in absolute units since only comparative measurements are required. The protocol is to block the reference beam somewhere in its path with a card and measure the object beam at the film location. The reference beam is then measured by blocking the object beam. The ratio of reference beam intensity to object beam intensity, measured at the film location, should be from about 3:1

to 10:1 for best results. If a meter is not available, then the beam strengths can only be judged by eye. If much hologram-making is anticipated, then one is well advised to invest in a meter or to make one using a photocell. The light levels will be very low, in photographic terms, if ordinary small laboratory lasers (the 5–50 mW variety) are employed.

7. Figure out an exposure time. If the meter has been calibrated, then exposure time can be calculated. If it has not been calibrated, then guess at an exposure. Write down what is used, as even a little data will reduce future guesswork and cut down the bill for photographic materials. If proper exposure must be determined by trial, then use small pieces of film or plate to do it. The stepwise exposure technique can also be employed. Just hold a card in front of the plate and move it in timed steps across the plate, multiplying the time in each step by 2 or 4 times. Holography is actually quite forgiving of exposure errors, especially if the film processing is monitored intelligently.

8. Mentally review the exposure sequence. Be sure the laboratory lighting is off. Block the laser beam at the laser using the shutter or a black card. Make sure that not much stray light is bouncing around. Take a film or plate from its holding box. Figure out which is the emulsion side. If film is used, the notch code will indicate the emulsion. When the film is held so that the edge notches are upper right, then the emulsion is facing you. Glass photoplate is not coded as is film. One way to detect the emulsion is to clamp the plate lightly between the lips. The side that sticks slightly is the emulsion side. If sanitation is a concern, then moisten finger and thumb and clamp them on opposite sides of the plate to see which side sticks. An alternative approach is to use a pocketknife to scratch the corner. If there is no resistance, then that side is the smooth glass.

9. It is a very good idea to use an indelible marker to mark one end of the emulsion side. Most holographers will hold the plate with emulsion side facing the body and write initials, date, and plate number across the top. The reason for doing this is so, after the plate is processed, the holographer can easily get it back into the holder in the same orientation it had when the hologram was recorded.

10. The plate or film is placed into the holder with the emulsion facing the object and the identifying code on the left.

11. The shutter is opened, or the card blocking the laser beam is removed, for the required exposure time. Use a timer, your watch, or just count seconds if the exposure is long enough so that small timing errors are not a significant fraction of the total. Very short exposures are best accomplished with a shutter in the setup.

12. Close the shutter or replace the masking card. Remove the film from the holder and place it in a dark box or envelope for transport to the processing area. The laboratory lighting can then be turned back up for safety.

13. Process the plate in a darkened room according to instructions for the film type. Do not use a safelight to any significant degree. When development is partly finished, the density can be judged periodically by looking through the plate at

a very dim and distant light or, for example, at a luminous clock face. As the transmittance approaches 50%, stop the development. If this 50% transmittance does not happen within the recommended development time, then extend the development until, hopefully, it does. Many holograms that were under exposed or overexposed can be rescued by monitoring the development and extending it or shortening it as appropriate. Fix and wash the film or plate as per instructions. Since this is not pictorial photography, the times may be shortened if the plate is not to be kept for the archives. Drying time is drastically shortened if the plate or film is bathed in methyl alcohol after washing.

14. The dry plate or film is placed back in the filmholder of the setup in the same orientation it had when the hologram was recorded. The object beam is blocked off and/or the variable beam splitter is adjusted to throw all the illumination into the reference beam, which is now the reconstruction beam. Place your eye behind the hologram so as to look through the hologram at the general location of the original object. A bright reproduction of the object should be visible. Check the parallax characteristics to see that it is a full three-dimensional image.

15. Record the details of the experiment in a notebook so that it may be repeated or the hologram can be reconstructed at a later date.

17.5 Hologram recording media

The main requirement for recording a hologram is that the medium be able to resolve and record a complicated grating structure that has a spatial frequency of 1,000 to 2,000 line/mm. Most photographic materials are not capable of resolving such fine structure. The grain size of the emulsion must be very small, of course. Typically, the requirement for small grain implies slow emulsion speeds. In addition to the severe resolution criterion, the film must respond to the wavelengths of common lasers used in holography. This spectral response requirement creates another problem, because most emulsion materials respond best at blue and violet wavelengths and poorly, without special treatment, at the red of helium–neon lasers.

For quite a long time after holography was invented, the only useful film was Kodak scientific plate 649-F, which was made for spectroscopy. It is very slow, so long exposures are required. Its resolution capabilities are outstanding, and it is still a good choice when fine grating structures are to be resolved.

Both Kodak and Agfa listings now include photographic emulsions that are designed specifically for holography at different wavelengths. These materials can be had, typically, in at least two sensitivities. The faster emulsions have, as expected, reduced resolution capability. The film should be matched to the application. Table 17.1 is a summary of photographic materials for holography, along with sensitivities, resolution limits, and

Table 17.1. *Holographic films and plates*

| | Exposure to achieve D = 1.0, (erg/cm²) | | | | | | | | Resolving power (lines/mm) @ TOC | | Emulsion | | | | |
	HeCd 3250	HeCd 4416	Ar 4880	Ar 5145	Nd:YAG 5320	HeNe 6328	Kr 6471	Ruby 6943	1000:1	1.6:1	Granularity @ D = 1.0 48 μm	6 μm	Contrast for plate/film (γ)	Thickness for plate/film (μm)	Development (min, developer)
Kodak															
Spectroscopic type 649-F plate film	—	500	800	800	1000	900	800	5000+	2000+	—	<5	<10	5/4	17/6	6–8, D-19
Holographic plate, type 120 (−0.2 or −01)/film, SO-173	—	500	—	—	—	400	400	400	2000+	—	<5	<10	5/4	6	6–8, D-19
Special plate, type 125 (−02 OR −01)/film, SO-424	20	80	50	50	100	—	—	—	1250	630	<5	13	4	7/3	6–8, D-19
High speed holographic plate, type 131 (−02 or −01)/film, SO-253	—	20–35	40–65	25–35	20–30	5–8	3.5–6	1000+	1250	800	<5	14	7	9	6–8, D-19
Technical pan film, 2415	—	0.4	0.8	0.8	0.7	0.4	0.3	—	320	125	8	—	1–3	7.5	6–8, HC-110(D) 4, D-19

Agfa–Gevaert															
8E56HD-AHI plate/film	—	350	600	350	300	—	—	—	2500+	—	—	—	—	4	7	6–8, D-19
8E75HD-NAH plate/film	—	400	—	450	150	150	100	2500+	—	—	—	—	3	7	6–8, D-19
10E56-NAH plate/film	—	60	30	20	20	—	—	1500+	—	—	—	—	7	7	6–8, D-19
10E75-NAH plate/film	—	60	—	120	60	20	20	1500+	—	—	—	—	4	7	6–8, D-19

Source: Courtesy of Newport Corp.

Definitions and notes:

Exposure: This measure of sensitivity is the radiant energy in ergs/cm^2 to produce a hologram with neutral density (D) of 1.0. Sensitivity varies with wavelength, as shown. In amplitude holograms, maximum diffraction efficiency occurs with D between 0.6 and 0.8. A slightly higher density is required with holograms that are to be bleached.

Resolving power: The emulsion resolution is obtained by photographing target objects with contrasts (TOC) of either 1000:1 (high) or 1.6:1 (low). This figure is most useful as a relative measure. Holograms can be obtained even if the fringes have spacings two or three times smaller than the resolving power obtained by this common test.

Granularity: These emulsion granularity figures are 1,000 times the standard deviation of film density about D = 1.0 when scanned by a microdensitometer having a circular aperture of either 48 μm or 6 μm. These values give a relative measure of granularity. In general, granularity should be small for holography. Grain size and sensitivity exhibit a somewhat reciprocal relationship.

Contrast: Contrast of photographic emulsions is usually reported in terms of the parameter γ, which is the slope of the characteristic curve for the emulsion. The characteristic curve is the plot of density (D) versus the log of the exposure (E); it is often called the H-D curve. Normal development is assumed; the γ can be modified by altering the development. The contrast differs slightly for film and plate, partly because of thickness differences.

Thickness: Emulsion thickness is not usually very important except when multiplexing holograms or making reflection holograms, in which case the thicker emulsions should be chosen. Note that the emulsion is usually thicker on glass plate.

Development: Normal development is in Kodak D-19 or dilute Kodak HC-110. The plates and films are fixed in Kodak Rapid Fixer and then washed. For quick drying, bathe the hologram in 50–70% alcohol for one minute and then blow dry. Kodak 120 and 131 carry a residual dye that can be removed by washing the holograms for 1–3 min in 75% methanol after normal water washing. Considerable latitude in film and plate processing is possible.

processing data. It is based on a useful chart that once appeared in the catalog of a commercial firm (Newport Corp., 1791 Deere Ave., Irvine, CA 92714).

A reuseable polymeric material for holography has been developed, and it has been marketed for some years by Newport Corporation. This medium uses a thermoplastic polymer layer atop a photoconductor and a glass insulating substrate. In use, an electric charge is first deposited on the combination. The plate is then exposed in the holographic setup using blue laser light (argon ion). Then, the plate is again subjected to an electric charge. Finally, heat is applied to soften the plastic so that the local charge variations will be converted to surface features. The hologram is then ready for viewing as with photoemulsion media. The hologram is erased by heating the plate so as to level the surface, after which it can be reused. These plates can be recycled up to 300 times, according to the manufacturer. The hologram images are relatively noise-free because of the lack of grain in the medium.

The only other optically sensitive materials available for large-aperture holography are the photoresists that have been mentioned in connection with moire measurements. Holograms have been made with these media. They are not very sensitive, and they respond only to the blue and violet end of the spectrum, so the range of lasers that can be employed is severely limited.

No electronic imaging devices have resolution capabilities that allow their use for even the crudest pictorial holography.

17.6 Double-exposure interferometry

The optical arrangements described in section 17.3 are suitable for interferometric measurements of displacement. Only minor additions are needed to adapt the procedure to the recording of double-exposure holographic interferograms according to the ideas presented in section 16.5.1. In particular, a second exposure period is added to the procedural sequence. The specimen is loaded or deformed between exposures, and nothing else can be disturbed during the entire time from the start of the first exposure to the end of the second.

The object must be mounted so it is stable, but so that the desired deformation or displacement can be applied between exposures. Typically, mechanical loading is done by applying a small, dead weight or by use of a micrometer screw. Remember that the deformation must be constant through the second exposure; any creep or load variation will cause the second image to go unrecorded.

These considerations also lead to the suggestion that only glass photoplate be used for interferometry. Flexible substrate material has been used with good success in double-exposure work if it is sandwiched between clear glass plates during the recording and viewing stages.

A critical fact is that the load or deformation must be of appropriate magnitude. Prediction of an acceptable displacement or load can be difficult. An approximate calculation that establishes a maximum displacement of 10 or so wavelengths maximum is a good start. If the structure response is truly unknown, no such calculation can be executed, and one must rely on intuition and some trial and error. The experiment can be repeated with different loads, once reasonable fringes are obtained, in order to calibrate the system.

Another important consideration is that the load or deformation must be applied expeditiously in the dark. Provision of stops on the load mechanism, or at least some tactile indicators, is a good idea. Rehearsal of the steps in applying the loads, first in the light and then in darkness, can save time and recording materials.

If quantitative results are sought, then the setup should be adapted to the problem so as to simplify interpretation, such as by converting it to the one-dimensional line-of-sight case. At least, the illumination and viewing angles must be measured. Then, fringe orders are established at the desired object points as viewed through the necessary number of small areas of the hologram. The sensitivity vector equations can then be used to determine the unknown displacements.

17.7 Real-time interferometry

The optical layouts given in section 17.2 also suffice for real-time measurements by holographic interferometry, as described in section 16.5.2. Only two refinements are needed. The first is that the specimen be held in a stable fixture that will allow small loads and deformations to be applied in a controlled and reproducible manner. This requirement is similar to that mentioned for the double-exposure technique, except it is less severe. The magnitude of the deformation can be adjusted while watching the fringes form, so it need not be chosen before the fact. Also, the adjustment need not be done in darkness. As with the other interferometry methods, the entire setup must not be changed or disturbed during the entire experiment, including the fringe-viewing stages.

The second refinement is critical to success. The hologram, after development, must end up in exactly the position it held during the exposure. The margin for error is only a few wavelengths, and even this error will cause an equal number of zero-state fringes to appear in the image. One implication is that only glass photoplate should be used for real-time work.

There are two convenient techniques for assuring that the photoplate will be properly located after development, fixing, washing, and drying. The first is to simply develop the plate in its holder after the exposure. The chemicals are brought to the plate. A basic approach is to use a holder that suspends the plate from above rather than supporting it from the bottom. The developer and fixer are prepared in plastic beakers or similar containers, which are strategically placed before darkening the laboratory. After exposure of the film, the beakers are raised from beneath to immerse the plate. A laboratory jack facilitates this maneuver and helps avoid bumping the container against the plate support.

A more sophisticated technique for developing in situ is to use a wet plate holder. These devices are available from suppliers or one can be built. The chemicals are poured or pumped into a glass tank containing the plate holder. After the processing stage is finished, the fluid is pumped out or else just drained from a tap at the bottom. Rather than trying to dry the plate in the tank after processing, the tank is often just filled with clean water and the holographic reconstruction is accomplished with a wet plate. At least one commercial holographic inspection system has used this approach. In fact, the chemicals were heated to reduce processing time, and the plates were not fixed beyond stopping development. The tanks were filled with water at the start, so both exposure and viewing were done with wet plates.

The second general method for assuring that the holographic interferogram is precisely located at its preprocessing position is to employ a real-time plateholder. Whether built or purchased, these devices incorporate some type of mechanical fixture to locate the plate. Three steel dowel pins serve nicely; the plate sits on two pins and one edge rides against the third pin. The plate is given a tap on the opposite edges to seat it, and a clamping device holds the front rim of the plate against a metal surface with an aperture of slightly less than the plate dimensions. This approach serves very nicely, and it tends to be less messy than bringing the photochemicals to the holographic table. The real-time plateholder is usually heavily built, and it is a good idea to clamp it down so that extracting and replacing the plate can be accomplished without moving the holder.

17.8 Time-average interferometry

This type of holointerferometry is nearly identical to recording and viewing a single exposure simple hologram. The theory for simple geometries is developed in section 16.5.3. The only difference is that some means of vibrational excitation of the specimen is required. Small electromagnetic drivers, piezo-electric exciters, or sound generators are employed, depending on the specimen and the frequency range sought. Arriving at the correct frequencies and excitation levels can be a problem. Preliminary work to establish modal frequencies and such will pay dividends. The resulting interferogram is similar to the two-exposure type in that the fringe-creating data are stored in the emulsion (Stetson 1974).

17.9 Holographic nondestructive testing

One of the significant areas of application of holography is in the field of nondestructive evaluation (NDE). Several diverse areas of use are described in Erf (1974). The great sensitivity and the ability to create three-dimensional interferograms of complex shapes are key factors in this development. Typically, only qualitative observations and comparisons of fringe patterns are adequate for the detection and rough characterization of manufacturing flaws or in-service damage. The holographic fringe patterns for a normal product are first established by testing several good samples. Then, samples with known or deliberately placed flaws are examined to establish fringe patterns and flaw sensitivities. Fringes from unknowns can then be compared with these calibration patterns. In many cases, it is necessary only to look for local discontinuities in the fringes. The patterns for a good sample usually show smoothly continuous fringes. For a damaged sample, the fringes are often irregular, broken, or grouped in the damage area.

Any one of the three interferometric techniques is suitable for nondestructive testing or inspection (NDI). The final choice depends very much on type of problem at hand. For static applications, the double-exposure method is preferred because it is easier and because it gives permanent documentation of the mechanical response in the form of a holographic plate. For situations in which the damage might be manifested by changes in the vibrational response, the time-average approach might be best.

Whatever the technique, the major problem is to figure out a loading scheme that will stress the object somehow to create a fringe pattern that will be sensitive to the types of flaws or damage that are of interest. Mechanical loading, thermal deformations, vibration, and air pressure are all to be considered.

Some examples will illustrate various approaches to loading for NDI/NDE. One of the earliest and most classic examples of NDI by holography was in the inspection of aircraft tires (Brown 1974). The tire was placed inside a chamber with a holography system set up to examine the interior. It was necessary only to change the pressure inside the chamber a small amount between exposures. The pressure causes any delamination or areas with incomplete fibers to bulge or deflate slightly because of trapped air or local stiffness discontinuities. The fringe patterns tend to group or show irregularities in the area of the delamination. Inspection of composite plates has been accomplished by heating the plate slightly between exposures. A lamp is one way to create enough heat transfer. Circuit boards can be tested by passing current through the conductors. Closed vessels are tested by low-level pressurization. The fringes will show irregularities around cracks, delaminations, or corroded areas.

Industry has not been quick to adopt holographic methods for routine inspection of consumer goods. Holography is not easy to adapt to the environment of the production plant with its noise, vibration, and dirt. It tends to be slow because processing of photographic materials is necessary. Except for the reusable polymeric hologram medium mentioned in section 17.5, the photoplates and chemicals tend to be too expensive for testing ordinary products. Inspection applications tend to be concentrated in industries that produce small quantities of special components whose failure would be deleterious to safety or cause financial loss. This picture is changing somewhat with the advent of inspection methods built around video image processing of coherent speckle fringe patterns, as discussed in the next part of this text.

17.10 Holographic photoelasticity

Photoelasticity, as described in Part II, has been established as a good method for determining principal stress difference in objects, in addition to its use in investigating material behavior. Other interferometry techniques serve to determine principal stress sum. Holography is basically an interferometric method, and one wonders if holography and photoelasticity might be combined to obtain a complete mapping of stresses in a deformed object.

Research has demonstrated that they can indeed be combined (Brčič, Powell, Der Hovanesian 1967; Fourney 1968; Vest 1979). The technique is an interesting and potentially useful example of hybrid interferometry. Even if the photoelastic data are not obtained simultaneously with the hologram, the holography does provide a quick method of obtaining principal stress sum so that stress separation can be accomplished. The method deserves a brief discussion for that reason alone. It also provides a first example of holography with a transparent object that affects primarily the phase relationships of the object beam.

The arrangement of optical elements for static photoelastic holography differs little from that used for ordinary holography. Light coming from a laser is naturally polarized, so the insertion of a $\lambda/4$ plate gives circular polarization for both reference and object beam. Figure 17.5 shows one setup for recording single- and double-exposure holographic photoelastic interferograms. Actually, it is better to use two circular polarizers, one of them between object and diffuser. The ground glass diffuser is added because the model is nearly transparent.

In general, the plane transparent model adds to the path length of the object beam. If the transparent model is birefringent, then the circularly polarized light behaves as we have discussed before. We expect the hologram to store all phase and amplitude information about the object, and this includes the isochromatic data as well as the interferometric data related to

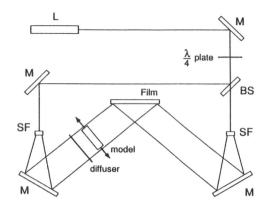

Figure 17.5. Optical arrangement for holographic photoelasticity.

strain sum. Holophotoelasticity is not seeing much use, and the subject will not be pursued further here.

17.11 Contour mapping

The determination of surface profiles in the form of fringe patterns that represent contour lines is another useful and interesting application of holographic interferometry (Varner 1974). One technique is to use a laser that is capable of generating coherent light at two wavelengths. Another uses an immersion tank in an ordinary setup for holographic interferometry, as illustrated in Figure 17.6. The object is placed in the tank, which must have at least one side of clear glass or plastic. Quantitative data interpretation is simplified if the object beam axis is normal to the transparent side of the tank and if a collimated object beam is employed. Another good idea is to include an object having known contour in order to simplify calibration. Enough water is put into the tank to cover the object. The system is allowed to come to thermal equilibrium so that the fringes will not be contaminated or eliminated by the effects of time-dependent thermal gradients. Either double-exposure or real-time interferometry can be used for creating the contour fringes. Assuming that the latter is selected, a hologram of the object in the immersion medium is made, and it is returned to the real-time plateholder after processing. The refractive index of the immersion water is then changed by mixing in a small amount of liquid having an index that is different from that of water. Alcohol or glycol can be used. The mixture is stirred gently and again allowed to stabilize. If the refractive index change is appropriate, then live interference fringes should be visible. A sample of a pattern that was obtained by this technique appears in Chapter 16.

The fringe order, which is proportional to object-beam path length change, is clearly a function of the geometry of the system, the thickness of the fluid traversed, and the change of refractive index of the medium. The sensitivity

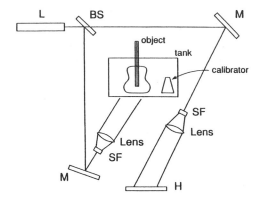

Figure 17.6. Arrangement for obtaining holographic contour fringes using refractive index change.

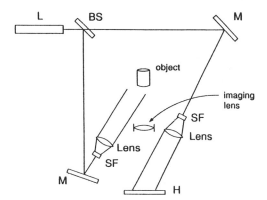

Figure 17.7. System for image plane holography.

can be calculated, or a simple object having known contour can be employed as a system calibrator. If near-normal incidence of both the object illumination and the viewing axis can be employed, then fringe interpretation is much simplified, as with line-of-sight observation of out-of-plane deformations.

If a trial-and-error approach is adopted as a measure to just obtain useful fringes, the difficult part is in inducing the correct change of refractive index. One should start with a small amount of the second liquid and increase the dose until fringes appear. The flexibility of the real-time method pays dividends in this context.

17.12 Image plane holographic interferometry

This technique, which seems not to be used very much, has great potential for strain and deformation measurement, especially with flat objects and thin phase objects, such as gas-filled channels. An imaging lens is inserted into the holographic system so as to create a real image of the test specimen at the film plane. Figure 17.7 is a schematic of such a system. Again, the interferometry can be done in live-fringe or frozen-fringe modes. One advantage of this image-plane technique is that the fringes are localized in the image of the object at the hologram plate. In fact, if double exposure is used, the fringes will usually be visible to the eye in white light, either transmitted through the hologram or in oblique reflection. This type of hologram has some of the properties of reflection holograms or rainbow holograms, such as are used on magazine covers. Reflection holograms have not been discussed in this book. Very briefly, the technique is to bring the object beams and reference beams to the holographic emulsion from opposite sides. The hologram grating fringes are then nearly parallel to the emulsion surface, and they serve as spectrally discriminating reflectors similar in function to multilayer interference filters. Light from a point source such as a penlight is coherent enough to reconstruct the holographic image.

References

Brčič, V., Powell, R. L., and Der Hovanesian, J. (1967). Application of holographic interferometry to photoelasticity. *Technika* (Yugoslavia), 22, 11: 233–8.

Brown, G. W. (1974). Pneumatic Tire Inspection. In *Holographic Nondestructive Testing*, Ed. R. K. Erf, Sec. 8.11. New York: Academic Press.

Erf, R. K., Ed. (1974). *Holographic Nondestructive Testing.* New York: Academic Press.

Fourney, M. E. (1968). Application of holography to photoelasticity. *Experimental Mechanics*, 8: 33–8.

Jones, R., and Wykes, C. (1989). *Holographic and Speckle Interferometry*, 2nd ed. Cambridge University Press.

Ranson, W. F., Sutton, M. A., and Peters, W. H. (1987). Holographic and Laser Speckle Interferometry. *Handbook on Experimental Mechanics*, Ed. A. Kobayashi. Englewood Cliffs, NJ: Prentice-Hall.

Smith, H. M. (1975). *Principles of Holography*, 2nd ed. New York: John Wiley and Sons.

Stetson, K. A. (1974). Holographic Vibration Analysis. In *Holographic Nondestructive Testing*, Ed. R. K. Erf, Ch. 7. New York: Academic Press.

Varner, J. R. (1974). Holographic and Moire Surface Contouring. In *Holographic Nondestructive Testing*, Ed. R. K. Erf, Ch. 5. New York: Academic Press.

Vest, C. M. (1979). *Holographic Interferometry.* New York: John Wiley and Sons.

Waters, J. P. (1974). Holography (Ch. 2) and Interferometric Holography (Ch. 4). In *Holographic Nondestructive Testing*, Ed. R. K. Erf. New York: Academic Press.

PART VII

Speckle methods
(*with contributions by C. Wykes, R. Jones,
K. Creath, and X. L. Chen*)

18

Laser speckle and combinations of speckle fields

Measurement methods based on the phenomenon known as coherent light speckle have become increasingly important in recent years. The development of electronic speckle pattern interferometry, in which the speckle patterns are acquired by a television system and then combined in a computer to create fringe patterns that can be displayed on a television monitor, has generated additional interest in speckle methods.

This chapter discusses the origins and nature of laser speckle, and it then goes on to describe the product of combining speckle fields in different ways. These notions are important to understanding the various methods by which speckle is employed in interferometry, as are described in subsequent chapters. Given the importance of electronic speckle techniques, the development assumes, where appropriate, that the recording of speckle irradiance is by means of an electronic detector rather than by a photographic emulsion.

Most of the concepts in this chapter have been treated exhaustively, lucidly, and creatively by Ennos (1975), Goodman (1975), Jones and Wykes (1983), and Vest (1979). The exposition that follows is synthesized primarily from these references, particularly the first three. In some cases, the words follow closely those of the the authors; their descriptions have become accepted as the classical standards. The references mentioned also carry extensive and useful bibliograpies.

18.1 The speckle effect

The invention of the laser created great anticipation among users of optics because it appeared to be the answer to a great many illumination problems. Here was a source that produced a beam of light that was intense, collimated, narrow, monochromatic, and coherent. Disappointment soon followed. Visual or photographic images of objects illuminated by a laser were covered with a grainy structure that severely limited the effective resolution.

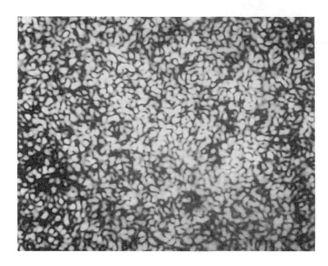

Figure 18.1. A magnified image-plane speckle pattern.

Photographs were greatly inferior to those made in noncoherent light. A magnified picture of a surface illuminated by coherent light, as offered in Figure 18.1, gives evidence of this effect.

Many of the photographs in this book, such as those from optical Fourier processing or moire interferometry, show this phenomenon, which is now known as "laser speckle." For some years, laser speckle was considered a nuisance that caused the laser to be less useful than expected. As understanding of the phenomenon was developed, it evolved from a problem into the basis of new measurement technologies. The new techniques, which are loosely grouped under the name "speckle methods" utilize speckle fields in combinations, as by double exposure, to create interference fringes that indicate displacement, strain, or motion. Now, measurement techniques based on speckle effects are probably at least as important, if not more so, than methods based on holography.

The causes of the laser speckle were recognized to lie in the coherence of the light and the minute roughness of any surface not ideally specular. Figure 18.2 illustrates the process of forming speckles in an exaggerated way for the simplest case.

The surfaces of most materials are rough on the scale of an optical wavelength (0.6 μm). Speckle can also be produced by transmission of coherent light through scattering objects such as ground glass, opal glass, or particles in liquid. When laser light is reflected or scattered from such a surface, the optical wave arriving at any receiving point consists of many waves, each arising from a different point of the illuminated surface. The path lengths traveled by these waves, from source to object point to the receiving point, can differ from zero to many wavelengths, depending on surface roughness and the geometry of the system. Interference of the dephased but

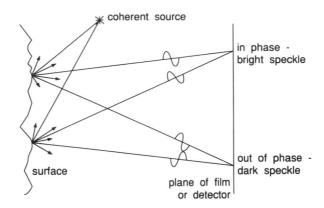

Figure 18.2. Speckles formed by interference of coherent light that is scattered or reflected from a surface.

coherent waves arriving at the receiving point will cause the resultant irradiance to be anything from dark to fully bright. The resultant of the waves arriving at a neighboring point will probably give a quite different brightness. This variation in resultant irradiance from one receiving point to another is the cause of laser speckle. This type is called "objective speckle." The second type of speckle will be introduced after some quantitative consideration of the first variety.

18.2 Types and sizes of speckle

18.2.1 Objective speckle

Figure 18.3 shows the parameters that are important in the formation of objective speckle. The maximum optical path lengths between the waves arriving at the image point depend on the random surface roughness and also on the maximum extent of the illuminated portion of the object in relation to the object–receiver distance. If the object distance is L and the illuminated area of the object has diameter D, then the statistical average of the path differences gives the average diameter S_{obj} of the speckle patches to be (Goodman 1976),

$$S_{obj} = 1.22 \frac{L}{D} \lambda \qquad (18.1)$$

18.2.2 Subjective speckle

If the optical system incorporates a lens to create an image of the object, then the diffraction limit of the lens must be considered. The theory presented in Chapter 10 suggests that a lens of finite aperture cannot transmit spatial

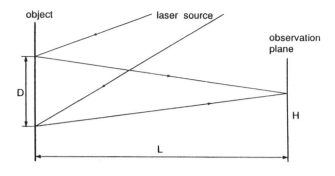

Figure 18.3. Parameters in the formation of objective speckle.

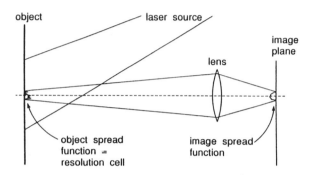

Figure 18.4. Formation of subjective speckle.

frequencies beyond a certain limit. This fact implies that there is a fundamental resolution limit for a given lens. Information within a patch or cell of this size on the object cannot be resolved. The resolution cell size defines the diffraction limit of any lens. At the image there will be a corresponding resolution cell whose size is related to the resolution cell at the object by the magnification factor of the system.

As a result of the diffraction limitation, the waves making up an image do not travel from object point to image point. Rather they go from object cell to image cell. The waves arriving at any given cell are added together. The irradiance within an image cell depends on the way in which the waves falling into the cell interfere with one another. Figure 18.4 suggests how this happens and shows the important parameters.

Another, and perhaps more precise, description of the formation of speckle in imaging systems looks at the brightness at any image point as the superimposition result of the coherent point spread functions for adjacent object points. The emphasis is a bit different, but the result is the same as before. Speckle created by imaging optics is referred to as "subjective."

In any case, the speckle size turns out to be the same as the resolution limit of the system, which we know is about the size of the Airy disc (see Chapter 10). The size S_{subj} of the individual speckles in this case is then related to the aperture ratio $F = $ focal length/aperture $= f/a$ of the lens (the $f/$ number) and the magnification M of the lens. The speckle size in the image is

$$S_{\text{subj}} \approx 1.22(1 + M)\lambda F \tag{18.2}$$

From simple lens theory, the speckle size on the scattering surface (object) is given by

$$S_{\text{subj}} \approx 1.22(1 + M)\lambda \frac{F}{M} \tag{18.3}$$

This is the same as the size of the resolution element on the object. Subjective speckle is the type that is used in most speckle metrology methods.

Note that any speckle observed by the eye, even if the eye is observing speckle cast on a screen, involves the imaging optics of the eye and so is properly identified as subjective speckle. Objective speckle can be observed by using it to expose a photoplate, for example, and then looking at the amplitude distribution on the developed plate.

In this book, only Gaussian speckle is considered. The term "Gaussian speckle" is derived from the fact that most real surfaces can be assumed to be rough in optical terms. This assumption results in a negative exponential probability density function for the intensity in the speckle pattern. Within the limits of the necessary conditions, the statistics of a Gaussian speckle pattern are independent of the nature of the scattering medium; in particular, the surface roughness does not influence the statistics provided that the surface roughness is greater than the wavelength and that a large number (N) of scatterers contribute to the intensity at any image point (patch). For most practical applications this is generally the case (Goodman 1975).

On the other hand, if only a few independent scattering areas are present, then the speckle statistics do contain information about the scattering medium. Such "small-N" speckle usually has non-Gaussian statistics, and hence the probability density function of intensity is usually not of the negative exponential type.

In the experimental measurement of the irradiance in a speckle pattern, the detector aperture must of necessity be of finite size. Hence the measured intensity is always a somewhat smoothed or integrated version of the ideal point-intensity, and the statistics of the measured speckle will be different from the statistics of the ideal speckle pattern.

18.3 Brightness distribution

The speckle is itself an interference phenomenon. The nature of the illuminated surface gives rise to two different classes of speckle patterns, regardless of whether the perceived speckle is subjective or objective. One class is called the "fully developed" speckle pattern; it develops only from interference of

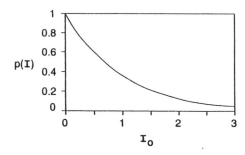

Figure 18.5. Probability density function of the brightness distribution of a fully developed speckle field.

light that is all polarized in the same manner. The speckle field itself will then be similarly polarized. Surfaces at which polarized light is singly scattered, such as matte-finished metal, generally give rise to polarized speckle fields, as do lightly scattering transmission elements such as ground glass. On the other hand, matte white paint surfaces or opal glass, into which the light penetrates and is multiply scattered, depolarize the light and thus do not generate a fully developed speckle pattern. The brightness distributions of the two classes of speckle patterns differ substantially (Ennos 1975).

The distribution of brightness of many different classes and types of speckle patterns has been thoroughly analyzed by Goodman (1975) and subsequently summarized in different ways by many others (e.g., Ennos 1975; Jones and Wykes 1983). Only the results that are essential to developing an understanding of speckle measurement methods are offered here. The so-called fully developed speckle pattern has a brightness distribution that is described by the following probability function,

$$p(I) = \left(\frac{1}{I_0}\right) e^{\left(\frac{-I}{I_0}\right)} \tag{18.4}$$

In this function, $p(I)$ is the probability that a speckle has brightness between the values I and $(I + dI)$, and I_0 is the average brightness. This relationship is plotted in Figure 18.5, and it shows that the most probable brightness for a speckle is zero, that is, there are more dark speckles in the field than speckles of any other brightness.

18.4 Coherent combination of speckle and uniform fields

In many holographic and speckle interferometry processes, a uniformly bright field of coherent radiation, which is the so-called reference beam, is added to the speckle field. The addition of the reference field will affect both the size and the brightness distribution of the speckle field. These effects are important in speckle metrology.

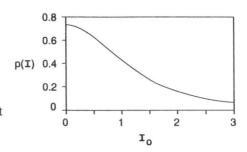

Figure 18.6. Brightness distribution for a coherent combination of speckle field and uniform field.

When a reference beam is introduced, the size of a speckle will approximately double (Ennos 1975). The reason for this involves the interference effect of adding a uniform strong wave to the speckle pattern in the direction of the optical axis. The size of a speckle without the addition of the reference beam corresponds roughly to the spacing of the interference fringes generated by waves coming from the opposite ends of a diameter of the speckle-forming pupil, if an imaging system is used. When the strong reference wave is introduced, the principal interference effects take place with respect to this central strong ray, so the maximum angle between interfering rays is halved. The equation for interference of two plane waves tells us that, when the angle between the beams is halved, the interference fringe spacing is doubled, as is the size of the speckle.

Burch (1970) analyzed the statistical distribution of brightness when the uniform field is added to the speckle pattern, for various ratios of irradiance of the two beams. A typical result is shown in Figure 18.6, which gives the distribution when the average speckle brightness is equal to the reference field brightness. The probability function for this case is

$$P(I) = \left(\frac{2}{I_0}\right) e^{-\left(1 + 2\frac{I}{I_0}\right)} J_0\left[2\left(2\frac{I}{I_0}\right)^{1/2}\right] \tag{18.5}$$

where J_0 is the Bessel function of zero order. This probability function does not differ appreciably from that of a single speckle pattern. Dark speckles will still predominate by far.

18.5 Coherent and incoherent mixing of speckle patterns

Some speckle interferometry applications require the mixing of two speckle patterns, as from two different scattering surfaces or else from one object illuminated with two different beams that might be coherent with one another. When this occurs, the size of the speckles does not change appreciably, but their brightness distribution might be altered, depending on whether the patterns are mixed coherently or not. For the case in which the two original

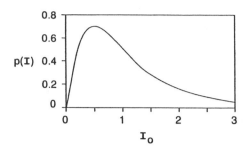

Figure 18.7. Probability brightness distribution for the incoherent mixing of two speckle fields.

speckle fields are brought together coherently, the result is a third speckle pattern, differing only in detail from its two component patterns but whose size and statistical brightness distribution remain unchanged. If the two original speckle fields are combined incoherently, then the brightness probability function becomes (Ennos 1975)

$$p(I) = 4\left(\frac{I}{I_0^2}\right)e^{\left(-2\frac{I}{I_0}\right)} \tag{18.6}$$

The brightness probability function for this case is plotted in Figure 18.7.

This result shows that there is zero probability of a dark speckle patch in the pattern. The highest probability is for a speckle of intermediate brightness. According to the authority just cited, this is not a surprising finding. If two speckle patterns of equal average irradiance are overlaid physically, there is a high probability that the bright areas of one pattern will be superimposed onto the dark areas of the other. Also note that surfaces that completely depolarize the light will give a speckle pattern having a brightness distribution that agrees with the preceding equation, because any two orthogonally polarized components of the scattered light are incoherent with one another.

Coherent and incoherent sums of speckle patterns are quite different from a statistical viewpoint. Under visual examination, the two are not easily distinguished.

18.6 Polarization effects

Because the sensing of phase information in interferometry relies on the formation of interferometric fringes formed by the constructive and destructive interference between the object and reference beams, the light in the two beams should have the same polarization. If a laser beam is divided and the two components are polarized orthogonally to each other, then they will not interfere. Most lasers are linearly polarized, which can in some instances cause problems in recording speckle patterns. Although both beams are derived from the same source, polarization changes can occur as a result of reflections.

Fortunately, most diffuse objects scatter light in randomly polarized fashion, and there will always be components having the same polarization as the reference beam.

If a polarization problem exists, it can be rectified by the following methods:

1. Change the surface properties of the object (e.g., paint it or spray with aluminum paint).
2. Rotate the plane of polarization in the reference or the object beam to match the other by inserting a half-wave plate into either beam and rotating it until the extinction angle is the same for each beam when tested with a polarizing filter.
3. Circularly polarize the light as it emerges from the laser by orienting a quarter-wave plate in the laser output beam.

18.7 Speckle pattern decorrelation

In the development and application of speckle measurement methods, the concept of speckle decorrelation is important. Basically, a speckle is a signature that is unique to a specific local microscopic surface element whose extent is dictated by the illumination or viewing aperture. The individual speckles do not change as the surface is moved; they move with the surface up to a point. Speckle metrology involves recording and/or combining two speckle patterns for two states of the specimen. In order for these methods to function, the two speckle patterns must stay correlated (in touch, marginally superimposed) with one another. If a speckle moves so far that its fundamental signature is changed, or if it moves so far that it cannot be compared with its first version, then correlation is lost and the measurement of its motion cannot be executed. This possibility is especially important when electronic detection with detectors or pixels of finite size must be employed.

The practical implication is that the possibility of speckle decorrelation places limits on the range of motion or deformation that can be recorded. More, certain unwanted motions can cause loss of correlation so that the motion of interest cannot be measured. The possibility of decorrelation limits the range of displacement for which speckle methods that rely on correlation can be utilized.

Note that a quite different situation arises with speckle photography methods (Chapter 19), which do not rely on correlation of speckles. In speckle photography, the speckles before and after deformation must form a pair that is separated by a small distance in order to form a sort of grating. The maximum limit on in-plane displacement for speckle correlation methods corresponds very roughly to the minimum displacement for speckle photography. The two methods are, therefore, complementary in terms of sensitivity and range.

This very fundamental aspect of speckle correlation metrology has been discussed by Jones and Wykes (1983). In their analyses, they point out that in speckle pattern correlation interferometry, it is assumed that neither the displacements that give rise to the phase variation causing the fringes nor other displacements not contributing to this phase variation significantly alter the random phase and amplitude of the speckles in the fringe observation plane. This approximation is, however, valid only when the displacements involved are less than some minimum value that depends on the kind of displacement involved and the viewing geometry. Larger displacements cause changes in the phase relationships that alter the fundamental speckle signature. For example, in the case of in-plane displacement interferometry, a body may undergo in-plane strain together with out-of-plane rotation. The latter can change the phase relationships within the speckles enough to reduce the contrast of the plane-strain fringes. Understand clearly that this type of decorrelation depends on alterations of *relative* phase relationships *within each speckle* during speckle movement. But, keep in mind that a uniform (nearly) gross change of phase for the entire speckle is what causes the desirable speckle fringes to form. For our purposes the decorrelation process just described is termed phase decorrelation. The degree of phase change within a speckle that can be allowed is debatable. As an upper bound, it seems that a phase change of 2π from edge to edge across a given speckle will result in total phase decorrelation.

Another important factor is that when the object deforms, the resolution elements also move. If the movement is large enough so that each displaced speckle does not appreciably overlap the corresponding original speckle position, then the original and final speckles cannot be correlated or compared. The maximum allowable motion of each speckle in speckle correlation interferometry is some fraction of the speckle diameter. If electronic detectors are used, then the allowable speckle movement is related to the detector size. If this movement is big enough to shift the speckle across a detector element, electronic or otherwise, at the image plane, then the detector will "lose memory" of the speckle that contains the displacement information. Memory loss is equivalent to decorrelation, and it is often a limiting factor in video-based speckle metrology (e.g., ESPI). Keep in mind that, in many electronic speckle setups, the speckle size is adjusted to be roughly equal to the detector size, in which case these two types of decorrelation by overly large displacements are equivalent. For our purposes, this type of decorrelation is called memory loss.

A significant implication of the memory loss idea is that the speckle size should be large for speckle correlation interferometry. This conclusion is opposite to that developed for speckle photography, which gives the highest sensitivity with small speckle size. Again, we find that the two methods are complementary.

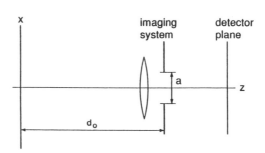

Figure 18.8. Schematic for analysis of speckle decorrelation.

The criteria to judge if there is decorrelation by phase change or by memory loss can be summarized as follows:

1. When the phase change across a resolution element is equal to or bigger than 2π, the corresponding individual speckle will become totally decorrelated in phase.
2. When the resolution element moves approximately a speckle diameter, there is decorrelation by memory loss. For electronic detection, when the speckle moves across and out of the detector element, the detector will lose memory of the information carried by the speckles.
3. As a rule of thumb, if the phase change and memory loss are less than about one tenth of the corresponding critical values calculated by the criteria just given, then the speckle is not decorrelated in phase or by memory loss.

The following estimations of limiting values of object motions for correlation to be maintained follow quite closely those of Jones and Wykes (1983), except that they did not consider the loss of memory that can take place when a detector of finite size (pixel or grain) is used at the image plane. We shall consider four different cases and determine whether phase decorrelation or memory loss is dominant for each.

Refer to Figure 18.8 for the configuration used in this analysis of speckle decorrelation. The illumination direction and the viewing direction are assumed perpendicular to the object surface. This assumption is often approximated in speckle interferometry. The detector can be, for example, a photometric device, a sensing element in a CCD video device, or the basic grain or the resolution cell in a photoplate image. For the moment it is most convenient to think of the sensor as an electronic element having small but finite dimension. Consider, in this illustrative treatment, only the case in which the speckle and the detector element have the same size. The effects of detector size are studied further in Chapters 21 and 22.

18.7.1 Out-of-plane translation

As the object is translated along the z-axis, the path lengths for the waves scattered from within a resolution element will change, causing a change of relative phase. From geometrical considerations of path length difference, the maximum value of this phase change will reach 2π across the resolution element when the object moves by an amount Δz_1 given by the following equation, after assuming $(x/d_0) \ll 1$ and neglecting the small quantities of second order:

$$\Delta z_1 = \frac{Mad_0^2}{1.22(1 + M)fx} \tag{18.7}$$

where d_0 is the object-to-lens distance
a is the aperture diameter
f is the focal length of the lens
M is the magnification of the imaging system
x is the distance from the optical axis to the center of the resolution element (the speckle) on the object surface

Also from geometric analysis, when the translation is

$$\Delta z_2 = 1.22(1 + M)\frac{fd_0\lambda}{Max} \tag{18.8}$$

the detector will lose memory of the corresponding resolution element, which means the resolution element will move out of the detector element or the original speckle area in the image. Remember, we have, for now, taken the detector size to be equal to the speckle size.

Consider a practical example to get in mind the relative magnitudes of the limiting motions based on these two criteria. If $M = \frac{1}{15}$, $d_0 = 500$ mm, $\lambda = 0.6328$ μm, $f/a = 16$ (a typical numerical aperture), and $x = 50$ mm, one obtains

$$\Delta z_1 = 16{,}009 \text{ μm} \qquad \text{and} \qquad \Delta z_2 = 1{,}976 \text{ μm}$$

In this case, the memory loss is found to be the limiting criterion for decorrelation, and one sees that the typical system has a large capability for correlation or memory retention in the z direction. Notice also that the tolerance to z-axis motion is very large near the center of the field in either case.

18.7.2 In-plane translation

As the object is translated in its plane, the speckle pattern is translated. It does not remain identical in form because the light scattered from a given point in the object is incident on the viewing lens at a different angle. Again, only geometric analysis of path length is needed. When the object is translated by

$$\Delta x_1 \approx \frac{Mad_0}{1.22(1 + M)f} \tag{18.9}$$

the phase change across the resolution element is 2π, and the speckle is therefore decorrelated.

When a body moves in its plane, the speckle pattern in the image plane also moves in proportion to the magnification of the viewing system. If a point on the object moves laterally by amount Δx_1, the corresponding point on the image will move by $\Delta x_1 M$. The speckle in the image plane will be decorrelated due to the memory loss when it moves by an amount equal to the resolution cell diameter S_{subj}. At the object, the memory loss occurs when the object is translated by

$$\Delta x_2 \approx 1.22(1 + M) \frac{f\lambda}{Ma} \tag{18.10}$$

Consider another practical example. If $M = \frac{1}{15}$, $d_0 = 500$ mm, $x = 50$ mm, $\lambda = 0.6328$ μm, and $f/a = 16$, the equations yield

$$\Delta x_1 = 1{,}601 \text{ μm} \quad \text{and} \quad \Delta x_2 = 198 \text{ μm}$$

Again the decorrelation from memory loss is more critical than is the decorrelation resulting from 2π phase change. Note that the decorrelation is uniform across the entire viewing plane in this case.

18.7.3 Out-of-plane rotation

When the resolution element is rotated by an angle θ about an axis lying in the object plane, the entire speckle pattern is decorrelated when θ attains the value

$$\theta = \frac{\lambda}{q} \approx \frac{Ma}{1.22(1 + M)f} \tag{18.11}$$

where q is the resolution element diameter, here equal to the speckle size. For $M = \frac{1}{15}$ and $f/a = 16$, one obtains

$$\theta = 0.18°$$

18.7.4 In-plane rotation

An in-plane rotation of the object gives rise to an equivalent rotation of the speckle pattern. A point at a distance R from the center of rotation moves by $R\alpha$ where α is the angle of rotation. The speckle is decorrelated when

$$\alpha \approx \frac{q}{R} = 1.22(1 + M) \frac{\lambda f}{aMR} \tag{18.12}$$

As an example, take $M = \frac{1}{15}$, $R = 50$ mm, and $f/a = 16$, to obtain

$$\alpha = 0.23°$$

Alternatively, for small in-plane or out-of-plane rotation, the movement of the resolution element could be analyzed directly by considering the motion to be a combination of in-plane and out-of-plane translations.

From the preceding analysis, the conclusion is that the memory loss effect is generally more important in causing speckle decorrelation than is the 2π phase change. Also, in-plane translation is more serious than out-of-plane translation.

18.8 Sensitivity vector

In some of the various forms of speckle metrology, the process yields a fringe pattern that, because it is a mapping of phase changes, is interpreted in terms of path length differences. The geometry of speckle systems, as of holographic systems, is often such that the conversion from path length information to the displacement vector for a point on the specimen is best accomplished by the use of vectors. This problem has been solved in general form for holographic interferometry; it is discussed in section 16.8, and it need not be repeated here.

References

Burch, J. M. (1970). Interferometry with scattered light. In *Optical Instruments and Techniques*, Ed. J. H. Dickson, pp. 213–29. Newcastle-Upon-Tyne: Oriel Press.

Ennos, A. E. (1975). Speckle Interferometry. In *Laser Speckle and Related Phenomena*, Topics in Applied Physics, Vol. 9, Ed. J. C. Dainty, Ch. 6. Berlin and New York: Springer-Verlag.

Goodman, J. W. (1975). Statistical Properties of Laser Speckle Patterns. In *Laser Speckle and Related Phenomena*, Topics in Applied Physics, Vol. 9, Ed. J. C. Dainty, Ch. 2. Berlin and New York: Springer-Verlag.

Goodman, J.W. (1976). Some fundamental properties of speckle. *Journal Optical Society of America*, 66, 11: 1145–50.

Jones, R., and Wykes, C. (1983). *Holographic and Speckle Interferometry*. Cambridge University Press.

Vest, C. M. (1979) *Holographic Interferometry*. New York: John Wiley and Sons.

19

Speckle photography

This chapter presents a remarkably simple and effective way to use the speckle effect in the measurement of displacements and deformations. It can give point-by-point or whole-field data, and the sensitivity can be made variable. The method can be extended to use noncoherent illumination, and an example of such an application is described. Certain versions of the technique are closely tied to moire and shearographic techniques. These parallels are noted because they provide valuable unifying insight.

19.1 Introduction

A direct and simple exploitation of speckle for engineering measurement is to use it as a microscopic marker of points on the surface of the object being studied. A single speckle is a unique signature derived from the local characteristics of a small area of the object surface and dependent on the geometry of the optical system and the numerical aperture of the illumination or viewing system. If a speckled image is created, then the speckle near a point in the image is uniquely identified with the corresponding point on the object. If the point on the object moves within certain limits, and if the optical system is not changed, then the speckle moves with the point, and the motion is apparent in the image. The speckle is not lost or reformed. If the speckle is recorded for two states of the specimen, then the displacement of the speckle corresponds to the local displacement of the surface.

Suggestions for the direct use of coherent light speckle in displacement metrology and contour mapping first began to appear in about 1968. Subsequent development was quite rapid and somewhat complicated with several different researchers working simultaneously but independently on differing approaches to similar problems. Given the curious history of unique and overlapping contributions, and because of the early and still-extant confusion in naming the various speckle methods, it seems worthwhile in this

Table 19.1. *Early history of speckle methods*

Approach	Described by	Date	Scope
Apparently earliest suggestions of displacement measurement by speckle interferometry	Groh Burch and Tokarski	1968 1968	One-dimensional, mention fatigue detection
Single illuminating beam method for vibration studies	Archbold et al. Eliasson and Mottier Burch Tiziani Köpf	1969 1971 1971 1971 1971–2	Mostly examinations of nodal patterns in vibrating bodies
Double illuminating beam method for displacement measurements and contour mapping	Brooks and Heflinger Archbold et al. Leendertz Butters Burch Hung and Der Hovanesian Adams	1969 1970 1970 1970 1971 1972 1972	Some quantitative results, but generally investigations of possibilities in speckle correlation and speckle-moire
Multiple aperture methods for displacement measurement	Duffy Hung et al.	1972 1974	Speckle-moire concept Shearing interferometer
Single aperture–single illuminating beam with optical Fourier processing for displacement measurements	Butters and Leendertz Archbold and Ennos Duffy Adams and Maddux	1971 1972 1972 1973	Speckle correlation Treat several methods, applications, and limitations Based on moire theory Pointwise observation of Young's fringes

instance to list the pioneering investigators. Table 19.1 is a summary of early progress based on published results (Cloud 1975). A review paper by Stetson (1975), the book by Jones and Wykes (1983), and the book edited by Dainty (1975) offer excellent reviews of progress, considerable scientific insight, and descriptions of several applications of the various techniques that are collectively called "speckle methods."

In the years since its inception, speckle photography, also called speckle interferometry and speckle-moire interferometry, has matured into a well-understood technique that offers a fruitful approach to a number of difficult problems in experimental mechanics. The method utilizes simple high-resolution photographs of a specimen that is illuminated with coherent light. The end result is a permanently stored whole-field record of interference fringes, which yields a map of displacements in the object. The fringe pattern is localized in the image of the object being studied, and two components of the displacement vector can be obtained individually and without ambiguity. Through minor changes of geometry of the optical system, the sensitivity of the measurement process may be varied continuously between wide limits; and these sensitivities are such that they fill the gap between standard interference methods and conventional moire techniques. The technique can be set up to measure only one displacement component at a time. In particular, it is naturally most responsive to in-plane motion and not sensitive to out-of-plane motion. Holographic interferometry has just the opposite primary sensitivities, so the two methods can be used to complement one another. In one of the several approaches to speckle-moire interferometry, the data are stored photographically in such a way that the sensitivity and range can be established after the experiment is completed. Because the approach is purely optical, there is no physical contact with the specimen. Only optical access is needed, and the recording station may be remote from the specimen site. Finally, the speckle method is much more tolerant of environmental vibrations than is holographic interferometry. The technique does have its shortcomings, and these will be discussed near the end of the chapter.

Given the advantageous characteristics just listed, speckle-moire interferometry seems to be well-suited for problems involving fragile specimens, surface phenomena, difficult specimen environments, or indeterminate or unusual magnitudes of displacement. Some particular examples of such problems include investigations of properties of foils, tests of biomaterials, measurement of deflection of shell and plate structures, and detection of onset of fatigue damage.

One complication that arises when we examine the literature is inconsistency in naming the method. This is partly reflective of a divergence of viewpoints regarding the particular interferometric process involved. It seems, for example, that several researchers did not realize that two-beam speckle interferometry can be seen as basically a moire method. The major cause of misunderstanding results from the use of the terms "speckle interferometry" to refer to two different techniques. A simple solution to the terminology problem was offered by Stetson (1975). If the technique involves incoherent recording of two speckle fields and if no reference beam is employed, then it is called "speckle photography." If a reference beam is present, then the speckle fields are still recorded incoherently, but the fields are being modified by the reference beam. In this case, the term "speckle interferometry" is used. Still, the two approaches are somewhat unified by a common body of theory, and optical interference is important in both because it is fundamental to the formation of speckle and in the recovery of data from recorded speckle patterns.

19.2 Single-aperture speckle recording

An optical system for recording speckle data on a photographic plate is shown in Figure 19.1. There is nothing special about this setup, which is an advantage of the method. Just illuminate the object with an expanded laser beam and take a picture. The spatial filter eliminates the usual spurious fringes, which would otherwise contaminate the speckle photograph, but not seriously so. The aperture of the recording lens governs the speckle size and thereby the basic sensitivity of the method. The lens numerical aperture must be chosen with an eye to the magnitudes of the displacements to be measured and the resolution capabilities of the film. For most purposes, a numerical aperture of $f/4$ to $f/11$ seems appropriate if working with holographic plate at roughly 1:1 magnification. Some trials might be called for in specific situations.

A typical double-exposure sequence is utilized, with the specimen being in a different state for each of the two exposures. The processed plate contains a speckled image of the specimen with no evident fringes. That the plate is a complex diffraction grating is readily ascertained by illuminating the plate in reflection or transmission with a point light source, such as a pen light. One will see several orders of Young's fringes, with spacing and direction dependent on the displacement of each area of the specimen between exposures. This diffraction pattern forms the basis of one method of displacement measurement. The penlight examination has been found to be a quick and simple test of success of the data recording phase of the experiment. The plate can be examined even while it is wet, and if diffraction performance is not good, the experiment can be repeated with only a few minutes loss of time.

It is highly recommended that glass photoplate be employed for speckle recordings, with holographic plate being the optimum choice. Film plate processing follows standard practice. Bleaching of the plate to convert it to a phase grating is a good idea when a low-power laser is used in data processing. Use of a simple bleach gives results with a 3-mW laser that otherwise could only be obtained with more than 10 mW. Best exposure of the plate seems to be that which gives an intensity transmittance of approximately 50% in the unbleached plate. Naturally, in double-exposure analysis, each exposure is one-half the total required.

Now consider from a physical viewpoint the structure that is recorded on the film. A single speckle pattern is a very fine random interference pattern of small patches which have the highest probability of being dark. The exposed film is a negative that will be dark in the areas of the bright speckles, so there will be a random array of relatively sparse dark spots in the negative

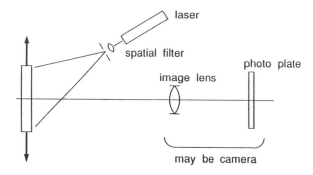

Figure 19.1. Single-aperture optical arrangement for photography of speckles.

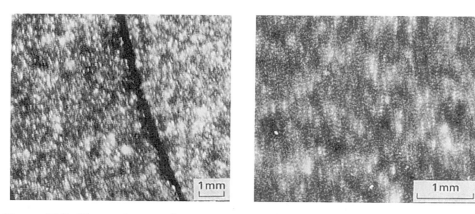

Figure 19.2. Photomicrographs at two magnifications of double-exposure laser speckle pattern.

image. The second exposure will add a similar array of dark spots that are shifted with respect to the first set by the displacement and deformation of the object. If the shift is not too large, but also if it is large enough so that speckles in the two systems do not overlap appreciably, then the resulting pattern will consist of a host of speckle pairs. The spacing between the two elements of any pair is indicative of the magnitude of the local deformation as seen at the film plane. The orientation of the speckles in a pair with respect to one another indicates the orientation of the local displacement as seen at the film plane. Figure 19.2 shows photomicrographs of just such a double-exposure speckle pattern. The pairing of the speckles can be seen.

With normal linear processing of the recording film, the dark patches resulting from bright speckles will be seen to have fuzzy boundaries instead of the sharp edges implied by most descriptions of the phenomenon. The reason is that the light distribution in a bright speckle is roughly that of the Airy disc or sinc function, which represents the diffraction function for a clear aperture (see section 10.4). The speckle patch has a bright center that rapidly falls to zero; then the brightness oscillates to create faint rings around the central spike.

Examination of many double-exposure patterns such as that just offered suggest that the speckle pairs seem to line up in such a way as to create the suggestion of "windrows." For a smoothly varying displacement field, the appearance is that of a rather poor moire grating. This behavior and similar observations for other speckle methods lead to the name "speckle moire interferometry." Most methods of speckle metrology can be explained in terms of geometric moire phenomena. Part of the reason for the grating structure evident in these speckle patterns lies in the way the irradiance distributions of adjacent speckles interact with one another when one set is shifted slightly relative to the other.

To begin to understand the technique mathematically, just describe the original speckle pattern as having some amplitude distribution that is a function of position in the film plane, as $A(x, y)$. The irradiance distribution at the film is just $I(x, y) = A^2(x, y)$. If the sensitivity function of the film is the same as was used in earlier chapters, then the transmittance of the film that would result from the first exposure is

$$T(x, y) = K_0 + K_1 I(x, y) \tag{19.1}$$

The basic premise of speckle photography is that the speckle distribution is not changed by the small deformations of the object. The speckle pattern moves with the surface. The amplitude distribution, and therefore the irradiance, is not changed except by a coordinate shift that is equal to the displacement. If u and v are the x and y components of the displacement of an individual speckle as seen at the film plane, then the new irradiance is just the old function with new coordinates, that is,

$$I(x + u, y + v) = A^2(x + v, y + v) \tag{19.2}$$

We have assumed that the illumination level is the same for both exposures. If it is not, then an extra constant must be inserted.

The new irradiance contributes another change of transmittance in the film. If linearity is assumed, which is reasonable physically, then the final transmittance after the double exposure and processing is

$$T(x, y) = K_0 + K_1[I(x, y) + I(x + u, y + v)] \tag{19.3}$$

The actual speckle distribution is not known in general, and in fact it is not of much interest. The problem is to somehow extract from this exposed film the displacements u and v of the specimen as evidenced at the plane of the recording film. If these can be obtained, then only geometric relations are needed to convert them to displacements at the specimen.

19.3 Point-by-point determination of displacement

Examination of Figure 19.2 convinces one that measurement of the displacements between speckle pairs with a measuring microscope is not a practical method. The structure is very complex, and the centers of the individual speckles are not easy to locate, partly because of interactions between the speckle fields that now are superimposed through double exposure. A better approach is to utilize the diffraction characteristics of this gratinglike structure to determine the speckle pair separations. There are two such techniques. The simpler one is discussed first.

Suppose that a slender beam of coherent light is passed through the double-exposure photographic negative. A screen is placed some distance away from the negative for observation of the diffraction pattern. Figure 19.3 is a schematic of such an arrangement.

Before resorting to a mathematical explanation of what is observed, let us consider it from a physical viewpoint, which is strengthened by our knowledge of diffraction by two simple structures. Assume that the displacement field is slowly varying relative to the cross section of the laser beam. Then the laser beam encounters an assembly of many speckle pairs. The spacing and relative shift of each speckle relative to its pair mate is the same for the whole assembly. In the average, the beam will be diffracted as by a simple grating that is confined within the area illuminated by the laser beam. Recall the intensity distributions that are created by diffraction at a clear aperture and at a grating consisting of equispaced apertures, as discussed in Chapter 10. One expects the result to be a broad on-axis sinc function that is modulated by a distribution of narrow sinc functions (delta functions in the limit) corresponding to the fundamental grating frequency and its harmonics.

This prediction is borne out by experiment. The diffraction pattern will be a "halo" of illumination that decreases in intensity away from its center. The halo is intersected by dark parallel fringes. The spacing of the fringes in the halo is indicative of the grating frequency, which is the spacing of the speckle pairs. The orientation of the fringes indicates the relative orientations of the speckle pairs. As suggested previously, these parameters at the film plane can be transformed to give the magnitude and direction of the displacement of the corresponding point (actually a small area) at the specimen surface.

Constructing a map of displacements over the extent of the image requires that arrays of points on the double-exposure specklegram be interrogated in turn. The fringe orientation and spacing at each point on the film are measured relative to specimen coordinates. All measurements on the film must be correlated with their corresponding points in the specimen. This process is rather tedious if many points must be examined. It is amenable to automation through the use of detectors or video image acquisition operated in conjunction with a computer.

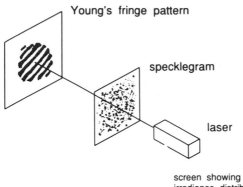

Young's fringe pattern

specklegram

laser

Figure 19.3. Point-by-point
interrogation of double-
exposure speckle
photograph
by a laser beam.

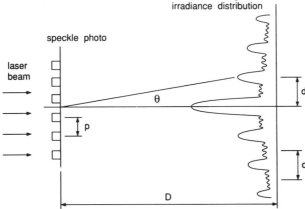

screen showing
irradiance distribution

speckle photo

laser
beam

θ

p

d

d

D

Figure 19.4. Relationship
between speckle pair
spacing and spacing of
Young's fringes.

This physical explanation based on diffraction by a grating can be extended to find the relationship between fringe spacing and displacement magnitude at the film plane. Figure 19.4 defines the variables involved. The speckle pair spacing is the grating pitch p; it is the displacement at the film. The distance from film to observing screen is D, and the spacing of the dark fringes at the screen is d, which is the same as the spacing of the bright areas. The grating equation gives

$$\sin\theta \approx \tan\theta = \frac{d}{D} = \frac{\lambda}{p}$$

and so,

$$p = \frac{\lambda D}{d} \tag{19.4}$$

This process of interrogating the double-exposure speckle photograph with a laser beam is, as mentioned before, actually one of diffraction at an aperture. Before leaving the topic, one should take a closer look to see if diffraction theory supports the expectations derived from physical reasoning based on grating behavior. A firmer grasp of the mathematics will be of advantage in understanding the whole-field method of fringe creation.

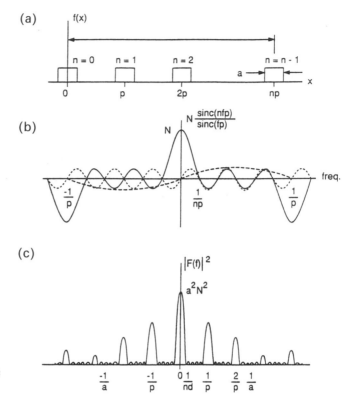

Figure 19.5. Fourier transform and irradiance distribution for an aperture containing a bar grating.

The far-field fringe pattern is the Fourier transform of the aperture transmittance function times a quadratic function of position and some constants, as was demonstrated in Chapter 10. It is illuminating to first look at the transform of a finite aperture that contains a grating of solid bars. This transform solution is well known. The spatial signal is shown in Figure 19.5a, its Fourier transform, in Figure 19.5b, and the irradiance distribution in Figure 19.5c. The irradiance distribution is closely matched by experimental observation, recognizing that the quadratic function is not incorporated, that the effect of the K_0 term in the transmittance is not included, and that the transform of the overall speckle distribution is also missing. Little of physical interest is gained by including these terms.

Further insight is gained by considering the Fourier transform of the tranmittance function in its general form as given in equation 19.3. Again, the constants and multipliers can be ignored because one is little interested in the precise irradiance distribution; the positions and meaning of the dark fringes are all that matters. For convenience, assume that there is only one component of displacement, u, at the film plane. This assumption is not restrictive as it implies a coordinate rotation with the x-axis lined up with the displacement. It allows the complete neglect of the y dimension. The parameter l specifies the angular deviation of the beam in transform space (frequency space) and replaces x_2/z_2 used in Chapter 10. The transmittance is substituted directly into the transform integral of equation 10.33 to obtain

$$U(l) = \int [I(x) + I(x + u)] e^{-iklx} \, dx \tag{19.5}$$

This integrand may be written as a convolution as follows,

$$U(l) = \int I(x) \otimes [\delta(x) + \delta(x + u)] e^{-iklx} \, dx \tag{19.6}$$

Now use the fact that the Fourier transform of a convolution of two functions is the product of the transforms of the functions. Let the transform of $I(x)$ be just $i(l/\lambda)$; it is not known. (Remember that the λ is required to establish the metric in transform space, and it was contained in the k in the integral.) The other transform pair for the second function in the integrand is

$$FT\{\delta(x) + \delta(x + u)\} = \cos \frac{\pi u l}{\lambda} \tag{19.7}$$

The complex amplitude in the transform plane is therefore

$$U(l) = i\left(\frac{l}{\lambda}\right) \cos \frac{\pi u l}{\lambda} \tag{19.8}$$

The irradiance distribution in the transform plane is

$$I(l) = \left[i\left(\frac{l}{\lambda}\right)\right]^2 \cos^2 \frac{\pi u l}{\lambda} \tag{19.9}$$

The first expression in this result is just the square of the Fourier transform of the original speckle pattern. It will be just another speckle pattern with a structure similar to that of the original. This speckle pattern is modulated by the $\cos^2$ fringes described by the second expression. The modulation pattern gives a bright center that is flanked by dark fringes. Dark fringes are attained whenever the argument of the function reaches an integer multiple of $\pi/2$. The distance between the dark fringes corresponds to a change in the argument of π radians, or

$$\frac{\pi u l}{\lambda} = \pi$$

Taking, as before, the fringe spacing as d and the distance to the viewing screen as D, then $l = d/D$ and the preceding equation reduces to the following for the speckle shift in the transform plane as caused by the specimen displacement

$$u = \frac{\lambda D}{d} \qquad (19.10)$$

This result is the same as that given in equation 19.4, since the displacement u is the same as the grating pitch p.

The quadratic function of position was ignored in this development, as was the effect of the finite aperture. Both of these will just shape the overall irradiance distribution of the diffraction halo. It is satisfying to find that the diffraction pattern is actually a modulated speckle pattern, as that is what is seen in the laboratory.

19.4 Spatial filtering for whole-field displacement fringes

The method just described requires point-by-point interrogation of the double-exposure speckle photograph. A favorable feature of optical methods in general is that they offer the possibility of parallel processing. In this instance, it is possible to spatially filter the data contained in the entire specklegram to obtain fringes in the image that are loci of points of equal displacement component. Figure 19.6 shows an optical spatial filtering system that can be used.

To thoroughly understand this technique for obtaining whole-field displacement fringes, let us resort to a physical line of reasoning. The discussion becomes a bit tedious, as does the mathematical treatment, because the process is fairly complex. Experience has shown that the resulting insight is well worth the effort expended. Not only does it aid in understanding the spatial filtering of specklegrams, but it also helps in comprehending Fourier transforms in terms of physical meaning.

First, review from a physical viewpoint the extent and nature of the data contained in a double-exposure specklegram. Assume that the specklegram is of an object in which the displacement varies in magnitude and direction over the surface. Then the spacing and relative orientations of the speckle pairs vary in the same way. One small area element of the speckle image has a gratinglike structure with a certain spacing and orientation. An adjacent area element has a slightly different grating spacing and direction. The whole image is composed of a multitude of small pieces of gratings, each one different, in general, from the others. In the limit, the image is made up of a continuously varying gratinglike speckle pair structure.

If the entire speckle photograph is illuminated by coherent light, all of these little grating components diffract light simultaneously. The result is a diffraction halo in the far field. From Chapter 10, the irradiance distribution in the diffraction halo is given by multiplying the Fourier transform of the

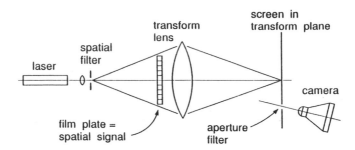

Figure 19.6. Simple optical Fourier transform spatial filtering system for extracting displacement fringes from double-exposure speckle photographs.

specklegram with a quadratic function of position. In general, there will be no recognizable structure in this halo because the specklegram is so complicated. It would be easy to dismiss the halo as just another speckle pattern. The physical understanding developed in Chapter 10 suggests, however, that the halo in the transform plane does indeed have meaning. Each particular area of gratinglike speckle pairs in the specklegram diffracts light to a particular set of points in the transform plane. Remember that position in the transform plane corresponds to spatial frequency and orientation in the signal plane.

Trying to view or utilize this spatial frequency and orientation information in the Fourier transform is obstructed by the problem that was pointed out in Chapter 10. For this large-aperture situation, the fully developed transform created under the Fraunhofer approximation is many kilometers away, and the near-field diffraction integral contains terms that contaminate the simple interpretation as a Fourier transform. The solution to the problem was presented in section 10.6. One merely places a lens in the system to, basically, bring the transform closer. To reiterate, diffraction at any aperture creates a Fourier transform in the far field, which is probably too far away to be of use for signal transparencies larger than a pinhole. A lens must be used to produce a Fourier transform in the usefully near field. The lens also changes the metric in the transform plane.

To facilitate this line of physical reasoning, visualize some simple examples. Regardless of where they are in the specklegram, any and all speckle-pair elements separated by a displacement vector of magnitude p and vertical orientation will diffract light to a series of bright bands (not dots), which are horizontal and which are separated at the viewing screen placed at the focus of the lens by fringe spacing $d = F\lambda/p$, this being just a different way of expressing equation 19.10. F is the focal length of the lens, and collimated illumination of the transparency is assumed. If the illumination is not collimated then the scaling factor is changed by a constant. Now, if l is the

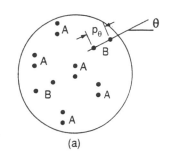

(a)

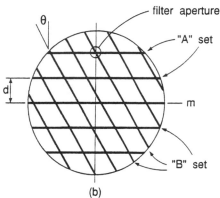

(b)

Figure 19.7. Relationship between speckle pairs in speckle photograph and fringes in the transform plane: (a) specklegram containing a few speckle pairs having two different orientations; (b) Fourier transform with fringes corresponding to the speckle pairs.

vertical distance in the transform plane, then there is a bright band wherever $l = nd = nF\lambda/p$, n being an integer. Inverting this relationship implies that at any bright band located at distance l from the origin in the diffraction halo, the corresponding speckle-pair separation or displacement magnitude in the specklegram is

$$p = \frac{nF\lambda}{l} \tag{19.11}$$

Figure 19.7 shows conceptually the relationship between a speckle photograph containing a few speckle pairs and the fringes that these cause in the Fourier transform plane. The preceding example corresponds to set A in the illustration. The illuminated areas are bands rather than dots because the diffracting area is small in the lateral direction. In fact, these are just Young's fringes for a given speckle pair, as described in the preceding section. A speckle pair, or the gratinglike structure formed by a small assembly of speckle pairs in a small area, which are separated by a distance slightly greater than p, will illuminate horizontal bands that are slightly farther apart. Speckle pairs oriented along an inclined axis (so its grating structure is perpendicular to that axis) will be mapped to bands that are inclined to the corresponding axis in the transform. See set B in Figure 19.7. An important fact is that these bright fringes will cross those created by the first two examples. The halo is formed by a multitude of bright diffraction bands all crossing one another.

Now that the information-carrying transform can be created, and now that the meaning of the transform is understood, the way to extract displacement fringes is clear. The concept of spatial filtering, as discussed in section 10.7, is put to good use. Suppose an opaque screen carrying a small hole is placed in the transform plane. Then the only light that can get through is that which carries information about areas of the specklegram having a given spatial frequency and orientation of the speckle pairs. All other information about all other areas having different spacing and orientation of speckle pairs is blocked.

Now a key point to finish the discussion. What if an inverse transform is made with another lens or the eye, using the light that gets through the hole? The inverse transform is an image of the original specklegram but with the spatial frequency content modified by the filter in the transform plane. The image is made only with the light that is available beyond the hole. That light comes from only those small areas in the specklegram that had a certain orientation and spacing of the speckle pairs. Therefore, only those areas of the specklegram image (specimen image) will be illuminated. The rest of the image will be dark.

Suppose that the spatial filter hole is placed at vertical distance l from the origin in the transform halo. The only areas of the specklegram that contribute light to that spot are those whose *vertical* separations of the speckle pairs are given by equation 19.11. In other words, they are loci of constant vertical displacement.

The difficult thing to see is that the vertical component of displacement is what is selected by putting the small filter aperture along the vertical axis of the transform halo. The explanation lies in the way that the Young's fringes in the transform tilt and cross one another. If the speckle pair displacement is tilted at angle θ with the horizontal, then their transform fringes are tilted at angle θ with the vertical, and they intersect the filter aperture at just the right distance to correspond to the correct vertical displacement. To show this, refer to Figure 19.7 and take p_θ as the displacement along the θ direction. The fringe separation at the transform plane in the corresponding direction is d_θ. Then the vertical locations of the fringes are $l = nd_\theta/\sin\theta$. The relationship used in developing equation 19.11 is again employed to write this as $l = nF\lambda/p_\theta \sin\theta$. But $p_\theta \sin\theta$ is just the vertical component of the original displacement vector, and it turns out to be $nF\lambda/l$, which is the same as given in equation 19.11.

A similar line of reasoning can be applied for any other position of the hole in the filter screen in the transform plane.

In summary, this spatial filtering of double-exposure speckle photographs yields fringes that are loci of constant displacement components. The particular component selected is established by the radial orientation of the spatial filter aperture in the transform plane. The sensitivity, which is reflected in the fringe spacing, is established by the amount of radial offset from the optical axis.

The resulting displacement fringes may be photographed by the camera that performs the inverse transform. The camera is placed behind the aperture so as to focus on the specklegram as seen through the transform lens. Any type of camera can be used. The spatial filter screen with its aperture can be attached to the front of the lens housing or mounted in an ordinary filter ring. That way the filter moves with the camera, and the aperture is always properly centered. The camera can be mounted on a swinging support so that locating the aperture in the desired position and aligning the camera is simplified. The focusing of the camera is best done in white light, because otherwise the low light intensities and speckle effects interfere with normal focus procedures. Just put a card with a sharp pattern on it in the specklegram holder and illuminate it with a desklamp to provide a good target for focus adjustment.

Figure 19.8 is an example of the flexibility of the method and the results that can be obtained (Cloud, 1975). Displacement fringes in a ring that is diametrally compressed are pictured for two sensitivities corresponding to two filter offsets. The pictures in the left-hand column give vertical displacements, and those in the right-hand column give horizontal displacements. It is worth noting that observation of fringe motion while moving the filter aperture gives exactly the same information as changing the load and watching the fringe motion in real-time. Clearly, the method yields a meaningful whole-field picture of displacements. The strain components can be obtained with appropriate differentiation of the displacement data. A crude but effective approach is to merely measure the spacing of the displacement fringes with a variable scale. Strain data that are more precise can be obtained with the manual or numerical plotting, curve fitting, and differentiation procedures that have been developed for moire strain analysis.

The relationships developed by means of physical argument can, of course, be derived analytically, starting with the expression for the Fourier transform of the specklegram, sampling a piece of it to represent spatial filtering, and taking the inverse transform of the piece. This mathematical treatment will not be undertaken here, as little additional insight is gained.

Little has been said about the relationship of speckle displacement in the speckle photograph and the actual object displacements. This conversion involves analytic geometry, and one is helped by keeping the setup simple.

A very important fact is that this form of speckle metrology is not sensitive to out-of-plane motion or tilt of moderate magnitudes, provided that the recording lens is sharply focused to image the object onto the film (Vest 1979). Low sensitivity measurements of out-of-plane motion can be extracted only because this motion changes the image size and therefore creates speckle pairs. Tilt of the object can be measured by recording the speckles in the

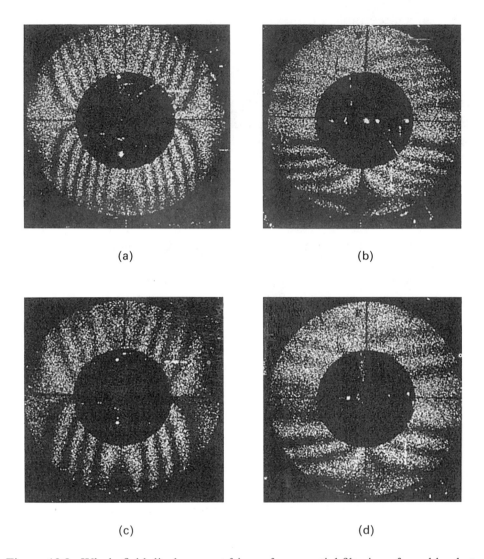

(a)

(b)

(c)

(d)

Figure 19.8. Whole-field displacement fringes from spatial filtering of speckle photograph of ring in diametral compression: (a) vertical displacement fringes at high sensitivity, (b) horizontal displacement fringes at high sensitivity, (c) vertical displacement fringes at low sensitivity, (d) horizontal displacement fringes at low sensitivity. Ring outer diameter is 88.8 mm; inner diameter is 44.4 mm; total deformation is 0.2 mm (Cloud 1975).

back focal plane of the recording lens rather than in the image plane. The insensitivity to motion out of the plane is a positive feature in many applications.

Time-average and real-time versions of speckle photography have been developed, but they will not be considered here. Most applications utilize the double-exposure technique.

19.5 Calibration of method

A most distinctive and attractive feature of speckle photography is its variable sensitivity. A related advantage in the case of the procedure discussed is that the sensitivity can be chosen *after* the data recording is completed. These gains come at a definite cost in that they tend to complicate the obtaining of quantitative displacement data.

As mentioned, the sensitivity is primarily a function of the focal length of the transform lens and the position of the filter in the image processor. Enough theory has been presented to allow one to calculate the fringe sensitivity for any optical setup. Such a procedure is not easy to follow in practice because of difficulties in establishing accurately the parameters involved, particularly filter offset. At best, such a procedure is time-consuming.

A simple, practical, and more direct procedure with a potential for good and consistent accuracy is to utilize a calibration specimen that exhibits a range of known displacements (Cloud, 1975). A speckle analysis of this calibrator serves to establish the sensitivity and errors for any given optical arrangement. The calibration procedure parallels the data recording and analysis procedure and may be carried out at any time. Knowledge of filter position, focal lengths, magnifications, and so on become irrelevant and unnecessary. The only restriction that cannot be violated is that the data processor shall not be disturbed between the calibration and the analysis of the unknown. An interesting aspect of this approach is that only one double-exposure photo plate of the calibration specimen is ever needed. It may be used at will to recalibrate the filtering setup regardless of rearrangement of apparatus. In fact, it has proven more practical to have on hand a series of calibration photo plates with different magnitudes and ranges of displacements.

A simple calibrator consists of a disc of aluminum that rotates about an axis through its center. Figure 19.9 pictures such a device. The surface of the disc is sandblasted for good diffusivity, and it carries fiducial markings to aid in identification, scaling, and focusing. The pivoted disc is rotated through a lever 154-mm long by means of a large micrometer head with a smallest scale division of 0.002 mm. Smallest readable rotation of the disc is 13 microradians. There is no backlash, and rotation is linearly related to micrometer reading over a range well beyond that expected in speckle work.

Typical fringes from a calibration sequence are shown in Figure 19.10. The figure shows fringe patterns corresponding to a disc rotation of 1.95 mrad for four different spatial filter positions as specified by parameter r. Notice that the fiducial marks in the image eliminate the need of calculating any

Figure 19.9. Calibrator for
speckle photography
(Cloud 1975).

scale factors, and distortions caused by a short-focal-length transform lens
can be eliminated easily. The positions of the fringes can be scaled directly
from the photos by whatever method gives the best compromise between
accuracy and speed.

Plots of fringe order as a function of displacement from these fringe
photographs are shown in Figure 19.11. The displacement sensitivities
obtained in this experiment vary from 9.5 to 21.3 μm per fringe. These
values are quite typical of those obtained with ordinary equipment and
holography film. The plots are linear within the investigated range of
±100 μm displacement.

19.6 Limitations

The positive features of displacement analysis by speckle photography tend
to draw attention from the price that must be paid in terms of film
requirements and loss of definition resulting from speckle effects in the fringe
photographs.

The upper and lower limits of displacement that can be measured by

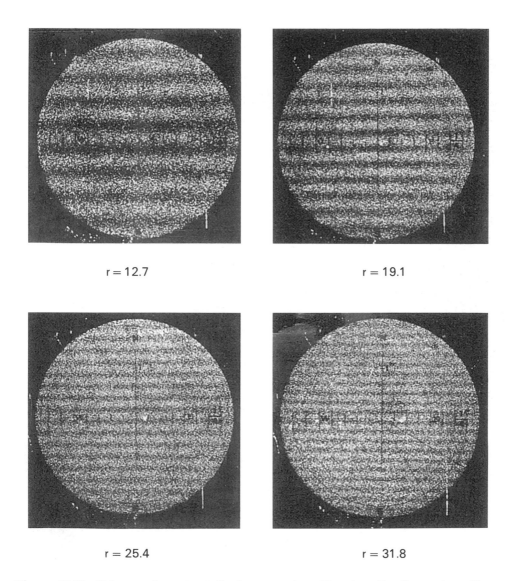

r = 12.7   r = 19.1

r = 25.4   r = 31.8

Figure 19.10. Fringes of constant displacement in calibration disc for various filter offsets: r is aperture offset, mm; disc diameter is 101.6 mm; disc rotation is 1.95 mrad; maximum displacement is 101 μm (Cloud 1975).

speckle photography have been given considerable attention (e.g., Archbold and Ennos 1972; Cloud 1975; Parks 1980). The sensitivity limit is fixed by the smallest speckle size, which in turn is established by the aperture ratio of the recording optics. This sensitivity limit takes on physical meaning if it is realized that, in the data-recording process, each exposure involves photography of a gratinglike structure (here random dots) consisting of all

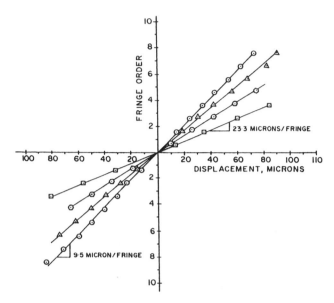

Figure 19.11. Fringe order as function of displacement obtained from whole-field spatial filtering of speckle photograph of calibration disc (Cloud 1975).

spatial frequencies out to the resolution limit of the system. For high sensitivity in measuring displacements, high-frequency gratings are required. Because grating frequencies are limited by film resolution and aperture, sensitivity is subject to the same limitations. This sensitivity limit is illustrated quite convincingly in the optical data-processing stage. The optical Fourier transform will be a halo disc whose finite diameter corresponds to the highest spatial frequencies on the input plate. There is no possibility of finding fringe data outside this limit because there is no energy there.

It appears that the only ways to gain sensitivity are to increase resolution capabilities of the recording film, to increase aperture, and to magnify the image of the specimen on the film plate. The last-mentioned approach creates a finer grating relative to specimen size.

A second limitation is imposed by the requirement that the filter aperture in the data processing system be small enough to pass fringe data resulting from a very narrow band of spatial frequencies. A practical rule of thumb seems to require that the aperture be approximately one-tenth the amount by which the aperture is displaced from the optical axis.

The best results are evident if a reasonable balance is maintained between sensitivity limit and displacement magnitudes. The sensitivity and range can be selected only within certain limits. The maximum sensitivity and the maximum range of displacements that can be measured correspond closely with those predicted by theory. The fact that these parameters can be chosen after the data-recording phase does greatly simplify the procedure, however.

The whole-field processing of speckle photographs is not economical of light. Only a small fraction of the light illuminating the speckle photograph passes through the spatial filter and ends up forming the image. The image tends to be dim unless a fairly large laser is employed, but much can be accomplished with lasers in the 5–20 mW power range.

19.7 Dual-aperture speckle photography

Many variations of the basic technique of speckle photography as just described have been developed. Space does not allow mention of many of them. A short discussion of one approach, known as the dual aperture technique (Duffy 1972), will be undertaken. The method gets around some of the problems with calibration, spatial filtering, and light loss that are inherent in the basic technique. More importantly, perhaps, it emphasizes the close relationship between speckle methods and moire. In fact, it is a variation in concept and methodology of the use of slotted apertures for moire analysis. Understanding of the basis of the method augments the discussion of the preceding sections on speckle photography. Finally, the concept of speckle shearography develops naturally from the idea of dual-beam speckle.

Figure 19.12 is a diagram of an optical system that can be used to record double-aperture double-exposure speckle photographs. This recording is the same as that for the usual speckle photography, except that a mask having two apertures equidistant from the lens center is placed in front of the lens or, preferably, inside the lens if it has several elements. The same results can be realized if two small, matched lenses are used for creating the speckle image on the film. The spacing of the two apertures establishes the sensitivity of the method, and the orientation of the axis connecting the two fixes the direction of the displacement component to which the system will respond.

The function of the dual aperture lens can be understood in the context of the ordinary method as discussed in sections 19.2 and 19.4 plus some knowledge of the optical transfer function of a lens and the concept of two-beam interferometry.

In the simplest view, the effect of having two apertures in the lens is to create two separate speckled images of the object. If the two image-forming beams are brought into exact registration at the image plane, then they will interfere. The new image speckle pattern, obtained by coherent addition of the two beams, will be modulated by interference fringes formed because the two beams come together at an angle. Figure 19.13 is a microphotograph of such a specklegram.

The pattern of speckles modulated by fringes forms a moire grating in the image. One such grating will be recorded for each exposure. Because the speckles move with the specimen surface as it is deformed, there will be a

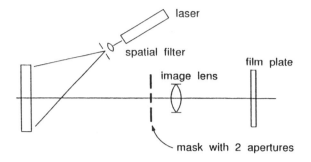

Figure 19.12. Apparatus for dual-aperture speckle photography.

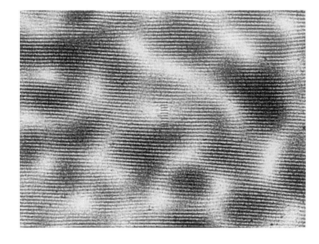

Figure 19.13. Microphotograph of speckle pattern for dual-aperture double-exposure laser speckle photography.

corresponding deformation of the grating. The speckle photograph so obtained will contain an ordinary double-exposure moire pattern, even though no specimen grating was used. The sensitivity, which depends on grating pitch, is fixed by the spacing between the recording apertures. The axis of the grating, which dictates the displacement component it will respond to, is established by the orientation of the axis between the apertures. If the apertures are on a horizontal axis, then the grating lines are vertical, and the measured displacement component is horizontal.

Because there is much less redundant information in this dual aperture image than there is for an ordinary specklegram, the displacement fringes are often visible without spatial filtering. If they cannot be seen in transmission, then try grazing incidence of light from a penlight or desklamp.

The best fringe patterns are obtained by spatial filtering, which is much less critical to perform than for ordinary speckle. Because of the narrow band of spatial frequencies present, the filtering is like that of superimposed moire gratings as discussed in Chapter 11. Instead of a complicated diffraction halo, most of the optical energy will be confined to two or four off-axis patches in the transform plane. The effect of filtering is mainly to reduce optical noise appearing in a faint halo. This spatial filtering is much more economical of light. The price is much less light at the film plane in the speckle recording stage.

The meaning and interpretation of the fringes is the same as for ordinary single-beam speckle photography.

A little thought suggests that this method and the previously discussed approach utilize the same ideas in differing order. In the dual-beam method, the filtering is done first and the imaging second. The reverse is true for the ordinary method. A gain is less critical spatial filtering and less trouble in quantitative interpretation. The price is less flexibility, because the measurement sensitivity is established at the time of speckle recording by the relative spacing and orientation of the two apertures.

More important, the relationships between speckle photography and moire methods are noteworthy and instructive. The entire theory for ordinary single-beam speckle photography can be worked out entirely from the moire idea for two apertures (Hung 1978). Each set of apertures in the entire imaging lens contributes its own set of moire gratings in the specklegram. The spatial filtering extracts the information contained in only one of these gratings from all the rest. The fact that the moire gratings are formed within speckles seems irrrelevant in this model. Actually, it is still important, as the speckles are what assures that the moire gratings move with the deformed surface.

If four or six apertures are used, then the data plate contains all information to determine three strain components. Spatial filtering must be used to separate them. This is the speckle version of the moire strain rosette.

This dual-beam technique is clearly a version of the use of slotted apertures to create moire gratings from somewhat random surface structures illuminated by white light (Burch and Forno 1975). The use of slotted apertures in multiplying moire sensitivity is described in section 11.5.

The dual-beam image is formed by combining two separate and complete images. One of the beams can be caused to be displaced slightly from its normal position by placing, for example, a wedge in front of one of the apertures. The result is that one image will be slightly displaced from the other. Interaction between these two coherent images forms the basis of the important technique known as speckle shearography, which can yield fringe patterns indicative of strain rather than displacement.

19.8 White light speckle photography and example application

The speckle methods discussed so far in this chapter have used the speckles formed in an image merely as fine markers. One is soon led to question whether it is necessary to create the speckles by coherent light. Could not a fine set of random dots on the surface of the object serve just as well? This

Figure 19.14.
Microphotograph of
single-exposure
noncoherent light speckle
pattern from sandblasted
aluminum disc illuminated
at a low angle by an
ordinary photographic
flashlamp (Cloud et al.
1980).

question was answered in the affirmative, particularly when Boone and DeBacker (1976) applied the speckle photography technique with a specially prepared retroflective surface that was illuminated with noncoherent (white) light. They also showed that placing special dots or materials on the surface was not necessary if the surface contained appropriate roughness features. When the surface is illuminated at a low angle, the peaks and valleys in the surface create shadows that can be used as a speckle pattern. At about the same time, Burch and Forno (1975) used similar procedures with slotted "tuned" apertures in a camera to perform moire analysis on large objects. Other aspects and applications of noncoherent light speckle were soon reported by several others (e.g., Asundi and Chiang 1982; Cloud et al. 1980; Cloud and Paleebut 1982; Cloud, Peiffer, and Radke 1980).

The recording of double-exposure noncoherent-light speckle patterns in the laboratory is very simple. The speckles are created by the low-angle lighting of the test surface, provided the surface has appropriate fine roughness features. Figure 19.14 is a photomicrograph of a single-exposure white light speckle pattern from a sandblasted aluminum surface. Notice the similarity to the laser speckle microphotograph of Figure 18.1. Because the speckles are uniquely related to local surface topography, they move with the surface as do laser speckles. One exposure is taken with the specimen in one state, and the second is recorded for a subsequent state. If the specimen is deformed or moved between exposures, then the result will be a specklegram that is analogous to that obtained by double-exposure laser speckle photography. Figure 19.15 shows photomicrographs at different magnifications of double-exposure speckle patterns, which were obtained by this method from a sandblasted aluminum disc that was rotated between exposures. The similarity of these patterns to those obtained in coherent illumination are obvious (see Fig. 19.2).

Extraction of displacement data from the speckle patterns parallels that for coherent speckle photography with one exception. Because the speckles are usually relatively larger than coherent light speckles, and because there are fewer of them, the whole-field spatial filtering technique does not, usually, produce very good results.

The theory of fringe formation for noncoherent light speckle photography has been developed, notably by Chiang and colleagues (Asundi and Chiang 1982; Chiang and Li 1985) and modified by Conley and associate (Conley 1986; Conley and Cloud 1991), among others. The development parallels that for coherent speckle, except that one must start with the point spread function for noncoherent light instead of the complex amplitude used for coherent light. The development is not reproduced here. The final result is the same as that given in section 19.3.

An interesting aspect of white light speckle is that it can be used on objects that, by virtue of size or location, cannot be illuminated with laser light. The technique of creating the speckle patterns from naturally occurring surface features causes it to be almost totally noninvasive. Problems in biomechanics (Cloud, Peiffer, and Radke 1980), for example, can be addressed in this way.

An example which illustrates the adaptability of the technique to field problems is taken from the area of geomechanics (Cloud and Conley 1983; Conley 1986; Conley and Cloud 1988; Conley and Cloud 1991). The motion of the surface of a portion of the Nisqually Glacier, Mt. Rainier National Park (Washington, USA) was measured using white light speckle. The specimen is about 6-km long and 800-m wide, it exists in a hostile enviroment, and it is not readily accessible for conventional measurements, especially in the steep icefall area that is of interest. Cameras especially constructed for the purpose were anchored on the valley walls to command a view of the upper elevations. From these locations, double-exposure images were acquired on holographic plate over intervals of 24 hours. Patterns were recorded when the sun was low so as to emphasize the undulations, sun cups, crevasses, and other surface features in the study area. The pictures were taken in pairs from the opposite valley walls so as to have a stereo-pair for data reduction. Figure 19.16 is a reproduction of one of these specklegram pairs.

Displacements in the film plane were gathered in the laboratory from these specklegrams by interrogation with a laser beam so as to produce Young's fringes, examples of which are reproduced in Figure 19.17. Not all image areas produce fringes because of a combination of atmospheric turbidity and

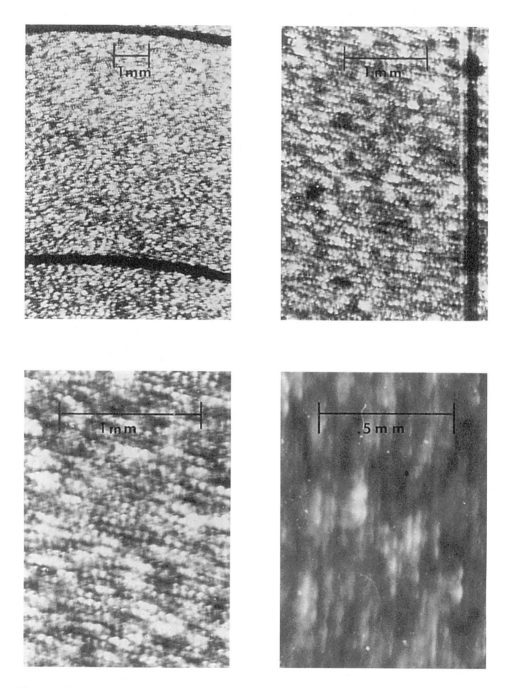

Figure 19.15. Photomicrographs at different magnifications of double-exposure speckle patterns from sandblasted aluminum disc illuminated at low angle by noncoherent light (Cloud et al. 1980).

Figure 19.16. Stereo pair of glacier specklegrams taken with cameras about 600 m apart (Conley and Cloud 1990).

 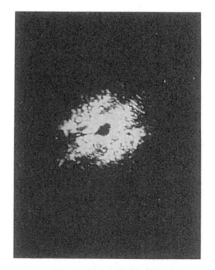

Figure 19.17. Young's fringes generated for two different locations on a glacier (Cloud and Conley 1983; reproduced by courtesy of the International Glaciological Society).

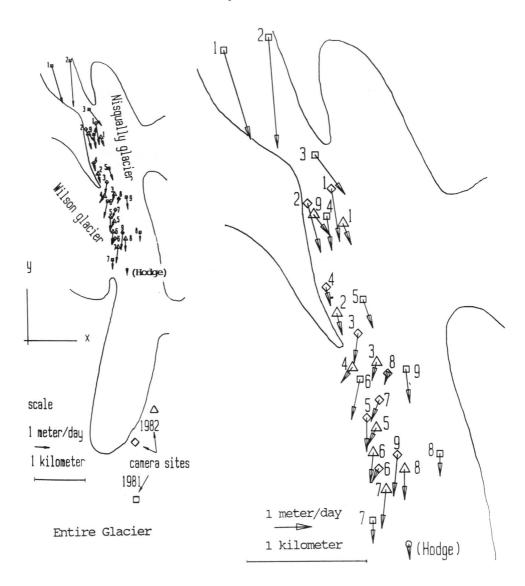

Figure 19.18. Glacier surface flow obtained during two field seasons using white light
speckle (Conley and Cloud 1988).

insufficient surface structure. The fringes that are obtained are noticeably
poorer than those obtained from better surfaces and by coherent speckle.
Those shown in Figure 19.17 are the first that were obtained from a glacier
surface, and subsequent results show improved fringe visibility.

The film plane displacements were corrected for camera motion and magnification, then projected onto the glacier using topographical map data and a computer algorithm similar to those used in photogrammetry. Figure 19.18 shows some results obtained in this difficult field application of white light speckle photography.

An interesting aspect of this study is that the sensitivity was found to be limited by atmospheric turbidity. Natural movement and time-dependent refractive index gradients of the air act to blur the image, preventing photography of detail below a certain size. The diffraction halo at fringe readout is therefore bandpass-limited and, in fact, serves as a measure of the finest speckle detail. This problem is seen in astronomical uses of speckle but had not been reported in terrestrial studies. It seems likely that similar effects would affect laser speckles recorded over long distances or in adverse environments.

References

Adams, F. D. (1972). *A Study of the Parameters Associated with Employing Laser Speckle Correlation Fringes to Measure In-Plane Strain.* Air Force Flight Dynamics Laboratory Tech. Report 72-20. Ohio: Wright Aeronautical Laboratories.

Adams, F. D., and Maddux, G. E. (1973). *On Speckle Diffraction Interferometry for Measuring Whole Field Displacements and Strains.* Air Force Flight Dynamics Laboratory Tech. Report 73-123. Ohio: Wright Aeronautical Laboratories.

Archbold, E., Burch, J. M., and Ennos, A. E. (1970). Recording of in-plane surface displacement by double-exposure speckle photography. *Optica Acta*, 17: 883–98.

Archbold, E., Burch, J., Ennos, A., and Taylor, D. (1969). Visual observation of surface vibration nodal patterns. *Nature*, 222: 263.

Archbold, E., and Ennos, A. E. (1972). Displacement measurement from double-exposure laser photographs. *Optica Acta*, 19, 4: 253–71.

Asundi, A., and Chiang, F. P. (1982). Theory and applications of the white light speckle method for strain analysis. *Optical Engineering*, 21, 4: 570–80.

Boone, P., and DeBacker, L. (1976). Speckle methods using photography and reconstruction in incoherent light. *Optik*, 44, 3: 343–55.

Brooks, R. E., and Heflinger, L. O. (1969). Moire gauging using optical interference. *Applied Optics*, 8, 5: 935.

Burch, J. M. (1971). Laser Speckle Metrology. In *SPIE Seminar-In-Depth on Developments in Holography*. Bellingham, WA: Int. Society Optical Engineers.

Burch, J. M., and Forno, C. (1975). A high sensitivity moire grid technique for studying deformation and strain in large objects. *Optical Engineering*, 14: 178–85.

Burch, J. M., and Tokarski, J. M. J. (1968). Production of multiple beam fringes from photographic scatterers. *Optica Acta*, 15, 2: 101–11.

Butters, J. (1970). Laser holography and speckle patterns in engineering metrology. In *Symposium on Advanced Experimental Technology in the Mechanics of Materials – Sept. 1970*. New York: Gordon and Breach (published as book in 1973).

Butters, J., and Leendertz, J. (1971). A double exposure technique for speckle pattern interferometry. *Journal of Physics E*, 4: 277.

Chiang, F. P., and Li, D. (1985). Diffraction halo functions of coherent and incoherent random speckle functions. *Applied Optics*, 24, 14: 2166–70.

Cloud, G. L. (1975). Practical speckle interferometry for measuring in-plane deformations. *Applied Optics*, 14, 4: 878–84.

Cloud, G., and Conley, E. (1983). A whole-field interferometric scheme for measuring strain and flow rates of glacier and other natural surfaces. *Journal of Glaciology*, 29, 103: 492–7.

Cloud, G., Falco, R., Radke, R., and Peiffer, J. (1980). Noncoherent-light speckle photography for measurement of fluid velocity fields. In *Proc. 24th International SPIE Technical Symposium: Applications of Speckle Phenomena*, Vol. 243, Ed. W. H. Carter. Bellingham, WA: International Society Optical Engineers.

Cloud, G. J., and Paleebut, S. (1982). Preliminary studies of strain within thick cracked specimens, on glaciers and on skin in vivo using multiple gratings and noncoherent speckle. In *Proc. VII International Conference on Experimental Stress Analysis*. Haifa, Israel: The Technion, Israel Institute of Technology.

Cloud, G., Peiffer, J., and Radke, R. (1980). Feasibility of noncoherent speckle photography for in vivo measurement of skin deformation. In *Proc. OSA Meeting on Holography and Speckle*. N. Falmouth, MA: Optical Society of America.

Conley, E. G. (1986). *Remote Sensing of Whole-Field Flow Rates of Glaciers and Other Large Bodies Using Interferometric Methods*. PhD dissertation. East Lansing: Michigan State University.

Conley, E.G., and Cloud, G. L. (1988). Whole-field measurement of ice displacement and strain rates. *Journal Offshore Mechanics and Arctic Engineering*, 110: 169–71.

Conley, E., and Cloud, G. (1990). White-light speckle for measuring geophysical surface motions. In *Proc. Hologram Interferometry and Speckle Metrology*. Baltimore, MD: Society Experimental Mechanics.

Conley, E. G., and Cloud, G. L. (1991). Resolution experiments using the white-light speckle method. *Applied Optics*, 30, 7: 795–800.

Dainty, J. C., Ed. (1975). *Laser Speckle and Related Phenomena*, Topics in Applied Physics, Vol. 9. New York and Berlin: Springer-Verlag.

Duffy, D. (1972). Moire gauging of in-plane displacement using double-aperture imaging. *Applied Optics*, 11, 8: 1728.

Duffy, D. (1974). Measurement of surface displacement normal to the line of sight. *Experimental Mechanics*, 14: 378.

Eliasson, B., and Mottier, F. (1971). Determination of the granular radiance distribution of a diffuser and its use for vibration analysis. *Journal Optical Society of America*, 60, 5: 559.

Groh, G. (1968). Engineering uses of laser-produced speckle patterns. In *Symposium on the Engineering Uses of Holography*, University of Strathclyde. Cambridge University Press.

Hung, Y. Y. (1978). Displacement and Strain Measurement. In *Speckle Metrology*, Ch 4. New York: Academic Press.

Hung, Y., and Der Hovanesian, J. (1972). Full-field surface-strain and displacement analysis of three-dimensional objects by speckle interferometry. *Experimental Mechanics*, 12, 10: 454.

Hung, Y., Hu, C., and Taylor, C. (1974). Surface measurements by speckle-moire and speckle-shearing interferometry. *Experimental Mechanics*, 14: 281.

Jones, R., and Wykes, C. (1983). *Holographic and Speckle Interferometry*. Cambridge University Press.

Köpf, U. (1971). Ein Koharent-optiches verfahren zur messung mechanischer schwingungen. *Optik*, 33: 517.

Köpf, U. (1972). Application of speckling for measuring the deflection of laser light by phase objects. *Optics Communication*, 5, 5: 347.

Leendertz, J. (1970). Interferometric displacement measurement on scattering surfaces utilizing speckle effect. *Journal of Physics E.*, 39: 214.

Parks, V. J. (1980). The range of speckle metrology. *Experimental Mechanics*, 20: 181–91.

Stetson, K. A. (1975). A review of speckle photography and interferometry. *Optical Engineering*, 14: 482–9.

Tiziani, H. (1972). Analysis of mechanical oscillations by speckling. *Applied Optics*, 11, 12: 2911.

Vest, C. M. (1979). *Holographic Interferometry*. New York, John Wiley & Sons.

20

Speckle correlation interferometry

This chapter describes a way of using speckles that seems more elegant than speckle photography. A reference beam or a second speckle pattern is coherently mixed with the object speckle. The brightnesses of the resulting speckles are very sensitive to object motion. Comparison of two such patterns through superimposition or digital processing yields fringes indicative of displacement. The photograph-based version of the method is not used as much as speckle photography owing to experimental difficulties. Study of the method is worthwhile since it is the basis of electronic speckle pattern interferometry, which is becoming increasingly important.

20.1 Introduction

We turn now to what might be considered a more sophisticated use of speckle information in measurement. Rather than use a speckle merely as a marker on the specimen surface, we utilize to some extent the phase information within a speckle and the coherent combination of speckle fields as the basis of measurement. Such an approach is properly interferometric in concept and execution, so the techniques in this class are usually lumped together under the terms "speckle interferometry" or "speckle correlation interferometry" as distinct from "speckle photography."

The somewhat confusing early history of these related but quite different techniques was presented in Chapter 19, so it will not be recapitulated here. The tutorial writings of Vest (1979), Ennos (1975), and Jones and Wykes (1983) are extremely useful from both the technical and historical viewpoints, as is the review by Stetson (1975).

One aspect of these techniques deserves special comment to preclude possible misunderstandings. The question that must be faced is, "Where, how, and why in the process is interference used to preserve, modify, or utilize phase information contained in the speckles?" After all, in the speckles themselves, the phase data are already converted to amplitude information by interference among the waves coming from various object points. Can some measure of the phase data, perhaps an average, within one speckle cell be recorded or preserved? In most of our studies, the interferometric combination of the waves coming from the specimen point in its initial and final positions occurs at the image plane. But, there is no direct way to preserve the speckle phase data from one specimen state for subsequent mixing with a later specimen state. Only the individual speckle irradiance can be recorded. Some speckle phase information can, however, be modified before it is converted to amplitude information by combining the speckle field with a reference beam, which, of course, might be just another speckle field. The phase data in the specimen or object speckles are, thereby, just converted to a speckle field that is different from the one existing in the image plane for the specimen by itself. The modification of the object speckle field assures that the phase changes seen as correlation fringes are properly related to the desired object displacement component. The mixing of speckle fields also increases the sensitivity to out-of-plane displacements with the appropriate apparatus.

In the end, these correlation methods sometimes seem more like geometric moire than coherent interferometry. Interference is used to create and modify the speckle patterns, which are recorded noncoherently. The recorded speckles for two specimen states are then combined noncoherently to create the fringes through what amounts to a geometric superimposition of random speckles.

The fringe patterns developed in speckle interferometry are best interpreted using the sensitivity vector that was discussed in section 16.11.

20.2 Basic concept

To fully understand the motive for modifying a speckle field by combining it with another field, a comparison with speckle photography is useful. Consider the speckle formed by the system shown in Figure 20.1, where an imaging lens creates a subjective speckled image, but a reference beam or reference speckle pattern is not used. This is just the speckle photography arrangement of Chapter 19.

We know that the speckles move with the object, but we also know that the speckle pattern does not change much with longitudinal (z-direction) motion of the specimen. Speckle photography is therefore not very sensitive to out-of-plane displacement. A fundamental reason for this behavior is that the speckle-producing wave groups are quite elongated (Vest 1979). The speckles are not really visible until they strike the film, but z-axis motion of the film will not change the speckles much, so the groups must be long. There is no

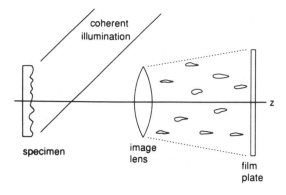

Figure 20.1. Conceptualization of appearance of speckle clusters when no reference pattern is used.

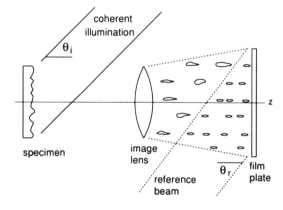

Figure 20.2. Use of reference beam to give sensitivity to z-axis displacement by speckle correlation.

interferometric combination of two independent waves at the film, so absolute average phase within a speckle is not important. This observation is supported by theoretical considerations.

Now modify the system by adding a reference beam inclined at angle θ_r, such as shown in Figure 20.2. In most practical forms of speckle interferometry, the setup is arranged so that the reference beam incidence angle is zero or nearly so. For those cases in which the reference impinges on the detector plane, the reference angle is actually not important because its phase does not change when the specimen moves. In other cases, the reference field is provided by a second beam that impinges on the object, and the angles of incidence of both object beams are important.

To facilitate understanding of what happens to the speckle, assume temporarily that the object and reference beam incidence angles are zero. The effect of the reference beam is to make the speckle very dependent on the phase difference between the reference and the object. This idea is represented in the illustration by showing the speckle-producing groups as short and periodic, instead of long and random as is the case for speckle photography. If the average phase in a speckle changes by one-half a wavelength because of specimen motion, then the speckle in the image changes from dark to light. The process is truly interferometric, and sensitivity to out-of-plane motion is equal to that of, say, holographic interferometry. This is true even though the comparison of the before and after speckles is a noncoherent combination.

The problem at this point is in somehow getting the two speckle images correlated so that fringes indicative of displacement can be created. It is no surprise that three modes have been devised: real-time, time-average, and double-exposure. Most speckle interferometric schemes can be adapted to either of the three modes of fringe observation. Some modes are better than others in a given application. In all examples of speckle correlation interferometry, at least those using photographic media, fringe visibility can be a serious problem.

Consider in general terms what might be seen in the real-time procedure with an interferometer as just described. A photographic plate of the speckle interference pattern is recorded, processed, and put back into its original position. It is a negative of the speckle, so bright speckles are dark on the plate. If the plate is brought exactly into registration, then all the bright spots line up with dark spots on the negative, and the whole image will be dark. The specklegram is acting as a filter mask.

Suppose the specimen is then shifted one-quarter wavelength toward the lens. A phase difference of one-half wavelength is induced in the object speckles. Upon mixing with the reference beam, all the formerly dark speckles are changed to bright, and the formerly bright ones become dark. The newly bright ones now line up with transparent patches on the film (formerly dark so unexposed), so the image becomes more-or-less bright. Another quarter-wavelength of motion causes a return to the original dark state. The image is alternately light and dark with successive $\lambda/4$ displacements of the object. This phenomenon is used to generate a fringe pattern indicative of out-of-plane motions of the object.

The remaining question has to do with the effects of in-plane motion (along the x-axis). Will not this motion create a fringe pattern as well? The speckles are known to move with the surface as it moves, and this fact is used in speckle photography. (For the setup having zero incidence angles, however, lateral motions on the order of several wavelengths will not cause phase shifts in the speckles with respect to the reference beam, so no fringes appear. Lateral motions on the order of the speckle size will cause the new speckle pattern to physically shift away from the original. This aspect of speckle motion is exploited in speckle photography. With larger motions, the speckles are said to be decorrelated because the old and new are not in registration. The criteria for decorrelation were analyzed in Chapter 18.

Notice that speckle correlation interferometry operates at a greater sensitivity than does speckle photography, and the two methods are complementary. For speckle interferometry, the speckle motions must be much smaller than the speckle size. The opposite is true for speckle photography. A related observation is that the effects of speckle size are different. Fine speckles are needed for the highest sensitivity in speckle photography. In speckle correlation interferometry, the sensitivity is largely fixed by the geometry of the setup and the wavelength. Larger speckles can be an advantage in that they extend the range of measurement before correlation is lost. If electronic recording is used, then size of the detector elements is also important in maintaining correlation, as will be discussed later.

Because speckle statistics predict low probability of bright speckles, the image is not really very bright. Fringe visibility in speckle correlation interferometry is therefore not very high.

We now turn to some practical implementations of the idea in configurations to measure specific motions.

20.3 Arrangement to measure out-of-plane displacement

A practical setup that is sensitive to out-of-plane displacement is shown in Figure 20.3 (Jones and Wykes 1983). Only the essential details are shown; the laser, spatial filters, and mirrors are left out. The object is illuminated by the object beam at an angle θ_i to the surface normal, and a speckled image is formed by the lens at image plane H. A reference wave, which usually has either spherical or planar wavefront and nominally normal incidence, is added to the image by the beam splitter. The purpose of the reference beam, as explained earlier, is to modify the object speckle by interferometric combination with a field having known reproducible phase. When the object is displaced out of its plane, there will be a change in all the speckles.

The change in the phase of the object beam relative to the reference beam is given by equation 16.44 with the viewing angle $\theta_v = 0$, so that this setup is sensitive to out-of-plane displacement. Note that the reference beam angle does not affect the result. The phase relationship is

$$\Delta\phi = \left(\frac{2\pi}{\lambda}\right)(1 + \cos\theta_i)\,\Delta z \tag{20.1}$$

Assume for the sake of convenience that the system is being operated in the real-time mode. The brightness of the image seen through the filter mask, that being the negative of the original speckle pattern, will vary as the specimen is shifted in the z direction. The image shows a dark fringe wherever $\Delta\phi = 2n\pi$, or in terms of displacement, there will be a dark fringe in the field wherever

$$\Delta z = \frac{n\lambda}{1 + \cos\theta_i} \tag{20.2}$$

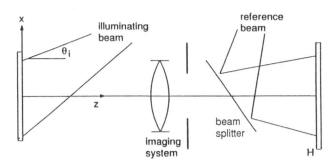

Figure 20.3. Arrangement
of speckle interferometer
sensitive to out-of-plane
displacement.

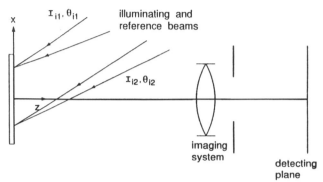

Figure 20.4. Speckle
interferometry with reduced
sensitivity for measuring
out-of-plane displacement.

Several applications of speckle interferometry, including electronic speckle interferometry, have been based on this setup. Note that, because of the use of an imaging lens in the object beam, the angle of viewing is not necessarily zero over the whole object field. Some attention is given this point in Section 20.5.

The sensitivity of the simple setup just discussed is often too high for practical problems. An arrangement that gives lower sensitivity is shown in Figure 20.4 (Jones and Wykes 1983). Two object illumination beams are utilized with different angles of incidence. Coherent mixing of two speckle fields is thereby attained, and a separate reference beam is not needed. Movement of the object will affect the phases of both object beams. The relation between phase and displacement is again calculated using the sensitivity vector (eq. 16.44) with the result,

$$\Delta\phi = \left(\frac{2\pi}{\lambda}\right)(\cos\theta_{i1} - \cos\theta_{i2})\,\Delta z \qquad (20.3)$$

where θ_{i1} and θ_{i2} are the angles of inclination of the two illuminating wavefronts to the surface normal of the object.

Viewing is in the normal direction. When the difference between θ_{i1} and θ_{i2} is small, $(\cos\theta_{i1} - \cos\theta_{i2})$ becomes small and the sensitivity reaches several millimeters per fringe. The maximum displacement still is limited by decorrelation and/or memory as explained earlier.

20.4 Arrangement to measure in-plane displacement

Rather than use an object beam and a reference beam for speckle correlation interferometry, two object beam speckle patterns can be combined coherently to make a new speckle pattern. The arrangement shown in Figure 20.5, which was described by Leendertz (1970) in his seminal paper on the subject of speckle metrology, gives fringes that are sensitive to in-plane displacement. Here the object lies in the xy-plane and is illuminated by two plane wavefronts, I_{i1} and I_{i2}, inclined at equal and opposite angles, θ, to the x-axis surface-normal. Again, no separate reference beam is needed at the image plane. The positive y-axis points out of the page, and the center of the viewing lens aperture lies on the z-axis.

For calculation of the fringe-displacement relation, let $A_1(x, y)$ be the first object beam at the film plane, and let $A_2(x, y)$ be the second. For any image point P, the irradiance from the combined beams is

$$I = A_1^2 + A_2^2 + A_1 A_2 \cos \gamma \tag{20.4}$$

where γ is the phase difference between the two waves contributing to point P in the image initial state.

When an element is displaced by a vector **d** having the three components d_x, d_y, d_z, irradiance becomes

$$I = A_1^2 + A_2^2 + A_1 A_2 \cos(\gamma + \Delta\phi) \tag{20.5}$$

where $\Delta\phi$ is the added phase difference caused by the motion.

To establish the relationship between phase change and displacement, the sensitivity vector (section 16.11) is used. The total phase change in beam 1 is calculated as

$$\delta_1 = (\mathbf{k}_3 - \mathbf{k}_1) \cdot \mathbf{d}$$

$$= \frac{2\pi}{\lambda} [\mathbf{k} - (-\sin\theta\,\mathbf{i} - \cos\theta\,\mathbf{k})] \cdot (d_x \mathbf{i} + d_y \mathbf{j} + d_z \mathbf{k})$$

$$= \frac{2\pi}{\lambda} [d_x \sin\theta + d_z(1 + \cos\theta)] \tag{20.6}$$

Similarly, the calculation for the total phase shift in beam 2 is

$$\delta_2 = (\mathbf{k}_3 - \mathbf{k}_2) \cdot \mathbf{d}$$

$$= \frac{2\pi}{\lambda} [\mathbf{k} - (\sin\theta\,\mathbf{i} - \cos\theta\,\mathbf{k})] \cdot (d_x \mathbf{i} + d_y \mathbf{j} + d_z \mathbf{k})$$

$$= \frac{2\pi}{\lambda} [-d_x \sin\theta + d_z(1 + \cos\theta)] \tag{20.7}$$

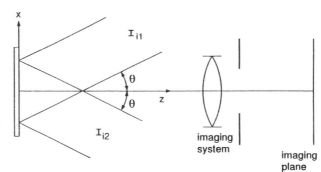

Figure 20.5. Arrangement of speckle interferometer sensitive to in-plane displacement.

The $\Delta\phi$ appearing in equation 20.5 is the difference of δ_1 and δ_2, which is

$$\Delta\phi = \left(\frac{4\pi}{\lambda}\right)d_x \sin\theta \tag{20.8}$$

Dark fringes will occur wherever

$$d_x = \frac{n\lambda}{2\sin\theta} \tag{20.9}$$

The relative phase of I_{i1} and I_{i2} is constant over planes lying parallel to the yz-plane, so the displacement components d_y and d_z lying in those planes will not introduce a relative phase change. This form of interferometer therefore allows an in-plane displacement distribution to be observed independently in the presence of out-of-plane displacements.

Similarly, y-axis illumination geometry, in which the object is illuminated at equal angles to the y-axis, will form a fringe pattern where

$$\Delta\phi = \left(\frac{4\pi}{\lambda}\right)d_y \sin\theta \tag{20.10}$$

To determine both in-plane components of the displacement vector, the fringe spacing measurements must be taken from both illumination geometry fringe patterns. Either the setup or else the specimen must be rotated 90° about its z-axis.

An alternative approach eliminates the need to rotate the specimen or the interferometer to obtain the two in-plane displacements (Jones and Leendertz 1974). The object lies in the xy-plane and is illuminated by plane wavefronts I_{i1} and I_{i2} propagating in the xz-plane and the yz-plane at equal angles θ to the surface normal. Viewing is in the normal z direction. The phase–displacement relationship is

$$\Delta\phi = \frac{4\pi}{\lambda}(d_x - d_y)\sin\theta \tag{20.11}$$

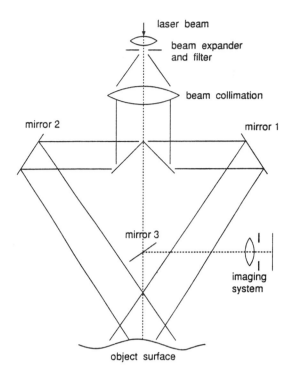

Figure 20.6. An alternative
arrangement of speckle
interferometer for
measurement of in-plane
displacement.

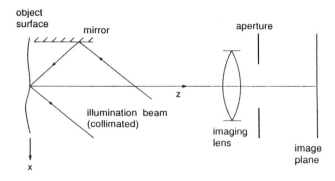

Figure 20.7. A simple
arrangement of speckle
interferometer for
measurement of in-plane
displacement.

Additional manipulation of the fringe data is required to obtain the separate
displacement components. The setup can also incorporate a third beam to
advantage. In practice it is difficult to utilize this orthogonal in-plane
approach for investigating an arbitrary strain field. In addition to the need
to differentiate the strain profile, or to know it beforehand, large optical
components are needed to examine large objects.

   Two practical in-plane sensitive setups are illustrated in Figures 20.6 and
20.7 (Jones and Wykes 1983).

20.5 Two special cases to aid in error analysis

To gain greater understanding about what a speckle metrology system really measures, we analyze two special cases – namely, pure out-of-plane and pure in-plane displacements – with the basic setup of the speckle correlation interferometer. These problems are particularly relevant to the development of electronic speckle pattern interferometry. In the analysis, diverging spherical illumination and viewing parallel with surface-normal are used. One question to be answered is what are the effects on the displacement measurement of variations of illumination angle over the field, as is inherent with divergent illumination. The other question is the extent to which in-plane displacement contaminates the measurement of the desired out-of-plane component.

20.5.1 Pure out-of-plane displacement

In this case, the displacement vector is parallel with the viewing vector, as shown in Figure 20.8. Here $\mathbf{d}$ is the displacement vector and $\mathbf{K} = (\mathbf{k}_1 - \mathbf{k}_2)$ is the sensitivity vector, where $\mathbf{k}_2$ is the viewing vector, $\mathbf{k}_1$ is the illuminating vector. Also, $\mathbf{k}_1'$ is parallel with $\mathbf{k}_1$. The $\mathbf{k}_1$, $\mathbf{k}_1'$, and $\mathbf{k}_2$ are the usual propagation vectors having magnitude $2\pi/\lambda$.

Now, the question is how the angle θ affects the measured displacement. From Figure 20.8 a relationship between the measured displacement d_m and the actual displacement can be derived. The result is

$$d_m = \mathbf{d} \cdot \mathbf{K} = \frac{(1 + \cos\theta)}{2} d \qquad (20.12)$$

where d is the actual displacement.

The relationship between measured d_m and actual d for angle θ less than $15°$ is plotted in Figure 20.9. There is a good approximation when θ is small. The difference can be ignored for many setups involving spherical waves, provided the specimen is not too large for the distances involved.

20.5.2 Pure in-plane displacement

Here, the displacement vector is normal to the viewing vector $\mathbf{k}_2$, as shown in Figure 20.10. In this case, the relation between measured and actual displacement is given by,

$$d_m = \mathbf{d} \cdot \mathbf{K} = (\sin\theta)d \qquad (20.13)$$

Note that the setup is a typical out-of-plane interferometer. The relationship between the measured displacement d_m and the actual displacement d for θ less than $15°$ is plotted in Figure 20.11. For small angles of illumination, the

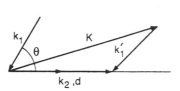

Figure 20.8. Vector relation-ships for pure out-of-plane displacement measurement.

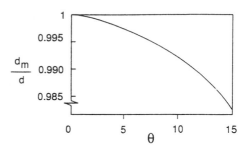

Figure 20.9. The relationship between d_m/d and illumination angle for pure out-of-plane displacement.

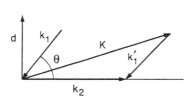

Figure 20.10. Vector relation-ships for pure in-plane displace-ment measurement.

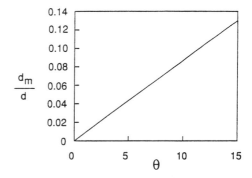

Figure 20.11. The relationship between d_m/d and illumination angle θ for pure in-plane displacement.

fringe pattern is not significantly affected by the in-plane component of displacement.

20.6 Recording speckle for correlation fringes

The techniques for performing speckle correlation interferometry parallel closely those for holographic interferometry. These are described in consider-able detail in Chapter 17 and will not be repeated here. Speckle correlation can be carried out in real-time, time-average, or double-exposure modes.

A critical factor is that the setup be stable to within true interferometric requirements. Speckle photography is tolerant of some vibration, but speckle interferometry is not. For real-time studies, the speckle plate must be returned exactly to its original position after processing. A solid interferometric plateholder must be used, or else the plate must be developed in place.

High-contrast photoplates are best for speckle correlation work. The emulsion does not need the resolution required for holography, especially if larger speckles are used. Quite ordinary black-and-white media are sufficient.

If the system is designed so that speckle size is large enough to be recorded by electronic means, then the possibility of doing away with photographic materials entirely is raised. This subject is discussed in Chapters 21 and 22.

20.7 Accessing speckle correlation fringes

Fringe contrast in speckle correlation interferometry is not nearly as high as it is in other types of interferometry, and some special efforts must often be exerted in order to see the fringes at all.

Real-time fringes are the most visible, although fringe visibility is still low, the theoretical maximum visibility being on the order of 14%. With care, visibility approaches this value, and the eye is rather good at extracting patterns at this level, provided overall brightness is not too high. One should experiment with dark-adapted eyes and low levels of object illumination when creating real-time speckle correlation fringes. Alternatively, heavy exposures of the photoplate are sometimes effective. With nonlinear film processing, fringe visibility can be boosted to the vicinity of 40% (Ennos 1975).

The double-exposure and time-average approaches are not as effective with speckle correlation as they are in holographic interferometry. The speckle fringes will have much lower contrast. The reason is that two speckle patterns are being incoherently added before film development. The developed plate is not acting as a filter mask as it is in real-time mode. Review of the probability density curve in section 18.5 for incoherent addition of speckle fields shows that the dark speckles have the lowest probability of occurring. The result is that areas where the patterns are uncorrelated will have only a few high-contrast speckles. Where the speckles are well correlated, however, the speckle pattern is that of two separate coherently combined patterns, meaning a formerly dark speckle is still dark. High-contrast speckles are formed in the correlated areas.

The result is that, in double-exposure and time-average patterns, the only real difference between correlated areas and uncorrelated areas is in the contrast of the speckles in these areas. To the eye, the appearance of the correlated areas is one of crispness as against the haziness in the uncorrelated portions. One does not see dark bands on a light background. There is not much gradation of brightness in the fringe pattern.

Some measures can be taken to optimize the viewing of double-exposure and time-average speckle correlation fringes because the fringe data are carried on the film plate. Looking directly at a plate that is backlit by softly diffused light is often effective. In some instances, particularly if heavy exposures are used, it is best to view the plate with grazing incidence of fairly bright and directional lighting, such as from the sun or a desklamp. Nonlinear exposure and processing of the plate can improve fringe contrast.

Another approach is to use coherent optical spatial filtering to improve fringe contrast. The speckle correlation photograph is placed in an optical Fourier processing system, as has been described in Chapters 10 and 11. The low-contrast uncorrelated regions will diffract light only poorly. The high-contrast correlated area will contribute most to the diffraction halo. The spatial filter is designed to take advantage of this fact. If the central portion of the transform is blocked, then the image is formed mostly with light from the correlated areas, and so on. This technique will consistently yield visible fringes from double-exposure speckle correlation photographs, and it is probably the best approach to use.

In electronic speckle interferometry, much of the difficulty with fringe visibility can be taken care of during the computer processing of the speckle images. Contrast enhancement, averaging, and digital filtering are all used to improve fringe rendition.

In summary, speckle correlation interferometry, after initial flurries of interest, tended to fall into disuse, largely because of stability requirements and the difficulty of obtaining adequate fringe patterns. The advent of electronic image acquisition and computer fringe processing changed all that. There is much interest in electronic speckle interferometry because it allows interferometric measurement of entire displacement fields without the necessity of processing photographic material.

References

Ennos, A. E. (1975). Speckle Interferometry. In *Laser Speckle and Related Phenomena*, Topics in Applied Physics, Vol. 9, Ed. J. C. Dainty, Ch. 6. Berlin and New York: Springer-Verlag.

Jones, R., and Leendertz, J. A. (1974). Elastic constant and strain measurements using a three beam speckle interferometer. *Journal of Physics E: Scientific Instruments*, 7: 653–7.

Jones, R., and Wykes, C. (1983). *Holographic and Speckle Interferometry*. Cambridge University Press.

Leendertz, J. A. (1970). Interferometric displacement measurement on scattering surfaces utilizing speckle effect. *Journal of Physics E: Scientific Instruments*, 3: 214.

Stetson, K. A. (1975). A review of speckle photography and interferometry. *Optical Engineering*, 14, 5: 482–9.

Vest, C. M. (1979). *Holographic Interferometry*. New York: John Wiley and Sons.

21

Electronic speckle pattern interferometry
(*with contributions by R. Jones and C. Wykes*)

The use of television image acquisition and computer image processing has revolutionized optical methods of metrology. A prime example is in the area of speckle correlation interferometry, which is discussed in this chapter. The implications of detector size are discussed, and limitations and advantages are outlined. An understanding of the material in Chapters 18 and 20 is strongly recommended.

21.1 Introduction

In spite of their obvious merits, holographic interferometry, speckle photography, and photograph-based speckle interferometry have not seen wide adoption by potential industrial and research users. The main reasons for this lack of acceptance seem to include the stability requirements, the necessity for photoprocessing, the requirements for postprocessing (such as image reconstruction and optical Fourier processing), and difficulties in fringe interpretation by persons not trained in optics. The processing and post-processing are particularly troublesome, since they increase the time required to complete a cycle of experiments.

For these reasons it is natural to investigate the use of television systems to replace photographic recording materials and to use electronic signal processing and computer techniques to generate interference fringe patterns. This technique is electronic speckle pattern interferometry (ESPI), although it is also called video holography, TV holography, or electronic holography (EH). The basic concepts of ESPI were developed almost simultaneously by Macovski, Ramsey, and Schaefer (1971) in the United States and by Butters and Leendertz (1971) in England. The latter group, especially, vigorously pursued the development of the ESPI technique in both theoretical and practical directions. Later, Løkberg and Høgmoen (1976) and Beidermann and Ek (1975) also undertook successful research and development in ESPI.

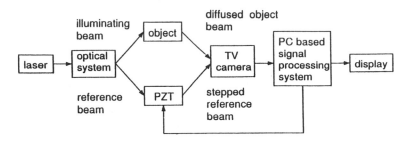

Figure 21.1. Block diagram of a generic system for electronic speckle pattern interferometry.

Most of the analysis presented in this chapter follows closely on the lucid exposition by R. Jones and C. Wykes (1983, 1989), who have been members of the active group at the University of Loughborough with J. Butters. Jones and Wykes, and their publisher, have kindly consented to the inclusion of some segments of one chapter of their fine book, with moderate changes and some reorganization. The author is most grateful.

For speckle pattern interferometry, the resolution of the recording medium used need not be high compared with that required for holography. Only the speckle pattern must be resolved, and not the very fine fringes formed by the interference of object and holographic reference beam. The speckle size is typically in the range of 5 to 50 μm, and it can be varied to some extent to suit the resolution limits of a television system. A standard television camera can be used to record the speckle pattern. The result is that video processing can be used to generate correlation fringes equivalent to those obtained by photographic speckle correlation, as discussed in the previous chapter.

The major feature of electronic speckle techniques is that they enable real-time correlation fringes to be displayed directly on a television monitor without recourse to any form of photographic processing, optical spatial filtering, or plate relocation. Furthermore, the vibration isolation is relaxed a bit because only $\frac{1}{30}$ sec is needed to record a frame of speckle pattern. No dark room is required for processing photo materials. A computer is used to control the entire process, to calculate the displacement, and to present results in graphical form. The advantages of this approach are readily apparent, but they are gained at some cost. Positive and negative features of the technique will be discussed later.

For orientation purposes, a block diagram of a generic system for ESPI appears in Figure 21.1.

In this chapter and the next, the optical and electronic aspects of ESPI are examined. Considered are optical setups, video processing, resolving power of the system, and phase measuring techniques. Some sample results are presented.

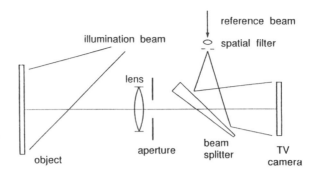

Figure 21.2. An in-line reference beam speckle interferometer that uses a beam splitter to combine the beams.

21.2 Optical setups for ESPI

There are two basic ways of combining the object and reference beams in an ESPI system for out-of-plane measurement. One of these, which is essentially the same as the basic speckle interferometer devised by Stetson (1970), is shown in Figure 21.2. It is a smooth reference beam system, because only an expanded laser beam serves as the reference.

In this system, the input reference laser beam is filtered and expanded by a spatial filter to remove extraneous optical noise and to create a spherical wavefront. This well-conditioned beam is particularly important for use in the addition mode (as explained later). The reference beam is then reflected onto the TV camera target by an optical-wedge beam splitter having a wedge angle of about 1°. The rear face of the beam splitter is antireflection coated in order to suppress secondary reflections. Another such wedge can be placed at the camera detector. These wedges reduce problems with multiple reflections of the speckle pattern. The source point (spatial filter) of the reference wavefront is made conjugate with the midpoint of the maximum focal range to approximately accommodate the two extreme focal conditions. Light scattered from the object is collected by the viewing lens, and an image is formed in the plane of the sensing plane of the video camera, where it interferes with the reference wavefront. A disadvantage of this system is that dust particles, which tend to collect on the beam splitter, together with any other small blemishes, act as light scattering centers. These cause reference beam noise and degrade the quality of addition fringes. In the subtraction mode of ESPI, this problem is not serious because the noise is the same for each of the images.

The second arrangement, also a smooth-reference type, which considerably reduces the reference beam noise, is shown in Figure 21.3. The laser reference beam in this setup is pre-expanded by lens L_1 and then focused down through a mirror having a small hole. The reference beam is aligned by translating the lens L_2 so that its focal point coincides with the hole. The small hole in the mirror does not degrade the image quality unless a significant fraction

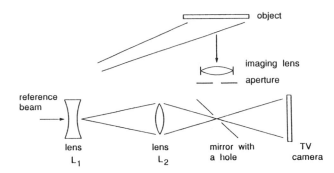

Figure 21.3. An in-line reference beam speckle interferometer.

of the light transmitted by the viewing lens is incident on the hole. The point of divergence of the reference beam and the center of the viewing lens are clearly not coincident in this arrangement. That this can degrade performance will be shown; but, because the tolerance along the optical axis direction is large, this layout can be made to function within appropriate limits.

21.3 Video system and other electronics in ESPI

The video system comprises the television camera, the signal processing and picture storage unit, and the display monitor. The aim of the system design is to obtain fringes of maximum visibility. Electronic noise and, in particular, high-frequency noise must be minimized. Because the cost of a laser increases approximately in proportion to the output power, it is important to optimize the sensitivity of the video system.

The most noise-sensitive part of the signal processing is at the first stage of video signal amplification, which is carried out at the camera head amplifier, so this section of the system must be carefully designed. Usually, DC-coupling is used to link the output signal to a high-impedance and low-noise device, which is then connected directly to the detector plate.

The CCD camera is used in modern ESPI systems because it has wide range of linearity, low noise at low light intensity, high signal-to-noise ratio, and peak spectral response at about 0.6 μm. The vertical resolution is determined by the number of scan lines, which is fixed when a standard video system is used. The frequency response of the system should be such that the horizontal spatial resolution is at least as good as the vertical resolution.

A video system converts an image formed on the detector plate of a television camera into an equivalent image on a television monitor screen. With light incident on the CCD array, a charge pattern will build up corresponding to the intensity distribution of the incident light. Charge is then transferred from collection site to collection site by changing potentials within what is essentially an insulator. The output will give rise to a voltage signal. A standard television system has 525 vertical lines and a complete scan is performed at a rate of 30 frames/sec (U.S. standard). For a standard video camera the active area of the camera is about 13×10 mm, so the image must be reduced to this size. After amplification and the addition of timing pulses, the camera signal is used to modulate an electron beam, which scans the screen of a television monitor so that the brightness of the screen is made to vary in the same way that the irradiance of the original image varies. Ideally, the brightness of the monitor should vary linearly with the irradiance of the original image. The exact relationship between the monitor and original image intensities is a complicated function of the electronic processing as well as the brightness and contrast controls of the television monitor. In the analysis of the video display of speckle correlation fringes, it will be assumed that (a) the camera output voltage is linearly proportional to the image irradiance and (b) the monitor brightness is proportional to the camera output voltage.

Correlation fringes in ESPI are observed by a process of video signal subtraction or addition, or by a combination of the two. In the subtraction process, the television camera video signal corresponding to the interferometer image plane speckle pattern (only an intensity signal, remember) of the undisplaced object is stored electronically. The object is then displaced and the live video signal, as detected by the television camera, is subtracted from the stored picture. The output is then high-pass filtered, rectified, and displayed on a television monitor where the correlation fringes may be observed live, that is, in real time. In this process, sophisticated software can be used to process the fringe data, calculate the desired information, and display it graphically on the screen. For the addition method, the light fields corresponding to the two states are added at the image plane of the camera. The television camera detects the added light intensity, and the signal is full-wave rectified and high-pass filtered as in the subtraction process. Again, the correlation fringes are observed on the television monitor.

21.4 Fringe formation by video signal subtraction

The detector plate of the camera is located in the image plane of the speckle interferometer. Under these conditions the output signal from the television camera, as obtained with the object in its initial state, is recorded in the computer's RAM memory. The object is then displaced, and the live camera signal is subtracted electronically from the stored signal. Those areas of the two images where the speckle pattern remains constant will give a resultant signal of zero, whereas changed areas will give nonzero signals. Figure 21.4 shows some examples of fringe patterns obtained with this approach. These were photographed directly from the television monitor of a commercial ESPI system (e.g., Nokes and Cloud 1992).

(a)

(b)

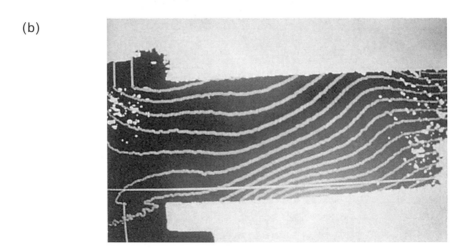

Figure 21.4. Fringe patterns showing displacements in a composite aircraft wing spar obtained by electronic speckle interferometry using video signal subtraction: (a) fringe pattern after digital filtering; (b) fringes traced and counted. Black-and-white reductions of photographs of false-color computer display.

To understand the formation of fringes, consider the intensities I_{before} and I_{after} (before and after displacement) as given by equations developed in the previous chapters,

$$I_{\text{before}} = I_R + I_O + 2\sqrt{I_R I_O} \cos \phi \qquad (21.1)$$

$$I_{\text{after}} = I_R + I_O + 2\sqrt{I_R I_O} \cos(\phi + \Delta\phi) \qquad (21.2)$$

where ϕ is the phase difference between the reference beam and the object beam before the displacement

$\Delta\phi$ is the phase change caused by the displacement

If the output camera signals V_{before} and V_{after} are proportional to the input image intensities, then the subtracted signal is given by

$$V_{\text{s}} = (V_{\text{before}} - V_{\text{after}}) \propto (I_{\text{before}} - I_{\text{after}})$$
$$= 2\sqrt{I_{\text{R}}I_{\text{O}}}\,[\cos\phi - \cos(\phi - \Delta\phi)]$$
$$= 4\sqrt{I_{\text{R}}I_{\text{O}}}\,\sin(\phi + \tfrac{1}{2}\Delta\phi)\sin(\tfrac{1}{2}\Delta\phi) \qquad (21.3)$$

This signal has negative and positive values. The television monitor will, however, display negative-going signals as areas of blackness. To avoid this loss of signal, V_{s} is rectified before being displayed on the monitor. The brightness on the monitor is then proportional to the absolute value of V_{s}, so the brightness B at a given point in the monitor image is given by

$$B = K|\sqrt{I_{\text{R}}I_{\text{O}}}\,\sin(\phi + \tfrac{1}{2}\Delta\phi)\sin(\tfrac{1}{2}\Delta\phi)| \qquad (21.4)$$

where K is a constant.

If the brightness B is averaged along a line of constant $\Delta\phi$, it varies between maximum and minimum values B_{max} and B_{min} given by

$$\begin{aligned} B_{\text{max}} &= K\sqrt{I_{\text{R}}I_{\text{O}}} & \Delta\phi &= (2n+1)\pi & n &= 0, 1, 2, 3, \ldots \\ B_{\text{min}} &= 0 & \Delta\phi &= 2n\pi & n &= 1, 2, 3, \ldots \end{aligned} \qquad (21.5)$$

The signals should be high-pass filtered to improve fringe visibility by removing low-frequency noise together with variations in mean speckle intensity.

21.5 Fringe formation by video signal addition

In this case, the two speckle patterns derived from the object in its two states are added together on the camera detector plate. The two images do not need to be superimposed simultaneously because a given camera has a characteristic persistence time (about 0.1 sec for a standard camera). The camera output voltage will be proportional to the added intensities if the time between the two illuminations is less than the appropriate persistence time. This technique is employed for the observation of time-averaged fringes when studying vibrations, and it is also used with a dual-pulsed laser.

When the two speckle patterns are added together, areas of maximum correlation have maximum speckle contrast. With diminished correlation, the speckle contrast decreases, as was pointed out in Chapter 20. It falls to a minimum but nonzero value where the two patterns are uncorrelated. This fact can be demonstrated by again calculating the intensity, which is proportional to camera output for the sum of the two input iluminations.

The voltage V_a is proportional to $I_{before} + I_{after}$ and is given by

$$V_a \propto (I_{before} + I_{after}) = 2I_R + 2I_O + 4\sqrt{I_R I_O} \cos(\phi + \tfrac{1}{2}\Delta\phi) \cos \tfrac{1}{2}\Delta\phi \qquad (21.6)$$

The contrast of the speckle pattern can be defined as the standard deviation of the intensity. For a line of constant $\Delta\phi$, this can be shown to be

$$\sigma_{RO} = 2\left[\sigma_R^2 + \sigma_O^2 + 8\langle I_R\rangle\langle I_O\rangle \cos^2 \frac{\Delta\phi}{2} \right]^{1/2} \qquad (21.7)$$

where $\langle I_R\rangle$ and $\langle I_O\rangle$ are the irradiances of the reference and object speckle pattern averaged over many points in the field, and

$$\begin{aligned}
\sigma_R &= \sqrt{\langle I_R^2\rangle - \langle I_R\rangle^2} = \langle I_R\rangle \\
\sigma_O &= \sqrt{\langle I_O^2\rangle - \langle I_O\rangle^2} = \langle I_O\rangle
\end{aligned} \qquad (21.8)$$

The σ_R and σ_O are the standard deviations of I_R and I_O. It is seen that σ_ρ varies between maximum and minimum values given by

$$\begin{aligned}
[\sigma_{RO}]_{max} &= 2\sqrt{\sigma_R^2 + \sigma_O^2 + 2I_R I_O} & \Delta\phi &= 2n\pi & n &= 0, 1, 2, \ldots \\
[\sigma_{RO}] &= 2\sqrt{\sigma_R^2 + \sigma_O^2} & \Delta\phi &= (2n+1)\pi & n &= 0, 1, 2, \ldots
\end{aligned} \qquad (21.9)$$

Though the contrast of the added intensities varies, the mean value along a line of constant $\Delta\phi$ is the same for all $\Delta\phi$, and is given by

$$\langle I_{before} + I_{after}\rangle = 2\langle I_R\rangle + 2\langle I_O\rangle \qquad (21.10)$$

When the sum of the two speckle patterns is directly displayed on the television monitor, the average intensity is constant, and the variation in correlation is shown as a variation in the contrast of the speckle pattern but not in its intensity. The DC component of the signal is removed by filtering, and the signal is then rectified. The resulting monitor brightness can then be considered to be proportional to σ_{RO}, thus,

$$B = K\left[\sigma_R^2 + \sigma_O^2 + 2\langle I_R\rangle\langle I_O\rangle \cos^2 \frac{\Delta\phi}{2} \right]^{1/2} \qquad (21.11)$$

The intensity of the monitor image varies between maximum and minimum values given by the two equations 21.9. Comparison of these with the two equations 21.5 shows that the fringe minima obtained with subtraction correspond to fringe maxima obtained using addition. The subtraction fringes are also found to have intrinsically better visibility than addition fringes because the subtraction fringe patterns contain zero intensity and addition fringes do not. However, when addition is used to observe the fringes in time-average vibration analysis, a video storage capability is not required. Addition mode is, therefore, attractive for modal analysis, even though fringe visibility is not the best.

21.6 Real-time vibration measurement

For vibration pattern measurement, the interferometer is operated in time-average mode. In this mode the images are being recorded and added together while the object is vibrating. Vibration fringes can, however, be produced by a time-average subtraction technique, which gives fringes better than those obtained by addition (Lu et al. 1989). Figure 21.5 shows an example of a fringe pattern obtained in time-average addition to show vibration modes in plates.

The analysis of time-average speckle follows closely that given for time-average holographic interferometry in Chapter 16. At a given instant t, the irradiance in the image plane is given by $I(t)$, where

$$I(t) = I_R + I_O + 2\sqrt{I_R I_O} \cos\left[\phi + \frac{4\pi}{\lambda} a(t)\right] \qquad (21.12)$$

The function $a(t)$ represents the position of a given point on the object at time t. The intensity is averaged over time τ to obtain

$$I_\tau = I_R + I_O + \frac{1}{\tau} 2\sqrt{I_R I_O} \int_0^\tau \cos\left[\phi + \frac{4\pi}{\lambda} a(t)\right] dt \qquad (21.13)$$

This average is evaluated for sinusoidal vibration, $a(t) = a_0 \sin \omega t$, and it can be assumed that $2\pi/\omega \ll \tau$, which means that the average is over several oscillations. The average reduces to

$$I_\tau = I_R + I_O + 2\sqrt{I_R I_O} \, J_0^2\left(\frac{4\pi}{\lambda} a_0\right) \cos \phi \qquad (21.14)$$

where J_0 is the zero-order Bessel function. This result is similar to that obtained for time-average holographic interferometry, as expected. The value of I_τ averaged over many speckle patterns is constant over the whole image, but the contrast of the speckle is seen to vary as the value of the J_0^2 function varies. When J_0^2 has a value of zero, the brightness varies only as I_O varies; whereas when J_0^2 has a maximum value, the brightness varies with the variation in I_O as well as with ϕ according to

$$2\sqrt{I_R I_O} \cos \phi \qquad (21.15)$$

Figure 21.6 shows the J_0^2 distribution that gives the irradiance distribution for time-average speckle fringes.

The correlation fringes thus observed map out the variation in a_0, the amplitude of the vibration across the object surface. The fringe minima correspond to the minima of the Bessel function. Table 21.1 shows the relation between the fringe order N and the vibration amplitude. Note that the relationship is not simply linear, so interpolation should not be used.

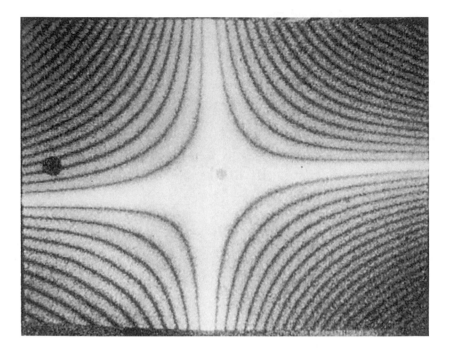

Figure 21.5. Vibration modal study of a plate obtained by time-average electronic speckle interferometry (courtesy of Karl Stetson).

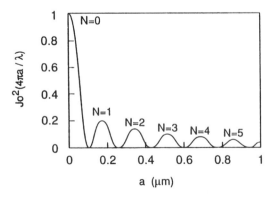

Figure 21.6. The form of the first few cycles of the J_0^2 function that defines the irradiance distribution of the time-average speckle fringes for a sinusoidally vibrating surface.

In the time-averaged mode, the phase information of the vibration components is lost owing to the averaging of the displacements over one or more vibration cycles. The sign of the time-varying displacement cannot, therefore, be obtained, so some of the information about the mode shape is lost. This limitation is circumvented if a stroboscopic technique is used (Høgmoen and Løkberg 1977). Another way to record the phase information is to use a dual-pulsed laser (Cookson, Butters, and Pollard 1978), or to move

Table 21.1. *Fringe order (N) vs.*
vibration amplitude (a) in time-average
ESPI mode

N (fringe order)	a (amplitude, μm)
0	0.00
1	0.19
2	0.35
3	0.51
4	0.67
5	0.83

the PZT mirror in the reference beam path at the same vibration frequency (Løkberg and Høgmoen 1976). More is to be said in the next chapter about the use of phase modulation techniques to determine the sign of the motion that is measured by ESPI.

21.7 Dynamic displacement measurement

When one takes a speckle picture in ESPI, it is necessary for all components to be stable to considerably better than a wavelength during the exposure time ($\frac{1}{30}$ sec). So, only static or slowly varying displacements can be measured using continuous lasers. With the use of a dual-pulsed laser, ESPI can be extended to the measurement of dynamic displacements. The two-pulse technique can also be used to measure large-amplitude periodic displacements. If, for example, the period of oscillation is 10^{-3} sec, a pulse separation of 10 μsec will enable 0.01 of the overall amplitude to be measured in a single recording. This procedure extends the range of vibration measurement to amplitudes approaching $10^3 \lambda$. These special techniques enhance the capability of ESPI, but they will not be studied here.

21.8 Optimization of light intensity

For a given type of TV camera, a certain minimum intensity is required to create a camera voltage output that can be detected above the background electronic noise. An increase in the incident intensity gives an increased output camera voltage until the intensity reaches the camera saturation level, beyond which the output voltage remains constant for any further increase in incident intensity.

The intensity of the speckle pattern varies randomly across the image; and to avoid losing information from the speckle pattern, the overall intensity should be below the saturation level of the camera for all of the useful picture area. If the mean value and the standard deviation of the speckle pattern are given by $\langle I_t \rangle$ and σ_t, then, when

$$\langle I_t \rangle + 2\sigma_t < I_{sat} \tag{21.16}$$

the intensity will be less than the saturation level of the camera for 95% of the image. Here I_{sat} is the saturation intensity of the camera.

When two speckle beams are used in a subtraction correlation system, the mean intensities of the two beams at the image plane should be equal, and the combined peak intensities should be approximately equal to the camera saturation intensity.

For the smooth reference beam case, experience suggests that best results are obtained when the reference beam intensity is set according to

$$\frac{1}{4} I_{sat} \leq \langle I_r \rangle \leq \frac{1}{2} I_{sat} \tag{21.17}$$

The value chosen depends on the signal-to-noise ratio and on the threshold set in the software. The sum of the intensities should be such that the peak intensities are near the saturation level of the camera.

21.9 Spatial resolution of the video system and its effect on ESPI

21.9.1 Camera resolution and speckle size

For experimentalists who happen to be trained in holography and similar processes, working with video-based systems requires some rethinking. The problem now is to get the spatial frequencies low enough, and to adjust the other measurement parameters to suit. The spatial resolution and dynamic range of video systems are considerably less than those of photographic and holographic emulsions. These factors affect the design of speckle pattern interferometers that use video systems to display the correlation fringes. A main objective, as will be seen, is to design the system in such a way that the speckle can be suitably resolved by the video system.

Let us be certain that we understand this problem. The ability of the video system to resolve fine detail in the image is limited. Thus, if a coarse black-and-white grid is imaged on the face plate of the TV camera, an equivalent grid will be observed on the monitor. As the spacing of the grid is reduced, the contrast of the grid observed on the monitor is reduced; and when the grid spacing is sufficiently small, no grid structure at all is seen. If the grid happens instead to be the more complex structure of a speckle pattern, the conclusion still holds. The video camera acts as a low-pass filter that severely limits the spatial frequencies that can be used in the speckle system, and the design of the system must accommodate this fundamental limitation.

The spatial resolution of a video system in the vertical direction is governed by the number of scan lines. A standard system having 525 scan lines and a detector active area of about 13×10 mm should have a minimum vertical resolution of approximately $20 \,\mu\text{m}$ on the face plate.

The resolution in the horizontal direction is governed by the temporal frequency response of the video system, which is determined by the electronic design of the system. Typically, the response is fairly uniform up to 4 MHz and falls to a low value at 10 MHz. The temporal frequency response is clearly related to the spatial frequency in the image. A sinusoidal intensity distribution in the image plane of spatial frequency $10^6/WNM$ mm^{-1} will give a sinusoidal output voltage of 1 MHz. Here, W = width of the active face plate area in mm, N = number of scan lines, and M = number of frames per second. For a standard TV system, 1 MHz corresponds to a horizontal spatial frequency of about 5 cycles/mm, or a line spacing of about $200 \,\mu\text{m}$.

A grid having a 20-μm spacing in the horizontal direction will give an output voltage at frequency 9.45 MHz. Because the frequency response is very low at 10 MHz, $20 \,\mu\text{m}$ can be taken as the upper limit of the spatial resolution in the horizontal direction. That is, the maximum horizontal and vertical spatial resolutions are about equal at $20 \,\mu\text{m}$ at the sensing plane for conventional television cameras. Incidentally, this limit is changing rapidly with improvements in television technology.

It was assumed in deriving equations 21.4 and 21.11, which describe the variation in monitor brightness to give subtraction and addition correlation fringes, that the random variations in I_R and I_O are fully resolved by the television camera; thus, the camera signal from a given point in the image is proportional to the irradiance at that point. When the points are not fully resolved, the output voltage is a function of the intensity of the image averaged over the resolution area of the camera; and the effect of this averaging is to reduce the standard deviation or contrast of the output voltage (Slettemoen 1977). Consider, for example, the case of two speckle patterns whose intensities have the same mean value and standard deviations. One of the patterns is fully resolved by the TV camera and the other is only partially resolved. The output camera signals will have the same mean values across the image, but the standard deviation of the signal will be less for the partially resolved pattern. If the speckles are sufficiently small, then the output signal will be uniform. We demonstrate now that the effect of this reduction in the standard deviation of the output signal is to reduce the visibility of the correlation fringes.

From equations 21.4 and 21.5, the correlation fringes are seen to arise from the variation with $\Delta\phi$ of the standard deviation of the subtracted or added signals; if the contrast of the individual signals is reduced, the standard deviation of the subtracted or added signals is also reduced, and the visibility of the correlation fringes is decreased. The camera output signal can be expressed as

$$V_{R,O} = V_O + (V_{corr})_{R,O} \qquad (21.18)$$

where V_O arises from the $(I_R + I_O)$ term
$(V_{corr})_{R,O}$ arises from the $2(I_1I_2)^{1/2}\cos\phi$ and $2(I_1I_2)^{1/2}\cos(\phi + \Delta\phi)$
terms

Since V_O remains unchanged when the object is displaced, it is a noise term. Consequently, the correlation fringe visibility is maximized when the standard deviation of $(V_{corr})_{R,O}$ is maximized. We can write the value of $(V_{corr})_{R,O}$ at a given point in the image as

$$(V_{corr})_{R,O} = k\langle\sqrt{I_R I_O}\,\cos(\phi + \Delta\phi)\rangle\Delta A \qquad (21.19)$$

where k is a constant
$\Delta\phi$ is the phase difference between reference beam and object beam
ΔA is the resolution area of the camera

If the variation of ϕ over the resolution area is very much less than 2π, then the value of $(V_{corr})_{R,O}$ for constant $\Delta\phi$ varies between $\pm[k\langle(I_1I_2)^{1/2}\cos\Delta\phi\rangle]_{max}$. When the value of ϕ varies by 2π or more over the resolution area, then the value of $(V_{corr})_{R,O}$ over the whole image is approximately zero. In this case, correlation fringes are not observed. The spatial frequency distribution of $\cos\phi$ must be such that the components of that distribution are at least partially resolved by the video system. Now, apply this understanding to two cases of special interest.

21.9.2 Case of two speckle beams

When the two wavefronts I_R and I_O are speckled, such as with the setup for measuring in-plane displacements, the spatial frequency distribution of the combination of the two beams will be similar to that of either I_R or I_O. The maximum spatial frequency of the distribution is determined by the diameter of the viewing lens aperture and is developed from equation 18.3 to give

$$\frac{1}{f_{max}} \approx 1.22(1 + M)\frac{\lambda d_o}{a} \qquad (21.20)$$

where d_o is the lens-to-image distance
M is the magnification of the lens
a is the lens clear aperture

If the distance from object to viewing lens is considerably greater than the lens-to-image distance, then, from the lens equation, d_o is approximately the focal length of the lens. The minimum speckle size S_{min} is then given by

$$S_{min} = \frac{1}{f_{max}} \approx 1.22(1 + M)\lambda F \qquad (21.21)$$

where F is the numerical aperture or f-number of the lens.

If the smallest spot resolved by the video system is 20 µm, corresponding to a spatial frequency of 50 lines/mm, then, to fully resolve the spatial frequencies arising from the speckle distribution, F must be greater than about 30. Quite a small aperture must be used if the scale of the speckle pattern is to be large enough to be fully resolved by the video camera.

If the numerical aperture of the lens is reduced below this value (the aperture is expanded), then the spatial frequency of the speckle pattern will increase, and it will no longer be fully resolved. However, the amount of light transmitted through the lens will increase, so the value of V_{corr} does not necessarily decrease. An exact analysis of the relationship between V_{corr} and the viewing lens aperture diameter is quite difficult. However, the optimum value can be readily found for a particular device simply by adjusting the aperture size until maximum speckle contrast is obtained.

21.9.3 Case of smooth reference beam

In this system, which is the classical speckle interferometer arrangement for out-of-plane measurement, I_O is a random speckle intensity and I_R is nominally uniform across the image plane. Figure 21.7 shows the essential geometry of the system.

The phase difference ϕ between I_O and I_R is found from consideration of the path lengths to be

$$\phi = \phi_s + \frac{2\pi l_{OP}}{\lambda} - \left[\phi_R + \frac{2\pi l_{O'P}}{\lambda} \right]$$

$$= \phi_s - \phi_R + \frac{2\pi}{\lambda}(l_{OP} - l_{O'P}) \qquad (21.22)$$

where   O is the center of the aperture of the viewing lens

O' is the point from which the reference beam appears to diverge

P is a point in the image plane (see Fig. 21.3)

ϕ_R is the phase of reference beam at O' and is constant across that beam

ϕ_s is the random phase of the object speckle at O

l_{OP} and $l_{O'P}$ are the distances from O and O' to P

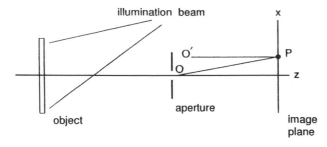

Figure 21.7. The general geometry of a smooth reference beam speckle interferometer.

It was pointed out in Chapter 19 that if ϕ varies by 2π or more within the resolution diameter of the resolution area of the camera, the correlation signal V_{corr} will be very small. The variation in ϕ_s is governed by the viewing lens aperture diameter, as in the case of the system with two speckle beams, and a suitable aperture ensures that it does not vary too rapidly. Clearly, the variation with P in the last term $\Omega = 2\pi(l_{\text{OP}} - l_{\text{O'P}})/\lambda$ must be such that Ω varies by considerably less than 2π over the diameter of the resolution area. If O and O' are coincident, Ω is zero, which is an optimum situation that, if possible, should be satisfied. O and O' are then said to be conjugate.

To determine the departure from conjugacy that can be tolerated, assume that the coordinates of O and O' are $(0, 0, l_{\text{O}})$ and $(\Delta x, 0, l_{\text{R}})$; the resolution diameter is x_{R}. When $\Delta x \ll x$ and $\Delta l = l_{\text{O}} - l_{\text{R}} \ll l_{\text{O}}$, the Pythagorean theorem yields

$$l_{\text{OP}} - l_{\text{O'P}} \approx \left(l_{\text{O}} + \frac{x^2}{2l_{\text{O}}}\right) - \left[l_{\text{R}} + \frac{(x - \Delta x)^2}{2l_{\text{R}}}\right] \tag{21.23}$$

Expand and neglect second-order terms of Δx and Δl to obtain

$$l_{\text{OP}} - l_{\text{O'P}} \approx \Delta l + \frac{x\,\Delta x}{l_{\text{O}}} - \frac{x^2\,\Delta l}{2l_{\text{O}}^2} \tag{21.24}$$

A lateral displacement Δx of O' with respect to O is seen to produce a linear variation in ϕ across the image plane. ϕ varies by 2π across an interval δx given by

$$\delta x = \frac{l_{\text{O}}\lambda}{\Delta x} \tag{21.25}$$

If this distance is to be resolved by the video system, δx should be considerably greater than x_{R}; that is,

$$x_{\text{R}} \ll \frac{l_{\text{O}}\lambda}{\Delta x} \tag{21.26}$$

or

$$\Delta x \ll \frac{l_{\text{O}}\lambda}{x_{\text{R}}} \tag{21.27}$$

The lateral departure from conjugacy must be less than that specified by equation 21.27. As an example, consider a lens-to-image distance of 100 mm and $x_R = 20\ \mu m$. The limit on lateral departure from ideal conjugacy is $\Delta x \ll 3.2$ mm.

A longitudinal departure from conjugacy of Δl gives rise to a variation in ϕ that varies as x^2. The maximum gradient of ϕ occurs at the edge of the image. Here, the distance δx in which ϕ varies by 2π is given by, taking the absolute value,

$$\delta x = \frac{\lambda l_0^2}{x\,\Delta l} \tag{21.28}$$

If this detail is to be resolved, we must have

$$\frac{\lambda l_0^2}{x\,\Delta l} \gg x_R \tag{21.29}$$

or

$$\Delta l \ll \frac{\lambda l_0^2}{x x_R} \tag{21.30}$$

Let x take its maximum value, which is equal to the half-width of the active area of the camera tube, typically about 5 mm. For a lens-to-image distance of 100 mm and $x_R = 20\ \mu m$, we obtain $\Delta l \ll 63$ mm.

To optimize the visibility of the correlation fringes, Δx and Δl should be made as small as possible. The results obtained show that displacements in the lateral direction give rise to variations of ϕ that are much more rapid than are given by equivalent displacements in the longitudinal direction. The lateral shift in particular should be minimized. It is not difficult in practice to do this. In addition, the size of the viewing lens aperture should be adjusted to give optimum speckle contrast.

Finally, note that because the speckle size is doubled owing to the interference with the smooth reference beam, the aperture needed to make the speckle size equal to the detector element size can be reduced to $F \approx 15$ for the example having $x_R = 20\ \mu m$.

21.9.4 Summary of resolution limitations

Because of resolution limitations, the use of a television camera to observe speckle correlation fringes requires the use of a small aperture in the viewing lens and results in relatively large speckles. This condition affects the performance of the interferometer in two ways.

First, the spacing of the correlation fringes must be greater than approximately 1/120 of a screen width. This conclusion follows because, when the fringe spacing approaches the minimum speckle size, the fringe visibility decreases; and it drops to zero when they are equal. A minimum ratio of fringe spacing to speckle size of about five is necessary to give reasonably visible fringes. Because the minimum speckle size is about 20 μm, the minimum fringe spacing is of the order of 10 μm at the camera. This figure corresponds to a fringe spacing of 1/120 of a camera detector plate of 12-mm width. Hence, a maximum 120 fringes of uniform spacing should, in theory, be observable. With a fringe pattern of variable spacing, the maximum number under optimum viewing conditions is found in practice to be about 50. If the number of fringes in one full screen is greater than 50, it will likely give rise to fringe counting errors.

Second, the use of a small aperture means that the system is not very efficient in light usage compared with the equivalent photographic speckle correlation system.

However, the speed and convenience of video processing compared with that of photographic processing generally outweighs these disadvantages. The limitations on fringe density, displacement range, and sensitivity can be partially overcome by the development of effective image-processing software. One also can use a laser with greater power. Most systems use a 10–15 mW laser. A higher power source such as an argon laser can be used for testing large objects.

21.10 Advantages and limitations of electronic speckle

Electronic speckle pattern interferometry has developed into an important technique, which has potential application in many areas of engineering and scientific measurement. Its good points are unique. The touting of its promise has led to confusion and disappointment among some who, unaware of the generic limitations of the approach, have purchased systems with unrealistic expectations. Electronic speckle, even though it is sometimes called video holography, is not an easy substitute for film-based holography. It is not even a substitute for photograph-based speckle methods. Rather, it is a strong addition to the arsenal of tools that the experimental analyst can bring to bear on a particular measurement problem.

21.10.1 Advantages

Here are listed the most practically significant positive features of electronic speckle interferometry:

1. Speed. Anyone who has performed film-based interferometry will appreciate the promise of whole-field interferometric fringe patterns fully developed in just a few seconds.

2. The method does not require the long-term, highly stable environment necessary for conventional interferometry.
3. The system can be used in brightly lit conditions; no dark room is needed during data acquisition or processing.
4. Because no film and no processing of film are needed, the material cost per experiment is very low, and potential safety problems are reduced.
5. People with little background in optics can operate the system.
6. Unique experimental capabilities cause the method to be the only one that can be employed in some situations.

21.10.2 Limitations

The main factors limiting the range of measurements that can be made using the various ESPI methods are now discussed.

Measurement sensitivity

ESPI can be used to give fringes that represent lines of either in-plane or out-of-plane displacement. The fringe sensitivity for an in-plane interferometer is calculated to be $\lambda/2 \sin \theta$, where λ is the wavelength of the light used and θ is the angle of incidence of the illuminating beam. Out-of-plane interferometers can give fringes representing constant displacements at intervals of the order of $\lambda/2$, or they can be desensitized up to about hundreds of micrometers. It is not possible to detect less than one fringe accurately with a conventional ESPI system, so these values represent the maximum sensitivities of the system.

We showed in section 21.9 that the fringe spacing on the TV camera must be greater than 1/120 of the screen width; this restriction limits the displacement gradient and also the total displacement that can be observed. The visibility of the fringes obtained using time-averaged ESPI to observe out-of-plane vibrations falls off rapidly with increasing vibration amplitude unless stroboscopic illumination or other techniques are used.

Object size limitations

The maximum area that can be inspected in one view is limited by the laser power available and the camera sensitivity. There is no reason why a larger area cannot be inspected if sufficient laser power is available, but the mechanical stability of the system and the coherence of the laser also limit the performance.

When in-plane measurements are being made, the illuminating wavefronts must be plane if fringes of uniform sensitivity are to be obtained. Unless a large collimating lens is used for large areas, the fringe pattern interpretation will be rather difficult. When the surface of the object being inspected is not flat, it can be shown that the fringes may be sensitive to out-of-plane movements.

To observe fringes on a small area of a large body, a relatively large deforming force must be applied, and this is likely to give rise to rigid-body translations, which cause speckle decorrelation and hence a reduction in fringe visibility. Large deformations and large forces must also be used when observing displacements to which the interferometer has little sensitivity. Decorrelation and memory loss (essentially decorrelation) caused by in-plane translations and rotations increase as the magnification of the viewing system is increased. Thus, decorrelation of the speckle, causing a reduction in fringe visibility, is likely when ESPI is used at high magnification. Fringes have been observed on an area of 0.05 mm^2 with an out-of-plane ESPI system (Jones and Wykes 1983).

Depth of field

Because the ESPI system must use a viewing system with a high F-number (small aperture), the depth of field of the system is large. The depth of field is usually not a restriction in ESPI.

Surface condition

It has been assumed throughout this discussion that the surface under examination does not change microscopically during the course of the measurement. If it is altered by oxidation, recrystallization, or other events, then decorrelation of the speckle will result, perhaps to a significant degree. Additionally, the nature of the scattering surface itself can lead to a much greater dependence of the speckle pattern on a tilt of the surface. For surfaces like abraded metal, the scattering from the surface is two-dimensional (only scattered once). When the light penetrates to some extent into the surface, the light is then scattered three-dimensionally within the material; thus, path changes resulting from a small tilt of the surface will vary in a random fashion from point to point. This occurs for such materials as paints, paper, and cardboard, and many organic substances. The net effect is that the tolerance on the allowable tilt is much narrower when a multiple-scattering surface is being studied than when the light is only scattered once.

Another consideration of practical importance relates to the type of surface required for measurement of in-plane displacement using double-illumination interferometry. Because of the high angles of incidence of the two illuminating beams, the surface being studied must be totally diffusing, with no enhanced scattering in the specular direction. For this reason, the object very often has to be coated with a matte white paint or powder.

21.10.3 Sources of error

Apart from the general limitations of the technique, several possible sources of error within the system might affect accuracy or, at least, might affect sensitivity. These and their contributions to measurement inaccuracy are listed here along with some possible corrective measures:

1. Frequency instability of the light source. This instability can cause λ to change slightly, but the effect on accuracy is very small.
2. Reference phase error. When phase-stepping is employed, as explained in the next chapter, this error involves the deviation for the ith shift or step between the actual and desired shift values (α_i). By carefully calibrating the phase-stepper used, this error can be minimized.
3. Detector nonlinearity. In technical manuals for detectors, the response of the detector is often described as a nonlinear power function such as I_r^r, with $0 < r < 1$. For CCD detectors, $r \approx 1$ and the nonlinearity is negligible. For vidicon tubes, $r \approx 0.7$ and the nonlinearity can be considerable. A CCD camera is commonly used in ESPI systems.
4. Amplitude instability of the light source. The light source intensity during recording of the frames can vary from one frame to another. This problem is minimized if the illuminating laser is sufficiently warmed up before collecting data. Chronic instability is cured by replacing the laser.
5. Environmental disturbances, vibration, and air turbulence. When looking at live correlation fringes, one can see the disturbances as fringe motion. If they do not affect the fringe pattern seriously, then no harm is done. If the fringes move too much, or if contrast is reduced because of vibration, then the system must be isolated by using air-supported tables and such devices as are used for holographic interferometry.
6. Noise in the detector output and quantization noise of the measured intensity. For a CCD camera, electronic noise is very small and usually negligible.

21.11 Example of application in nondestructive inspection

Nondestructive evaluation is potentially a major application of electronic speckle pattern interferometry. Figure 21.8 is a schematic of a system that has been used for that purpose and others (Cloud, Nokes, and Chen 1993; Nokes and Cloud 1992).

As explained for holographic interferometry, nondestructive evaluation requires some means of disturbing the object so as to highlight the displacement field in the vicinity of flaws. Several loading methods have been used, including gravity, vacuum, vibrations, thermal soak, and transient heating. Figure 21.9 shows a fringe pattern from the video monitor for a composite plate that has a small delamination produced by impact. The loading was by

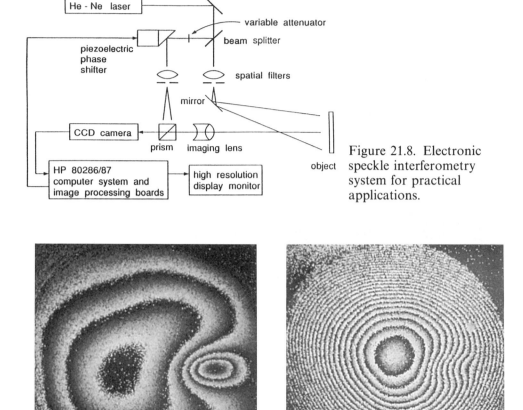

Figure 21.8. Electronic speckle interferometry system for practical applications.

Figure 21.9. Electronic speckle fringe patterns obtained by transient heating of a composite plate with delamination and different edge conditions; black-and-white reductions of photos of false-color computer displays.

transient thermal pulse. These fringes are not optimized by high degrees of digital processing, but the flaw is quite readily visible.

Digital reduction and processing of electronic speckle facilitates production of various schemes for presenting the data in meaningful ways. Figure 21.10 is a two-dimensional plot of out-of-plane displacements for a given axis in the specimen surface. The axis is taken through the damaged area of the specimen.

Figure 21.11 is another way of plotting the data automatically from the electronic speckle system. It is an isometric plot in which the damage zone is clearly visible.

Figure 21.10. Two-dimensional plot of displacement for given axis through the damage zone of a composite plate; traced from computer printout.

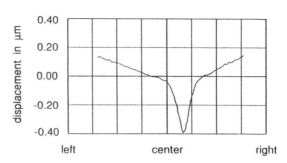

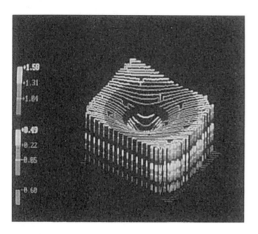
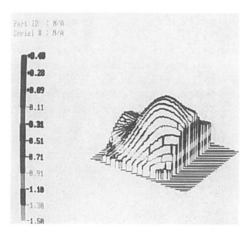

Figure 21.11. Isometric computer plots of displacements in composite plates by electronic speckle interferometry; black-and-white reductions of photos of false-color computer displays.

Electronic speckle pattern interferometry is a promising and dynamic field of research, both in the development and in the application of the technique. The same seems true of electronic speckle shearography, which is not discussed in this book. It seems doubtful that the full potential of either of these techniques has yet been realized.

References

Beidermann, K., and Ek, L. (1975). A recording and display system for hologram interferometry with low resolution imaging devices. *Journal Physics E: Scientific Instruments*, 8: 571.

Butters, J. N., and Leendertz, J. A. (1971). Holographic and video techniques applied to engineering measurements. *Journal Measurement and Control*, 4, 12: 349–54.

Cloud, G., Nokes, J., and Chen, X. (1993). Combined interferometric techniques for assessment of damage and verification of repair in composite structures. *Proc. 1993 ASM–ESD Conference on Advanced Composites*. Dearborn, MI: American Society for Metals.

Cookson, T. J., Butters, J. N., and Pollard, H. C. (1978). Pulsed lasers in electronic speckle pattern interferometry. *Optics and Laser Technology*, 10, 3: 119–24.

Høgmoen, K., and Løkberg, O. J. (1977). Detection and measurement of small vibrations using electronic speckle pattern interferometry. *Applied Optics*, 16: 1869–75.

Jones, R., and Wykes, C. (1983). *Holographic and Speckle Interferometry*. Cambridge University Press.

Jones, R., and Wykes, C. (1989). *Holographic and Speckle Interferometry*, 2nd ed. Cambridge University Press.

Løkberg, O. J., and Høgmoen, K. (1976). Use of modulated reference wave in electronic speckle pattern interferometry. *Applied Optics*, 15: 2701–4.

Lu, B., Yang, X. Abendroth, H., and Eggers, H. (1989). Time-average subtraction method in electronic speckle pattern interferometry. *Optics Communications*, 70, 3: 177–80.

Macovski, A., Ramsey, S. D., and Shaefer, L. F. (1971). Time-lapse interferometry and contouring using television systems. *Applied Optics*, 10: 2722–7.

Nokes, J., and Cloud, G. (1992). The use of high frequency dynamic information in the evaluation of damage in fiber reinforced composites. In *Proc. VII International Congress on Experimental Mechanics*. Bethel, CT: Society for Experimental Mechanics.

Slettemoen, G. A. (1977). Optical signal processing in electronic speckle pattern interferometry. *Optics Communications*, 23: 213–16.

Stetson, K. A. (1970). New design for laser image, speckle interferometer. *Optics and Laser Technology*, 2, 4: 179–81.

22

Phase shifting to improve interferometry
(*with contributions by K. Creath*)

It seems fitting to close this text by discussing a technique that can be used to improve the precision, convenience, and usefulness of all varieties of interferometry. The basic idea is to insert into one of the optical paths a device that will provide known phase shifts. By doing this a few times, the exact phase at a point in the unknown can be deduced from only intensity measurements at that point. There is no need to map fringes and interpolate between them. The origins of the idea seem to date from the early days of photoelastic interferometry when simple compensators were used to improve measurements of birefringence. The approach has gained popularity with the advent of powerful microcomputers and improved electronic imaging technology. It is especially useful in electronic speckle pattern interferometry, so the idea and examples are presented in that context. Keep in mind that existing and potential applications are much broader in scope.

22.1 A perspective

When first getting involved in the use of phase shifting to enhance interferometry, whether electronic speckle or any other variety, it is very easy to become mired in the mathematics and the various computer algorithms that are promoted. One then loses sight of the essential simplicity of the concept, its universality, and its utility. Some background is in order, and then the basic concept will be presented in order to efficiently learn about the phase-shifting approach.

First, let us retrace briefly what we have been doing in collecting and interpreting interferometric fringe data. With few exceptions, one being pointwise birefringence measurements using a compensator or polarization (e.g., Cloud 1968), it is usual that a picture of a fringe pattern is created. This picture is a map of a warped wavefront, which is in turn indicative of something useful, such as deflection of a structure, the shape of a surface, the

477

strain in a surface, or the stress state in a transparent object. To interpret these patterns, one counts the fringes and multiplies the fringe order by some factor depending on wavelength, geometry of the setup, and possibly some material properties. One then is able to interpolate between the fringes to obtain the variable of interest anywhere in the field.

This process is subject to some difficulties, although it has worked well enough for many years. It tends to be slow and tedious. Sometimes there are not enough fringes to allow valid interpolation. Often special measures, such as the introduction of bias fringes (actually a sort of phase shift), the insertion of a compensator (another phase shifter), or optical Fourier processing, are used to improve results. Still, the main reason that the direct attack with a fringe pattern has worked so well lies in the unmatched ability of the human eye–brain combination to locate the center of a fuzzy dark patch (a fringe), even when it is contaminated with optical noise.

It is natural, with the advent of television, computer image acquisition, and potent personal computers, that the fringe patterns be acquired digitally. The next step was to process them digitally using computer processes that were analogous to what the human did. This approach has not worked very well. The computer is not able to dependably map fringe patterns and number the orders. It has trouble finding fringe centers or even the edges. If the fringes are highly convoluted or broken, or if singular points are present, then the fringe maps are suspect, and the operator must intervene. Patterns that are contaminated by noise cause catastrophic difficulties; this includes the speckle noise that is prevalent with coherent optics. The problem is not that the computer is deficient; it is just being used in the wrong way.

22.2 The basic idea of phase-shifting interferometry

A better approach is to use the video–computer system to do what it does best. The video camera gives a good quick map of brightness, intensity, or irradiance; and this map can quickly and accurately be put in digital form and stored in a computer. The computer can do computations on these digital images and display the results pictorially. The problem, then, is to somehow get all the needed information from no more than the original maps of local image brightness. This is where the phase-shifting idea enters the picture, so to speak. Do not, however, entertain the notion that the use of phase shifting is a response to computer technology. The idea has been used for a long time in many different ways, such as in the context of photoelasticity.

To efficiently gain understanding of the concept, consider a simple inter-ferometric thought experiment. Refer to Figure 22.1 and suppose the objective

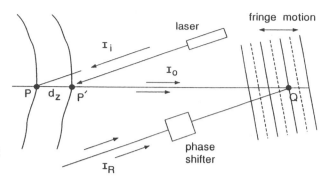

Figure 22.1. Simple
interferometric experiment
to explain phase shifting.

is to measure the out-of-plane displacement of a single point P on an object, using coherent light. Leave speckle notions out of the picture for now. The object is illuminated with coherent light striking it close enough along the surface normal so that geometric factors can be ignored. A scattered object wave passes from the point P to the single detector at Q. We shall use only narrow beams in order to avoid use of an imaging lens in this conceptual experiment. To create an interferometer, a reference beam, also close along the optical axis, is caused to fall on the single detector. Of course, a system of near-vertical interference fringes is set up in the vicinity of Q, and these are represented by the solid lines. Assume for the moment that a dark fringe passes through the detector, so the irradiance indicated by the detector is close to zero, but probably not exactly zero because of imperfect balancing of the beams and various sources of noise. Call this minimum detector output I_1.

Now, suppose that the point P is displaced through distance d_z along the optical axis, that is, toward Q. The standing interference fringes will be forced to move relative to the detector, because the object wave will have undergone a phase shift given by $\phi_2 = 4\pi d_z/\lambda$ (remember the extra factor of 2 in the total path length difference from source to object to detector). The detector will yield a new output I_2.

The problem is to deduce, if possible, the phase ϕ_2, and therefore the distance d_z, from the measurements of intensity that we have, namely I_1 and I_2. Figure 22.2 shows the problem in pictorial form. The mathematical relationship between detector output for this simple case is just

$$I_2 = \frac{I_{\max} + I_1}{2} + \frac{I_{\max} - I_1}{2}\cos\phi_2 \qquad (21.1)$$

This single equation cannot be solved with the information at hand. Neither ϕ_2 nor $I_{\max}$ is known.

The way out is to take another measurement with a known phase shift. Just introduce into the reference path a calibrated device that will give a

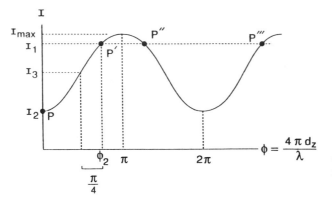

Figure 22.2. Plot of detector output as a function of phase shift for the simple experiment.

phase shift of, say, $-\pi/4$. A third detector output I_3 is recorded. A second equation can then be written as

$$I_3 = \frac{I_{\max} + I_1}{2} + \frac{I_{\max} - I_1}{2}\cos\left(\phi_2 - \frac{\pi}{4}\right) \tag{22.2}$$

The problem is reduced to two equations in the two unknowns mentioned earlier. The ϕ_2 is extracted and converted to d_z.

Now extend the thinking a bit. Suppose there is a whole array of detectors, in other words a video camera, in the plane containing Q. A lens to image the object into this plane would also need to be in place. The same process is completed for each one of the detectors, probably by computer since there are a great number of them. The result will be a phase map indicating the position and shape of the displaced or deformed object.

This basic conceptual experiment does not, of course, tell the whole story. It would be a rare thing if the detector initially were inside a dark fringe. The implication is that the first measurement of irradiance would not give directly the minimum intensity. Anyway, the measurement on the undeformed object is probably being subtracted from the subsequent measurements so that all the data are in terms of phase difference. Either situation would leave three unknowns: $I_{\min}$, $I_{\max}$, and ϕ_2. Three observations on the deformed object would be required: the initial one and two others with two values of known phase shift. The problem is more complicated than is the thought-experiment, but the concept and procedure are the same.

More serious difficulties are created if the displacement of the object might be greater than one-half wavelength or one full wavelength. Such situations are represented in Figure 22.2 by imagining that the point P might move to P″ or P‴. These are separate problems. Displacements greater than $\lambda/2$ but less than λ giving $\pi < \phi < 2\pi$, are handled by being careful with the signs in the equations. Larger displacements require a step known as phase unwrapping, which is discussed later in this chapter. The process is analogous to determining the nearest whole isochromatic order in photoelasticity before adding on the fractional order, which is actually what is determined in phase shifting.

The precision of phase-shifting techniques has been demonstrated to be 10 to 100 times greater than that obtained by computerized counting of fringes. It is also quite simple to carry out. With modern computers and phase shifters, it is also very quick.

With the basic ideas understood, the mathematics and techniques can be pursued. There has been a great surge of interest in phase-shifting interferometry in recent years, and many fine papers on the subject have been produced. The most useful major review of the technique, with an extensive bibliography, is by Creath (1988). An updated treatment has been published (Creath 1993). Many portions of the technical exposition given below follow closely Creath's authoritative treatment, which has become a standard. They are used with permission of author and publisher, for which the writer is most grateful. A paper by Creath (1990) also carries a long list of references.

This general approach, wherein known phase shifts are used in detecting the deformation or displacement, is called phase-measurement interferometry (PMI), phase-stepping interferometry, or else phase-shifting interferometry. Also, the discussion is within the context of Electronic Speckle Pattern Interferometry, although the ideas and techniques apply equally to other interferometric methods.

22.3 Methods of phase shifting

There are many ways to determine the phase of a wavefront. For all techniques, a temporal phase modulation, which is a time-dependent relative phase shift between the object and reference beams in an interferometer, is introduced to perform the measurement. By measuring the speckle pattern intensity as the phase is shifted, the phase of the wavefront can be determined with the aid of a computer-controlled electronic processing system.

Phase modulation in an interferometer can be induced by moving a mirror, tilting a glass plate, moving a grating, rotating a half-wave plate or polarizer, or using an acousto-optic or electro-optic modulator (Hu 1983; Shagam and Wyant 1978; and Wyant 1975). Phase shifters such as moving mirrors, gratings, tilted glass plates, or polarization components can produce continuous as well as discrete phase shifts between the object and reference beam.

The most common and straightforward phase-shifting technique is the placement of a mirror mounted on a piezo-electric transducer (PZT) in the reference beam (Wyant 1982). Many brands of PZTs are available to move a mirror linearly over a range of several micrometers. A high-voltage amplifier

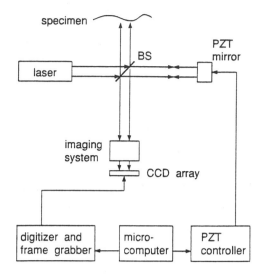

Figure 22.3. Schematic of electronic speckle interferometer using piezo-electric-mounted mirror as phase shifter.

is used to produce a linear ramping signal from zero to several hundred volts. If there are nonlinearities in the PZT motion, they can be accounted for by using a programmable waveform generator. If a phase-stepping technique is preferred to a continuous modulation, any calibrated PZT can be used because only discrete voltage steps are needed.

Figure 22.3 is a schematic of an ESPI setup that incorporates phase shifting of the reference beam. The emphasis is on the phase shifting, so some of the components are not represented. All required devices were shown in Figure 21.8. The light from a He–Ne laser is split into object and reference beams by a beam splitter. The object beam is further expanded by an objective lens and filtered by a pinhole to illuminate the object with a clean diverging beam. Diffusely scattered light from the object is collected by an imaging lens and imaged on the detector plate of a TV camera. The reference beam is also expanded and filtered through a spatial filter to interfere with the speckle image of the object on the detector plate of the TV camera.

To determine the phase information, a mirror mounted on a driver device incorporating a piezo-electric crystal (PZT) is placed in the reference beam path to artificially change the optical path difference between reference and object beams. The light intensity of the interference speckle image pattern for each mirror location of interest is sampled at each point (pixel) to yield a series of digital pictures. Each sample point is quantized to, typically, 256 discrete gray levels. The digital pictures are stored in a digital frame memory. Further processing depends on the application and the calculation algorithm. Usually, the phase shape of the test surface is sought at each point, because it is geometrically related to the wavefront phase that is calculated from the series of patterns obtained with known phase shifts.

If the detector elements are several times larger than the smallest speckles in the pattern, the modulation that is generated by changing the phase of the reference beam is reduced. This reduction is offset to some extent by the increased light level on the detectors, so there is a tradeoff between light level and modulation. The best setting can be determined by experiment.

22.4 Phase-measurement algorithms

Techniques using phase shifting and computer reduction to determine phase all have some common denominators. These techniques shift the phase of one beam in the interferometer with respect to the other beam and measure the intensity of an interference pattern at many different relative phase shifts. To make these techniques work, the interference pattern must be sampled correctly to obtain sufficient information to reconstruct the wavefront. The detected intensity modulation as the phase is shifted can be calculated for each detected point to determine if the data point is good (sufficient modulation).

Fringe modulation is a fundamental problem in all phase-measurement techniques (Creath 1985). When a fringe pattern is recorded by a detector array, there is an output of discrete voltages representing the average intensity incident on the detector element over the integration time. As the relative phase between the object and reference beams is shifted, the intensities read by the detector element should change. Because any array samples the image over a finite number of discrete points, the maximum image spatial frequency that can be unambiguously transformed is the Nyquist frequency, at which each sensing element corresponds to either a maximum or a minimum in the image. If we have at least one detector element for each speckle, the Nyquist frequency condition will be satisfied. However, if there is an entire fringe containing many speckles over the area of the detector element, there will be no modulation.

Figure 22.4 illustrates three different extreme cases that are useful in developing an understanding of this sampling problem. Case (1) of Figure 22.4 indicates a highly sampled condition, in which speckle size is much larger than the detector element; this gives highly detailed modulation output. Case (2) represents the sufficiently sampled case, in which the speckle and the detector element have almost the same size and the output has high modulation. Case (3) shows the undersampled case, in which the speckles are much smaller than the detector element and the output has low modulation. The detector element size influences the recorded fringe modulation, whereas the detector spacing determines if the wavefront can be reconstructed without phase ambiguities.

Many different algorithms for the determination of wavefront phase have been published, and others are appearing at a rapid rate. Some techniques

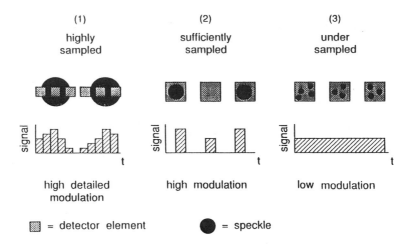

Figure 22.4. Camera detector array used to detect speckle and resulting modulation for cases in which: (1) detector element is much smaller than speckle; (2) detector element and speckle are approximately equal in size; (3) speckle is much smaller than detector element.

change the phase by a known amount between intensity measurements, whereas others integrate the intensity while the phase is being shifted. The first is the so-called phase-stepping technique, and the second is usually referred to as an integrating-bucket technique. A minimum of three measurements is necessary to determine the phase because there are three unknowns in the interference equation:

$$I = I_R + I_O + 2\sqrt{I_R I_O} \cos \phi \qquad (22.3)$$

The unknowns are the reference beam irradiance I_R, the object beam irradiance I_O, and the phase difference ϕ between the reference beam and object beam. The equation can also be put in terms of an average irradiance and the modulation level. The phase shift between adjacent irradiance measurements can be anything between 0 and π as long as it is constant (for the phase-stepping method) or linear in time (for the integration method).

In general, N measurements of the intensity are recorded as the phase is shifted. For the general technique the phase shift is assumed to change during the detectors' integration time, and this change is the same from data frame to data frame. The amount of phase change from frame to frame may vary, but it must be known by calibrating the phase shifter or by measuring the actual phase change. Unless discrete phase steps are used, the detector array will integrate the fringe intensity data over a change in relative phase of Δ. One set of recorded intensities may be written as (Greivenkamp 1984)

$$I_i(x, y) = \frac{1}{\Delta} \int_{(\alpha_i + \Delta/2)}^{(\alpha_i - \Delta/2)} I_0(x, y)[1 + \gamma_0 \cos(\phi(x, y) + \alpha(t))] \, d\alpha(t) \qquad (22.4)$$

where
$I_0(x, y)$ is the average intensity at each detector point (DC component)
γ_0 is the modulation of the fringe pattern
α_i is the average value of the relative phase shift for the *i*th exposure
$\phi(x, y)$ is the phase of the wavefront being measured at the point (x, y)
α is the time-dependent phase shift

The integration over the relative phase shift Δ makes this expression applicable for any phase-shifting technique. After integrating this expression, one finds the recorded intensity to be

$$I_i(x, y) = I_0(x, y)\left\{1 + \gamma_0 \left(\frac{\sin \dfrac{\Delta}{2}}{\dfrac{\Delta}{2}}\right) \cos[\phi(x, y) + \alpha_i]\right\} \qquad (22.5)$$

It is important to note that the only difference between integrating the phase and stepping the phase is a reduction in the modulation of the interference fringes after detection. If the phase shift is stepped a small amount ($\Delta \approx 0$) and not integrated, one obtains

$$\frac{\sin \dfrac{\Delta}{2}}{\dfrac{\Delta}{2}} = \mathrm{sinc}\, \frac{\Delta}{2} = 1 \qquad (22.6)$$

Clearly, phase stepping is a simplification of the integrating method. At the other extreme, if $\Delta = 2\pi$, there is no modulation of the intensity. Because this technique relies on a modulation of the intensities as the phase is shifted, the phase shift per exposure needs to be between 0 and π.

For a total of N recorded intensity measurements, the phase can be calculated using a variety of methods. A least-squares approach is useful to illustrate a typical procedure. This type of approach can be found in the literature (e.g., Greivenkamp 1984). First, equation (22.5) is rewritten in the following form:

$$I_i(x, y) = a_0(x, y) + a_1(x, y) \cos \alpha_i + a_2(x, y) \sin \alpha_i \qquad (22.7)$$

where

$$a_0(x, y) = I_0(x, y) \tag{22.8}$$

$$a_1(x, y) = I_0(x, y)\gamma_0 \frac{\sin \dfrac{\Delta}{2}}{\dfrac{\Delta}{2}} \cos \phi(x, y) \tag{22.9}$$

$$a_2(x, y) = I_0(x, y) \frac{\sin \dfrac{\Delta}{2}}{\dfrac{\Delta}{2}} \sin \phi(x, y) \tag{22.10}$$

The unknowns of this set of equations are $I_0(x, y)$, γ_0, and $\phi(x, y)$. The least-squares solution to these equations is

$$\begin{bmatrix} a_0(x, y) \\ a_1(x, y) \\ a_2(x, y) \end{bmatrix} = A^{-1}(\alpha_i)B(x, y, \alpha_i) \tag{22.11}$$

where

$$A(\alpha_i) = \begin{bmatrix} N & \sum \cos \alpha_i & \sum \sin \alpha_i \\ \sum \cos \alpha_i & \sum (\cos \alpha_i)^2 & \sum \cos \alpha_i \sin \alpha_i \\ \sum \sin \alpha_i & \sum \cos \alpha_i \sin \alpha_i & \sum (\sin \alpha_i)^2 \end{bmatrix} \tag{22.12}$$

and

$$B(\alpha_i) = \begin{bmatrix} \sum I_i(x, y) \\ \sum I_i(x, y) \cos \alpha_i \\ \sum I_i(x, y) \sin \alpha_i \end{bmatrix} \tag{22.13}$$

The matrix A must be calculated and inverted just once because it is dependent only on the phase shift. The phase at each point in the speckle pattern is determined by evaluating the value of B at each point and then solving for the coefficients a_1 and a_2. The final result is

$$\tan \phi(x, y) = \frac{a_2(x, y)}{a_1(x, y)} = \frac{I_0\gamma_0 \dfrac{\sin \dfrac{\Delta}{2}}{\dfrac{\Delta}{2}} \sin \phi(x, y)}{I_0\gamma_0 \dfrac{\sin \dfrac{\Delta}{2}}{\dfrac{\Delta}{2}} \cos \phi(x, y)} \tag{22.14}$$

This phase calculation assumes that the phase shifts between measurements are known and that the integration period Δ is constant for every measurement.

Besides a reduction in intensity modulation resulting from the integration over a change in phase shift, the finite size of the detector element also contributes a reduction in intensity modulation, as was discussed. To make reliable phase measurements, the incident intensity must modulate sufficiently at each detector point to yield an accurate phase. The recorded intensity modulation can be calculated from the intensity data using the equation,

$$\gamma(x, y) = \gamma_0 \frac{\sin \dfrac{\Delta}{2}}{\dfrac{\Delta}{2}} = \frac{\sqrt{a_1(x, y)^2 + a_2(x, y)^2}}{a_0(x, y)} \tag{22.15}$$

This expression can be used to determine if a data point will yield an accurate phase measurement or if it should be ignored.

Because the problem has three unknowns, it would seem that it is always best to use three phase steps. There are other possibilities for which the calculations are simplified or improved. Let us look at some specific choices, beginning with the most obvious.

22.4.1 Three-step technique

Because there are three unknowns in equation 22.3, a minimum of three sets (exposures) of recorded fringe data are needed to reconstruct a wavefront; the phase can then be calculated from a known phase shift of α_i per exposure. Two common choices for the phase shift value are $\Delta = \pi/2$ and $\Delta = 2\pi/3$.

(a) If $\Delta = \pi/2$ and three shifts are employed ($N = 3$) then $\alpha = \pi/4$, $3\pi/4$, and $5\pi/4$. The three intensity measurements can be expressed as (Creath 1988)

$$I_1(x, y) = I_0(x, y)\left(1 + \gamma \cos\left[\phi(x, y) + \frac{1}{4}\pi\right]\right) \tag{22.16}$$

$$I_2(x, y) = I_0(x, y)\left(1 + \gamma \cos\left[\phi(x, y) + \frac{3}{4}\pi\right]\right) \tag{22.17}$$

$$I_3(x, y) = I_0(x, y)\left(1 + \gamma \cos\left[\phi(x, y) + \frac{5}{4}\pi\right]\right) \tag{22.18}$$

When discrete steps are used, $\gamma = \gamma_0$, and when the phase is integrated over a $\frac{\pi}{2}$ phase shift per frame, $\gamma = 0.9\gamma_0$. The phase at each point is then simply,

$$\phi(x, y) = \arctan\left(\frac{I_3(x, y) - I_2(x, y)}{I_1(x, y) - I_2(x, y)}\right) \tag{22.19}$$

The intensity modulation is

$$\gamma(x,\ y) = \frac{\sqrt{[I_1(x,\ y) - I_2(x,\ y)]^2 + [I_2(x,\ y) - I_3(x,\ y)]^2}}{2I_0} \qquad (22.20)$$

(b) If a phase shift of 120° $(2\pi/3)$ is used, $\alpha_i = -2\pi/3,\ 0,$ and $+2\pi/3;$ and $\Delta = 2\pi/3.$ We have for this case,

$$\phi(x,\ y) = \arctan\left(\frac{\sqrt{3}(I_3(x,\ y) - I_2(x,\ y))}{2I_1(x,\ y) - I_2(x,\ y) - I_3(x,\ y)}\right) \qquad (22.21)$$

For the integration method, $\gamma = 0.83\gamma_0,$ and the detected intensity modulation is

$$\gamma(x,\ y) = \frac{\sqrt{3[I_3(x,\ y) - I_2(x,\ y)]^2 + [2I_1(x,\ y) - I_2(x,\ y) - I_3(x,\ y)]^2}}{2I_0} \qquad (22.22)$$

22.4.2 Four-step technique

This is a common algorithm for phase calculations (Wyant 1982). In this case $\alpha_i = 0,\ \pi/2,\ \pi,$ and $3\pi/2;$ and $\Delta = \pi/2.$ $\gamma = \gamma_0$ for the discrete four-step technique and $\gamma = 0.9\gamma_0$ for the integration technique. Note that integrating the phase produces a very small effect for a phase shift of $\frac{\pi}{2}$ per exposure. Thus, linearly ramping the phase shifter while taking measurements makes more sense than stepping and waiting for the reference beam to settle down. Using calculations similar to those for three-step technique, the phase at each point for the four-step method is

$$\phi(x,\ y) = \arctan\left(\frac{I_4(x,\ y) - I_2(x,\ y)}{I_1(x,\ y) - I_3(x,\ y)}\right) \qquad (22.23)$$

and the recorded modulation can be calculated from

$$\gamma(x,\ y) = \frac{\sqrt{[I_4(x,\ y) - I_2(x,\ y)]^2 + [I_1(x,\ y) - I_3(x,\ y)]^2}}{2I_0} \qquad (22.24)$$

One nice thing about this technique is that by using pipe-line image processing (using look-up tables), the noise caused by the speckles can be removed, which gives noticeable fringe visibility improvement (Creath 1990).

22.4.3 Carré technique

In the previous equations the phase shift is known either by calibrating the phase shifter or by directly measuring the amount of phase shift each time it is moved. Carré (1966) devised a technique of phase measurement that is independent of the amount of phase shift. It assumes that the phase is shifted by a constant value α between consecutive intensity measurements to yield four equations:

$$I_1(x, y) = I_0(x, y)\left[1 + \gamma \cos\left(\phi(x, y) - \frac{3}{2}\alpha\right)\right] \qquad (22.25)$$

$$I_2(x, y) = I_0(x, y)\left[1 + \gamma \cos\left(\phi(x, y) - \frac{1}{2}\alpha\right)\right] \qquad (22.26)$$

$$I_3(x, y) = I_0(x, y)\left[1 + \gamma \cos\left(\phi(x, y) + \frac{1}{2}\alpha\right)\right] \qquad (22.27)$$

$$I_4(x, y) = I_0(x, y)\left[1 + \gamma \cos\left(\phi(x, y) + \frac{3}{2}\alpha\right)\right] \qquad (22.28)$$

The phase shift is assumed to be linear with time. From these equations, the phase at each point can be calculated by

$$\phi(x, y) = \arctan \frac{\sqrt{[3(I_2 - I_3) - (I_1 - I_4)][(I_2 - I_3) + (I_1 - I_4)]}}{(I_2 + I_3) - (I_1 + I_4)} \qquad (22.29)$$

For this technique, the intensity modulation is

$$\gamma = \frac{1}{2I_0} \sqrt{\frac{[(I_2 - I_3) + (I_1 - I_4)]^2 + [(I_2 + I_3) - (I_1 + I_4)]^2}{2}} \qquad (22.30)$$

This equation assumes that α is near $\frac{\pi}{2}$. An obvious advantage of the Carré technique is that the phase shifter does not need to be calibrated. It also has the advantage of working when a linear phase shift is introduced in a converging or diverging beam where the amount of phase shift varies across the beam.

22.5 Comparison of phase-measurement techniques

In general, the integration methods give the same results as the phase-stepping methods except in the case of nonlinear phase-shift errors, where the integration method is superior. The Carré algorithm is the best to use where phase-shifting errors are present, and the four-step technique is the best for eliminating effects caused by second- and third-order detection nonlinearities.

22.6 Phase unwrapping

Because of the nature of arctangent calculations, the equations presented for phase calculation are sufficient for only a modulo π calculation. To determine

Table 22.1. *Determination of the phase modulo 2π*

Numerator	Denominator	Adjusted phase	Range of phase values
$\sin\phi$	$\cos\phi$	—	—
Positive	Positive	ϕ	$0-\frac{1}{2}\pi$
Positive	Negative	$\pi - \phi$	$\frac{1}{2}\pi-\pi$
Negative	Negative	$\pi + \phi$	$\pi-\frac{3}{2}\pi$
Negative	Positive	$2\pi - \phi$	$\frac{3}{2}\pi-2\pi$
0	Anything	π	π
Positive	0	$\frac{1}{2}\pi$	$\frac{1}{2}\pi$
Negative	0	$\frac{3}{2}\pi$	$\frac{3}{2}\pi$

the phase modulo 2π, the signs of quantities proportional to $\sin\phi$ and $\cos\phi$ must be examined. For all techniques but Carre's, the numerator and denominator give the desired quantities (Creath 1985). Table 22.1 shows how the phase is determined by examining the signs of these quantities after the phase is calculated modulo $\frac{\pi}{2}$ using absolute values in the numerator and denominator to yield a modulo 2π calculation.

Once the phase has been determined to be modulo 2π, the phase ambiguities due to the modulo 2π calculation can be removed by comparing the phase difference between adjacent pixels. For reliable removal of discontinuities the phase must not change by more than π between adjacent pixels. As long as the data are sampled as described in the sampling requirements, the wavefront can be reconstructed.

For the Carré technique, simply looking at numerators and denominators is not sufficient to determine phase modulo 2π (Creath 1985). The process is somewhat complicated and will not be discussed here.

Any phase unwrapping needs some reference point, that is, a point to start. If the reference point is a fixed point, say a point in the fixture that holds the object being tested, then the calculated phase change will be an absolute value. Otherwise, the calculated phase change will be a relative value.

To calculate the displacement, the phases before and after displacement are first subtracted from each other. These phase values have been unwrapped before the subtraction.

$$\Delta\phi(x, y) = \phi_2(x, y) - \phi_1(x, y) \tag{22.31}$$

Of course, if there is interest only in the surface shape, then there is no need to measure the deformed stage of the object, and the phase subtraction need not be performed. Only the phase of the stable object is measured, and the phase is unwrapped.

22.7 Converting phase change to displacement

Once the phase change is known, the corresponding displacement can be determined from the phase information. The surface displacement d_z at the location (x, y) is

$$d_z(x, y) = \frac{\Delta\phi(x, y)\lambda}{2\pi(\cos\theta_i + \cos\theta_v)} \tag{22.32}$$

where λ is the wavelength of illumination;

ϕ_i and ϕ_v are the angles of illumination and viewing with respect to the surface normal

The illumination and viewing directions will affect the measured value. The detailed relationships between actual and measured displacement as a function of geometry of the setup were discussed in previous chapters.

References

Carré, P. (1966). Installation et utilisation du comparateur photoelectrique et interferentiel du Bureau International des Poids et Mesures. *Metrologia*, 2, 1: 13–23.

Cloud, G. L. (1968). Improvement in use of photometric methods of measurement of birefringence. *Experimental Mechanics*, 8, 3: 138–41.

Creath, K. (1985). Phase shifting speckle interferometry. *Applied Optics*, 24: 3053–8.

Creath, K. (1988). Phase measurement interferometry methods. In *Progress in Optics XXVI*, Ed. E. Wolf, pp. 349–93. Amsterdam: Elsevier Science Publishers.

Creath, K. (1990). Phase measurement techniques for nondestructive testing. In *Proceedings: Hologram Interferometry and Speckle Metrology*. Bethel, CT: Society for Experimental Mechanics.

Creath, K. (1993). Temporal phase measurement methods. In *Interferogram Analysis*, Ed. D. Robinson and G. Reid, Chap. 4. Philadelphia: Institute of Physics.

Greivenkamp, C. J. (1984). Generalized data reduction for heterodyne interferometry. *Optical Engineering*, 23: 350–2.

Hu, H. Z. (1983). Polarization heterodyne interferometry using a simple rotating analyzer. 1: Theory and error analysis. *Applied Optics*, 22: 2052–6.

Shagam, R. N., and Wyant, J. C. (1978). Optical frequency shifter for heterodyne interferometers using multiple rotating polarization retarders. *Applied Optics*, 17: 3034–5.

Wyant, J. C. (1975). Use of an AC heterodyne lateral shear interferometer with real-time wavefront correction systems. *Applied Optics*, 14: 2622–6.

Wyant, J. C. (1982). Interferometric optical metrology, basic systems and principles. *Laser Focus*, May 1982: 65–71.

Author index

493

Subject index